T0400778

MATHEMATICS RESEARCH DEVELOPMENTS

BROWNIAN MOTION

THEORY, MODELLING AND APPLICATIONS

MATHEMATICS RESEARCH DEVELOPMENTS

Additional books in this series can be found on Nova's website
under the Series tab.

Additional E-books in this series can be found on Nova's website
under the E-books tab.

MATHEMATICS RESEARCH DEVELOPMENTS

BROWNIAN MOTION

THEORY, MODELLING AND APPLICATIONS

ROBERT C. EARNSHAW
AND
ELIZABETH M. RILEY
EDITORS

Nova Science Publishers, Inc.
New York

For permission to use material from this book please contact us:
Telephone 631-231-7269; Fax 631-231-8175
Web Site: http://www.novapublishers.com

Library of Congress Cataloging-in-Publication Data

Brownian motion : theory, modelling and applications / editors, Robert C. Earnshaw and Elizabeth M. Riley.
p. cm.
Includes index.
ISBN 978-1-61209-537-0 (hardcover)
1. Brownian motion processes. I. Earnshaw, Robert C. II. Riley, Elizabeth M.
QA274.75.B76 2011
530.4'75--dc22

2010054273

Published by Nova Science Publishers, Inc. † New York

CONTENTS

PREFACE

Brownian motion is the seemingly random movement of particles suspended in a fluid or the mathematical model used to describe such random movements, often called a particle theory. This new book presents topical research in the study of Brownian motion, including Markov chains and Brownian motion models in insect ecology; applications of multifractional Brownian motion to internet traffic; enhanced thermal conductivity of nanofluids using the Brownian motion-based model; Brownian dynamics simulation of surfactant micellar microstructure and rheology and the development of the fractional Brownian motion and its applications on coastal dispersion modeling. (Imprint: Nova Press)

Chapter 1 – Maruyama introduced the formal representation $db = w(t)\sqrt{dt}$ to relate the Brownian motion $b(t)$ and the white noise $w(t)$. It is shown that if one combines this modelling with fractional calculus based on the modified Riemann-Liouville derivative, together with integral w.r.t. $\sqrt{dt}$, one can arrive at a stochastic calculus which is quite similar to the Itô's one. This result could contribute to support the engineering mathematics viewpoint in accordance of which the elemental basic stochastic process would be the standard white noise instead of the Brownian motion. The main interest of this approach is that it applies to the equation $db(t,\alpha) = w(t)(dt)^{\alpha}$ therefore a new framework which gives a sound significance to this generalized Maruyama's notation. It is shown that this order parameter α has direct meaning in terms of subjectivity via observation with informational invariance as well as in coarse-graining phenomena, therefore the use of models in which it is time varying. By using the fractional Taylor's series, the definition of fractional stochastic differential equation is revisited in the engineering mathematics framework, and further modeling results are provided. Lastly, the authors show how these results could apply to fractional differential geometry, where white noises seems to be quite relevant as far as the authors are then dealing with curves which are continuous everywhere but nowhere differentiable. In all the article, the basic stochastic process is the Gaussian white noise, instead of the Brownian motion.

Chapter 2 – This chapter discusses some applications of the fundamental theory of probability laws and stochastic processes such as the Brownian motion in an attempt to describe the time evolution of certain developmental events during the life cycle of an insect. In describing several ecological phenomena, deterministic and stochastic models are widely applied in which the theory of certain processes play an important role. One such process is the Wiener process which was introduced by N. Wiener (1923) as a mathematical model for the random movement of colloidal particles first observed by a British Botanist, R. Brown (1828), but has now applications in several fields of science. Markov processes are considered as one of the most important and fundamental concepts in today's probability theory having several applications in ecology and other fields. Here the authors are interested in the time evolution of certain developmental events in the life cycle of an insect and therefore most values are treated as discrete states in a Markov process. Fundamental theoretical results are first outlined and the mathematical approach that was applied is presented. The work proceeds by expressing the transition probabilities of stationary Markov chains by means of matrices and probability laws. The method is further illustrated by using real data sets applied on periodic phenomena in the biology of an insect such as diapause induction, diapause termination as well as insect seasonality in temperate climates. The applied stochastic models are able to give part of the realization of the above physiological processes. The time evolution of the developmental events is generated as a scaling limit of a random walk with stationary independent increments while, in the discussion of the examples based on original and simulated data, improved estimates are computed. The presented approach provides a relatively flexible method for handling discrete-valued biological data and modeling their time evolution. In particular, it appears quite useful for detecting abrupt or steady changes in the structure and the time evolution of the process as a result of seasonal and long term variations of environmental factors such as temperature and photoperiod. Although the models are implemented under the Markov assumptions for stationary independent increments, the work also discusses the presence of non stationary processes and memory effects, and how they can be assessed, both by change of deviance as well as graphically, via periodogram plots of residuals. In fact, the presented Brownian motion arises as a limit of many discrete stochastic processes (displacements of the particles) in much the same way that Gaussian random variables appear as a limit of other random variables through the central limit theorem. In this context the time evolution of several biological random variables can follow certain probability laws. Finally, fundamental for a potential field application of such models, is also a biological interpretation, which is explained by clearly understanding causality and association to several external factors.

Chapter 3 – Measuring fluctuations of the Internet traffic (traffic for short) is a challenge and complicated issue in computer communication networks. There are two types of fluctuations of traffic. One is local irregularity or local roughness and the other long-range dependence (LRD). The former is a local behavior of traffic at small time scales while the latter a global property of traffic at large time scales. In the world of computer science, people often use the terms of small time scaling phenomenon and the large time scaling one to represent the fluctuations of traffic locally and globally, respectively. The authors follow this preference in this chapter without confusions.

Tools to separately describe the scaling phenomena of traffic are desired in several aspects of computer communications, such as performance analysis of network systems. As for this tough issue, standard fractional Brownian motion (fBm) and or its increment process, fractional Gaussian noise (fGn), may be never enough even they are on an interval-by-interval basis, because two types of scaling phenomena may not be characterized by the Hurst parameter H singly. Though the multifractional fBm (mBm) that is indexed by the local Hölder function $H(t)$ may be used to capture the scaling phenomena of those two but $H(t)$ describes two types of scaling phenomena on a point-by-point basis only. In addition, $H(t)$

linear relates to the time-dependent fractal dimension $D(t) = 2 - H(t)$. To overcome the limitation of $H(t)$, the authors propose to describe the local irregularity of traffic, i.e., small time scaling phenomenon, by using $D(t)$ on a point-by-point basis and to measure the LRD of traffic by using $E[H(t)]$ on an interval-by-interval basis. Since $E[H(t)]$ is a constant within an observation interval while $D(t)$ is random in general, they are uncorrelated with each other. Thus, the authors proposed method can be used to separately characterize the small time scaling phenomenon and the large one of traffic in the global sense, providing a new outlook or tool to investigate the fluctuations of traffic.

Chapter 4 – Closed-form analytical solutions for the pricing of options are available only in a limited number of simple cases. Beyond them, in more complex settings, the value of options is typically driven by several sources of risk. Other methods need to be developed in order to solve these cases. This paper develops several methods for pricing American and Bermudan options. In these options the exercise time is not restricted to the maturity of the options: rather, they can be exercised at any time prior to maturity or at pre-specified times before the expiration date. A number of numerical examples are shown.

First the authors address the analytical solution for a perpetual American option with two sources of risk. Both sources determine the value of a project, which serves as the underlying asset of an American option.

Second, the authors assess American and Bermudan options by means of multidimensional binomial lattices. A contribution of this chapter to the relevant literature is that in order to avoid negative probabilities the lattices are deployed to represent the behaviour of the corresponding futures contracts (whose stochastic equation has no drift), and futures prices are then used to estimate the spot price in a risk-neutral world. In this part I draw the optimal exercise boundary, i.e. the boundary between the "invest" region and the "wait" region. The authors also show the solution to an optimal control problem.

Another section is devoted to the Monte Carlo method. The authors obtain the price of path-dependent and American-style options, in the latter case using Least Squares Monte Carlo method along with the optimal exercise boundary under multiple sources of risk.

Chapter 5 – Nanofluids are a new class of heat transfer fluids which are engineered by dispersing nanometer-sized solid particles in conventional fluids. This is a rapidly emerging interdisciplinary field where nanoscience, nanotechnology, and thermal engineering meet. Since the novel concept of *nanofluids* was coined in 1995, this research topic has attracted tremendous interest from researchers worldwide due to their exciting thermal properties and potential applications in numerous important fields. Although research works have shown that nanofluids exhibit significantly higher thermal conductivity compared to their base fluids, the underlying mechanisms for the enhancement are still debated and not thoroughly understood. Despite considerable theoretical efforts devoted to the development of model for the prediction of the effective thermal conductivity of nanofluids, there has been little agreement among different studies and no widely accepted model is also available due to inconclusive heat transfer mechanisms of nanofluids. Nevertheless, fundamental understanding of the underlying mechanisms and development of a unanimous theoretical model are crucial for exploiting potential benefits and applications of nanofluids. In this chapter, a new and improved Brownian motion-based model is introduced for the prediction of the enhanced thermal conductivity of nanofluids. In addition to the Brownian motion of nanoparticles, this model also takes into account several other important factors such as particle size and interfacial nanolayer that contribute to the enhancement of the effective thermal conductivity of nanofluids. The conventional kinetic theory-based Brownian motion term has been renovated using effective diffusion coefficient concept. The present model shows reasonably good agreement with the experimental results of various aqueous nanofluids and gives better predictions compared to classical and other recently developed models. Besides providing a brief review on theoretical studies and various heat transfer mechanisms of nanofluids, details of the present model development and its validation with the experimental results are also discussed in this chapter.

Chapter 6 – As increasing the demand for electronic application of polymeric materials, a polymer with good electrical conductivity has been desired and investigated intensively these days. Carbon nanotubes (CNTs) are widely used as one of the high-performance modifiers because of the unique properties such as high stability, high electric and thermal conductivity, and large aspect ratio. Therefore, the properties of polymer composites containing CNTs have been studied intensively over the last decade for applications in electrostatic dissipation, electromagnetic interference shielding, and radio frequency interference shielding. In this chapter, the effect of Brownian motion of CNTs in a polymer matrix on the distribution in a molten state and the electrical conductivity in a solid state is explained in detail.

It is found that CNT-filled polymer composites without post-processing annealing procedure could be a thermodynamically non-equilibrium system even for the sample prepared by a compression-molding, in which conductive network formation depends on the processing temperature and residence time. Further, post-processing annealing can change the distribution state of CNTs and thus the electrical and mechanical properties. The redistribution process of CNTs at the post-processing annealing is known as dynamic percolation, which was firstly reported by the research group of Sumita. When a composite is heated at a temperature above the melting point of a polymer matrix, it is expected that CNTs move to a stable condition by means of Brownian motion, and finally form a continuous network if CNT content exceeds a critical value.

Furthermore, the thermally activated transfer process of CNTs between surfaces of immiscible polymer pairs is investigated. Compression-molded sheets of polypropylene (PP) containing 20 wt% of CNTs are laid on another polymer sheet that is immiscible with PP. It is found that CNTs immigrate from PP to the surface of the immiscible polymer such as polycarbonate (PC) during heat treatment by diffusion. The piled sheets are easily separated each other after cooling owing to the immiscible nature. The formation of a thin CNT-rich layer on the surface of the separated PC sheet produces electrical conductivity. Since CNT transfer is attributed to Brownian motion, heating conditions such as temperature and duration time are responsible for the amount of transferred CNTs. This transfer is influenced by the dispersion state of CNTs in the composite, the size of CNTs and their compatibility, and the chemical structure of the second polymer. Moreover, this method can be used to improve surface properties while minimizing CNT content in the bulk of the polymer composite and could be a feasible process to integrate CNTs into various devices.

Chapter 7 – Three-dimensional Brownian dynamics simulation was conducted for dilute micellar surfactant solution under a steady shear flow and a steady uniaxial elongational flow. The rodlike micelle in the surfactant solution was assumed as a rigid rod made up of beads that were lined up. Lennard-Jones potential and soft-sphere potential were employed as the inter-bead potentials for end-end beads and interior-interior beads, respectively. The motion of the rodlike micelles was determined by solving the translational and rotational equations for each rod under hydrodynamic drag force, Brownian force and inter-rod potential force.

Velocity Verlet algorithm was used in the simulation. The micellar microstructures and the rhelogical properties of the surfactant solution at different shear rates and elongation rates were obtained. The micellar network structure was formed at low shear or elongation rates and was destructed by high shear or elongation rates. The computed shear and elongational viscosities and the first normal stress coefficient showed shear thinning characteristics. The relationship between the rheology and the microstructure of the surfactant solution was revealed. The effect of surfactant solution concentration on the micellar structures and rheological properties was also investigated.

Chapter 8 – The methods for generating a Brownian and simple random walks are introduced. An improved fractional Brownian motion(fBm) model has developed and compared with the Mandelbrot fBm model.

A new particle tracking technique is introduced where non-Fickian diffusion is generated by employing fBm. The modelling of pollutant dispersion in an idealised coastal bay is carried out and the differences between a traditional Brownian particle tracking model and the new fBm particle tracking model within the coastal bay are compared in detail.

The results show that the fBm particle tracking model gives more flexibility in controlling the spreading of the diffusing cloud and, moreover, it is closer to reality (in that non-Fickian diffusion may be modelled).

Simulation of real observed data from Numthumbrian coastal waters is then carried out and an improved accelerated fBm method is developed for application in large water bodies such as coastal waters and the ocean surface.

The authors simulation results and the HR Wallingford simulation results are compared and tested with the observed data sets. The novel and practical fBm particle tracking model may become a useful engineering tool for the prediction of contaminant spread in environmental fluid flows.

Chapter 9 – In this paper the authors consider plasma confinement in tokamaks and stellarators as the problem of differential game that will solve the Grad-Shafranov equilibriums. Turbulence can show deterministic and stochastic regimes.

Chapter 10 – Transport processes of nanoparticles in gas flows have been considered in many distinguished papers [1 and references therein]. To mention that practically important process of Brownian deposition of nanoparticles on the wall was investigated in [1 – 9 and references therein]. At present time considerable attention is attracted to the specific of Brownian diffusion of nanoparticles in non-isothermal gas flows in the flow reactors. This interest is due to the ability to control movement and deposition of nanoparticles by changing the temperature of the reactor wall [3, 6-9]. The objective of the authors research is to investigate by the methods of mathematical modeling of the interference of thermophoresis and Brownian diffusion on the movement of nanoparticles with a laminar gas flow in a flow cylindrical reactor when the inlet temperature of the flow and the wall one are very different. The structure of the paper is the following. The mathematical model is described in the chapter 1. The semi-quantitative analysis based on the Galerkin method is given in next chapter. Then the authors present results of numerical simulation. Finally the authors summarized the authors main results and conclusions.

Chapter 11 – The paradigmatic model of a random process with continuous paths is the classical Brownian motion or Wiener process. By this the authors understand a continuous-time Gaussian Markov process with independent increments and continuous sample paths.. The history of Brownian motion as a mathematical model started in 1900 when L.Bachelier proposed modeling the fluctuations in stock prices by Brownian motion. Arigorous mathematical proof of the existence of such a process was first given by N. Wiener in 1923, by introducing convenient measures on an appropriate infinite-dimensional functional.

Chapter 12 – In this chapter, starting from fractional Brownian motion of unique parameter H ($0 < H < 1$), the authors propose two variantswhich extend the classical fBm. First, the piece-wise fractional Brownian motion model of parameters *Ho*, *Hi* and □presents two spectral regimes: it behaves like fBm of parameter *Ho* for low frequencies $| f | <$ □and like fBm of parameter *Hi* for high frequencies $| f | >$ □. Second, the nth-order fractional Brownian motion enables the H parameter to be within the range $]n-1,n[$, where n is any strictly positive integer. The properties of these two processes are investigated. Special interest is given to their increments, which extend fractional Gaussian noises. Synthesis of realizations of such processes is discussed. Finally, an application of such processes for the characterization of trabecular bone radiographs is presented. The overall objective of this application is to provide methods helping in the early diagnosis of osteoporosis for elderly people.

Chapter 13 – After Einstein's theoretical explanation of Brownian motion in 1905, and a few years later by Langevin, and then experimentally corroborated by J. Perrin, an amazing number of works on the same topic has been broadly extended to other branches of science. Nowadays, the theory of Brownian motion continues to be of great interest in physics, chemistry, biology, and actually finds interesting applications in astronomy, economy, physiology, medicine, and many other fields. After these pioneer studies, the problem of Brownian motion in an electromagnetic field was and continues to be an interesting subject of research in other physical contexts.

Chapter 14 – The authors develop an analysis of Brownian motion representation from the quantum to the diffusion MRI level. The authors then present an example of an application of Brownian motion computer simulations to diffusion MRI. The purpose of this work is to provide a general description of the mathematical and physical tools involved, and is aimed at graduate student level and above. The focus of the text is on the physics of Brownian motion, and so there is a large amount of equations.

In: Brownian Motion: Theory, Modelling and Applications ISBN: 978-1-61209-537-0
Editors: R.C. Earnshaw and E.M. Riley

Chapter 1

WHITE NOISE CALCULUS, STOCHASTIC CALCULUS, COARSE-GRAINING AND FRACTAL GEODESIC: A UNIFIED APPROACH VIA FRACTIONAL CALCULUS AND MARUYAMA'S NOTATION

Guy Jumarie*

Department of Mathematics, University of Québec at Montréal, Downtown Station, Montréal, Qc, Canada

ABSTRACT

Maruyama introduced the formal representation $db = w(t)\sqrt{dt}$ to relate the Brownian motion $b(t)$ and the white noise $w(t)$. It is shown that if one combines this modelling with fractional calculus based on the modified Riemann-Liouville derivative, together with integral w.r.t. $\sqrt{dt}$, one can arrive at a stochastic calculus which is quite similar to the Itô's one. This result could contribute to support the engineering mathematics viewpoint in accordance of which the elemental basic stochastic process would be the standard white noise instead of the Brownian motion. The main interest of this approach is that it applies to the equation $db(t,\alpha) = w(t)(dt)^{\alpha}$ therefore a new framework which gives a sound significance to this generalized Maruyama's notation. It is shown that this order parameter α has direct meaning in terms of subjectivity via observation with informational invariance as well as in coarse-graining phenomena, therefore the use of models in which it is time varying. By using the fractional Taylor's series, the definition of fractional stochastic differential equation is revisited in the engineering mathematics framework, and further modeling results are provided. Lastly, we show how these results could apply to fractional differential geometry, where white noises seems to be quite relevant as far as we are then dealing with curves which are continuous everywhere but nowhere differentiable. In all the article, the basic stochastic process is the Gaussian white noise, instead of the Brownian motion.

* E-mail: jumarie.guy @ uqam.ca

Keywords: Modified Riemann Liouville derivative, fractional Taylor series, white noise calculus, Itô's calculus, Maruyama's notation, fractional noise, subjectivity, relative observation, coarse-graining phenomenon, differential geometry

Mathematics subject classification: 60G22, 60J65, 28A80, 26A33, 53-XX

First Part : Motivation for Fractal Modeling

1. Introduction

The present paper originates from three remarks.

(First remark). Formally, one can generalize the dynamical equation

$$dx = f(x,t)dt$$

in the form

$$dx = f(x,t)(dt)^{\alpha}, \quad 0<\alpha<1, \tag{1.1}$$

where α is customarily referred to as Hurst's parameter, and all the question is of knowing whether we can put in evidence in a simple manner clear practical significances for α. And fortunately, "yes we can". It appears that subjectivity via observation with informational invariance and coarse-graining phenomenon in time can be suitably described by this model.

(Second remark) A few decades ago Maruyama proposed the formal equation

$$db(t) = w(t)\sqrt{dt} \tag{1.2}$$

to relate the Brownian motion $b(t)$ with the Gaussian white noise $w(t)$, and which allows us to write the standard Lévy-Itô-Doob stochastic differential equation in the form

$$dx = f(x,t)dt + g(x,t)w(t)\sqrt{dt}\ . \tag{1.3}$$

which can be formally manipulated as if $w(t)$ were a continuous function.

(Third remark) There is an issue in the scientific community about which process is the basic elemental process in the theory of stochastic processes. Is it the Brownian motion or on the contrary is it the white noise $w(t)$? Following Wiener a long time ago, engineers and physicists dealt with white noses more or less formally (sometimes with mistakes), until the advent of the Brownian motion theory which culminated with the Itô's stochastic calculus. Recently Hida [12,13] came back on the basic role of white noise in the study of stochastic processes, and the debate is now running on again. The present work can be thought of as a synthesis expanded around these remarks. We shall show that if one considers the above equation (1.1) in the framework of fractional calculus based on modified-Riemann Liouville derivative 23] and fractional difference, combined with integral w.r.t. $(dt)^{\alpha}$, then one can recover the famous Itô's lemma. In other words, one could start with the Gaussian white noise

to expand a theory of stochastic differential equations. The advantage of this point of view is that its extension to $(dt)^{\alpha}$, $0<\alpha<1$, is straightforward. The article is organized as follows. We shall first bear in mind the essential of the fractional calculus, mainly the revisited definition of fractional derivative, our new fractional Taylor's series and some basic formulae of fractional analysis, in the framework of the so-called modified Riemann-Liouville definition. Then we shall give a short background on observation with informational invariance and coarse-graining phenomenon to show how they introduce the Hurst's parameter α in quite a natural way and how we arrive at a fractional calculus on coarse-grained increment $(dt)^{\alpha}$. Then we shall bear in mind the keys of Itô's mean-square approach therefore we shall manipulate the equation by using result of fractional calculus . By this way, we shall arrive at the Maruyama notation, and we shall combine it with fractional calculus of order α. Then we shall exhibit the relation between Maruyama's notation and white noise calculus and we shall outline the relation with stochastic fractional dynamics. Lastly, we shall show how this framework could be used in an approach to differential geometry on fractal manifolds where the Gaussian white noise should play a role of importance We believe that all the matter can be dealt with by using Gaussian white instead of Brownian motion.

The theoretical framework of this paper is engineering applied mathematics, and for instance, we shall manipulate Brownian motion without explicitly referingreferring to probability spaces and filtration.

2. Background on Modified Riemann-Liouville Derivative

2.1. Fractional Derivative via Fractional Difference

Definition 2.1 Let $f:\Re\rightarrow\Re,\, x\rightarrow f(x)$, denote a continuous (but not necessarily differentiable) function, and let $h>0$ denote a constant discretization span. Define the forward operator $F_w(h)$ by the equality (the symbol := means that the left side is defined by the right one)

$$F_w(h)f(x) \;:=\; f(x+h); \tag{2.1}$$

then the fractional difference of order α, $0<\alpha<1$, of *f(x)* is defined by the expression

$$\begin{aligned}\Delta^{\alpha} f(x) \;&:=\; (F_w-1)^{\alpha} f(x)\\ &=\sum_{k=0}^{\infty}(-1)^k\binom{\alpha}{k}f[x+(\alpha-k)h],\end{aligned} \tag{2.2}$$

and its fractional derivative is defined as the limit [21,22,23]

$$f^{(\alpha)}(x) \;:=\; \lim_{h\downarrow 0}\frac{\Delta^{\alpha}[f(x)-f(0)]}{h^{\alpha}}. \;\blacksquare \tag{2.3}$$

Remark This definition is slightly different from the Grunwald-Letnikov's one, and which reads

$$f^{(\alpha)}(x) = \lim_{h \downarrow 0} h^{-\alpha} \sum_{k=0}^{(x-\alpha)/h} (-1)^k \binom{\alpha}{k} f(x-kh) \cdot$$

In addition, this equation (2.2) is not new and can be found in the literature, see for instance [36,37]. *Lemma 2.1* With the definition (2.1) above, the fractional derivative of order α, $0<\alpha<1$ of a constant K is zero. ■ This definition is close to the standard definition of derivative (calculus for beginners), and as a direct result, the α-th derivative of a constant, $0<\alpha<1$, is zero..

2.2. Modified Riemann-Liouville Derivative

Proposition 2.1 (Riemann-Liouville definition revisited). Assume that $f(.), x \to f(x)$, is a $\Re^+ \to \Re$ function. According to the definition 2.1, its fractional derivative of order α, $\alpha<0$, can be defined by the expression [21,22,23]

$$f^{(\alpha)}(x) := \frac{1}{\Gamma(-\alpha)} \int_0^x (x-\xi)^{-\alpha-1} f(\xi) d\xi, \quad \alpha<0. \tag{2.4}$$

For positive α one will set

$$\begin{aligned} f^{(\alpha)}(x) &:= \left(f^{(\alpha-1)}(x)\right)', \quad 0<\alpha<1, \\ &= \frac{1}{\Gamma(1-\alpha)} \frac{d}{dx} \int_0^x (x-\xi)^{-\alpha} \left(f(\xi)-f(0)\right) d\xi \end{aligned} \tag{2.5}$$

and

$$\begin{aligned} f^{(\alpha)}(x) &:= \left(f^{(\alpha-n)}(x)\right)^{(n)}, \quad n \le \alpha < n+1, \quad n \ge 1 \\ &= \frac{1}{\Gamma(1-(\alpha-n))} \frac{d^{n+1}}{dx^{n+1}} \int_0^x (x-\xi)^{-(\alpha-n)} \left(f(\xi)-f(0)\right) d\xi \,. \blacksquare \end{aligned} \tag{2.6}$$

Proof. A way to obtain this result is to show that the Laplace's transforms of (2.3) and (2.5) are the same, and to this end one notices that the Laplace transform of (2.2) is

$$\begin{aligned} L\{\Delta^\alpha (f(x)-f(0))\} &= \sum_{k=0}^{\infty} (-1)^k \binom{\alpha}{k} e^{-(k-\alpha)hs} L\{f(x)-f(0)\}. \\ &= (1-e^{-hs})^\alpha L\{f(x)-f(0)\} \\ &= h^\alpha s^\alpha \left(L\{f(x)\} - s^{-1} f(0)\right), \end{aligned}$$

therefore, on letting $h \downarrow 0$,

$$L\{f^{(\alpha)}(x)\} = s^{\alpha} L\{f(x)\} - s^{\alpha-1} f(0). \tag{2.7}$$

Further remarks and comments

(i) Notice the difference between (2.4) and (2.5). The second equation involves the constant $f(0)$ whilst the first one does not. We shall refer to this new fractional derivative so defined as to the *modified Riemann Liouville derivative* [21,22,23].

(ii) Remark that our basic definition is the definition (2.1), equation (2.3), and that the triple (2.4-6) is provided only as a useful alternative which will be helpful for practical calculus. The second definition refers to $\Re^{+} \rightarrow \Re$ functions, mainly because the equivalence between these two definitions is obtained by using the Laplace's transform.

(iii) The Liouville-Djrbashian-Caputo definition of fractional derivative [5,7] is the expression

$$f^{(\alpha)}(x) := \frac{1}{\Gamma(1-\alpha)} \int_0^x (x-\xi)^{-\alpha} f'(\xi) d\xi, \quad 0 < \alpha < 1,$$

of which the Laplace's transform is also (2.7). It follows that if $f(x)$ is differentiable, then the two definitions are quite equivalent, otherwise, when $f(x)$ is not differentiable, it may happen that the two definitions yield different results.

(iv) By using the modified definition, one can so define the fractional derivative of a function even when it is not differentiable, as it is the case with the white noise, for instance, and this is of paramount importance for us here.

(v) It is of order to notice that this definition is different from Kolwankar and Gangal's [26,27] despite it results in a similar framework.

As an alternative to the expression (2.6), we could have selected to set

$$f^{(\alpha)}(x) = D^{\alpha-n}\left(D^{n} f(x)\right), \quad n < \alpha < n+1$$

$$= \frac{1}{\Gamma(1-(\alpha-n))} \frac{d}{dx} \int_0^x (x-\xi)^{-(\alpha-n)} \left(f^{(n)}(\xi) - f^{(n)}(0)\right) d\xi,$$

but we shall delete this approach, in order to comply with the principle of increasing order of derivation which states that if we have to calculate D^{a+b} by using D^{a} and D^{b} serially in any order, then we shall set

$$D^{a+b} := D^{\max(a,b)} D^{\min(a,b)}.$$

Notice also that generally one has $D^{a} D^{b} \neq D^{b} D^{a}$

There is a large literature on fractional calculus, and for further details one can see for instance [2-5,8,9,11,20-28,33,35-40,43].

2.3. Taylor's Series of Fractional Order

A generalized Taylor expansion of fractional order which applies to non-differentiable functions (F-Taylor series in the following) reads as follows [22,23]. *Proposition 2.1* Assume that the continuous function $f:\Re \to \Re, x \to f(x)$ has fractional derivative of order $k\alpha$, for any positive integer *k* and α, $0<\alpha\leq 1$, then the following equality holds, which is

$$f(x+h) = \sum_{k=0}^{\infty} \frac{f^{(\alpha k)}(x)}{(\alpha k)!} h^{\alpha k}, \quad 0<\alpha\leq 1, \tag{2.8}$$

where, and this is of paramount importance, $f^{(\alpha k)}(x) := D^{\alpha} D^{\alpha} .....D^{\alpha} f(x)$ is the derivative of order αk of *f(x)*, and with the notation

$$\Gamma(1+\alpha k) =: (\alpha k)!,$$

where $\Gamma(.)$ denotes the Euler's gamma function.■

Indication on the proof. If $F_w(h)$ denotes the forward operator, clearly

$$F_w(h) f(x) := f(x+h),$$

then one can show that it satisfies the fractional differential equation

$$D_h^{\alpha} F_w(h) = F_w(h) D_x^{\alpha},$$

D_x is the derivative operator with respect to x, of which the solution is

$$F_w(h) = E_{\alpha}(h^{\alpha} D_x^{\alpha}),$$

where $E_{\alpha}(u)$ is the Mittag-Leffler function defined by the series.

$$E_{\alpha}(u) = \sum_{k=0}^{\infty} \frac{u^k}{(\alpha k)!}.$$

The first two terms of this series (say the fractional Rolles formula) had been obtained by Kolwankar and Gangal who work with the Cantors set [26,27].

This series is useful only when $f(x)$ is non-differentiable at x, for instance, it applies to the Mittag-Leffler function $E_{\alpha}(x^{\alpha})$ at $x=0$, but it does not hold when $x \neq 0$. Moreover it is different from other fractional Taylor's series previously obtained in the literature [38].

A heuristics is as follows. One first notices that this fractional Taylor's series holds for the Mittag-Leffler function, and then, it is sufficient to consider the special class of those functions which can be uniformly approximated by a sequence of Mittag-Leffler functions.

2.4. Fractional Taylor's Series for Multivariable Functions

Assume that $f(x,y)$ is non-differentiable with respect to x; then we can write the series

$$f(x+h,y) = \sum_{k=0}^{\infty} \frac{h^{k\alpha}}{(k\alpha)!} D_x^{k\alpha} f(x,y),$$

and then, on assuming now that the fractional derivatives with respect to x are themselves non-differentiable w.r.t. y, the fractional Taylor series w.r.t. y yields

$$f(x+h,y+l) = \sum_{k=0}^{\infty} \frac{h^{k\alpha}}{(k\alpha)!} \sum_{k=0}^{\infty} \frac{l^{r\alpha}}{(r\alpha)!} D_y^{r\alpha} D_x^{k\alpha} f(x,y)$$

$$= \sum_{k=0}^{\infty}\sum_{r=0}^{\infty} \frac{h^{k\alpha} l^{r\alpha}}{(k\alpha)!(r\alpha)!} D_y^{r\alpha} D_x^{k\alpha} f(x,y). \tag{2.9}$$

3. Basic Formulae for Fractional Derivative and Integral

3.1. Fractional Derivative of Compounded Functions

The equation (2.8) provides the useful differential relation

$$d^{\alpha} f \cong \Gamma(1+\alpha) df, \quad 0<\alpha<1, \tag{3.1}$$

or in terms of fractional difference, $\Delta^{\alpha} f \cong \alpha! \Delta f$, *which holds for non-differentiable functions only.*

Corollary 3.1 The following equalities hold, which are

$$D^{\alpha} x^{\gamma} = \Gamma(\gamma+1)\Gamma^{-1}(\gamma+1-\alpha) x^{\gamma-\alpha}, \quad \gamma>0, \tag{3.2}$$

or, what amounts to the same (we set $\alpha = n+\theta$)

$$D^{n+\theta} x^{\gamma} = \Gamma(\gamma+1)\Gamma^{-1}(\gamma+1-n-\theta) x^{\gamma-n-\theta}, \quad 0<\theta<1,$$

$$\left(u(x)v(x)\right)^{(\alpha)} = u^{(\alpha)}(x)v(x) + u(x)v^{(\alpha)}(x), \tag{3.3}$$

$$\left(f[u(x)]\right)^{(\alpha)} = f_u^{(\alpha)}(u)(u'_x)^{\alpha}. \tag{3.4}$$

$u(x)$ is non-differentiable in (3.3) and (3.4) and differentiable in (3.5), $v(x)$ is non-differentiable, and $f(u)$ is differentiable in (3.4) and non-differentiable in (3.5). ■

Indication on the proof. The proof of these results is based on the equality (3.1) which holds for non-differentiable functions only. For instance, for (3.3), one has successively

$$d(uv) = udv + vdu,$$

therefore, on assuming that both $u(x)$ and $v(x)$ are not differentiable,

$$\alpha! d(uv) = u(\alpha! dv) + v(\alpha! du),$$

that is to say

$$d^{\alpha}(uv) = ud^{\alpha}v + vd^{\alpha}u,$$

therefore the result.

The formula (3.4) is a direct consequence of the equality

$$\frac{d^{\alpha}f(u)}{dx^{\alpha}} = \frac{d^{\alpha}f}{du^{\alpha}}\left(\frac{du}{dx}\right)^{\alpha}.$$

Again regarding (3.4), let us point out that an incorrect argument would be to write

$$\frac{d^{\alpha}f(u)}{dx^{\alpha}} = \frac{d^{\alpha}f(u)}{du}\frac{du}{dx^{\alpha}} = \frac{df(u)}{du}\frac{\alpha! du}{dx^{\alpha}} = \frac{df(u)}{du}\frac{d^{\alpha}u}{dx^{\alpha}}$$

but when we use the equality $d^{\alpha}f = \alpha! df$, we implicitly assume that $f(u)$ is non-differentiable with respect to u, in such a manner that the derivative df/du turns to be meaningless.

According to the αth-derivative $D^{\alpha}u$ of u with respect to u itself, one has the equality

$$\frac{d^{\alpha}u}{du^{\alpha}} = \frac{1!}{(1-\alpha)!}u^{1-\alpha}, \tag{3.5}$$

therefore the relation

$$(du)^{\alpha} = (1-\alpha)! u^{\alpha-1} d^{\alpha}u, \tag{3.6}$$

which provides (on dividing both sides by $(dx)^{\alpha}$)

$$(u'_x(x))^{\alpha} = (1-\alpha)! u^{\alpha-1} u_x^{(\alpha)}(x). \tag{3.7}$$

This equality is to be understood as follows. If $u'(x)$ exists then $u_x^{(\alpha)}(x)$ also exists and both are related by (3.7). But $u_x^{(\alpha)}(x)$ may exist, whilst $u'(x)$ is not defined, in which case (3.7) fails to apply. This is the case for instance for a self-similar function with a Hurst parameter lower than the unit.

Remark that the equality (3.6) is always satisfied whichever the assumptions on the derivatives, and it is only when we divide by $(dx)^{\alpha}$ that we begin to get trouble with the significance of (3.7).

Corollary 3.2. Assume that $f(u)$ is differentiable with respect to u, and that $u(x)$ also is differentiable, then one has the equality

$$\left(f[u(x)]\right)^{(\alpha)} = \frac{1}{(1-\alpha)!}\left(f(u)\right)^{1-\alpha}\left(\frac{df}{du}\right)^{\alpha}\left(\frac{du}{dx}\right)^{\alpha} . \blacksquare \quad (3.8)$$

Proof. It is a direct combination of equations (3.4) to (3.7).

Corollary 3.3 Assume that both $f(u)$ and $u(x)$ have fractional αth -derivative, then one has the equality

$$\left(f[u(x)]\right)^{(\alpha)} = (1-\alpha)!u^{\alpha-1}\left(f(u)\right)^{1-\alpha} f_u^{(\alpha)}(u)u_x^{(\alpha)}(x) . \blacksquare \quad (3.9)$$

Proof. One has successively

$$\frac{d^\alpha f(u(x))}{dx^\alpha} = \frac{d^\alpha f}{d^\alpha u}\frac{d^\alpha u}{dx^\alpha} = \frac{d^\alpha f}{du^\alpha}\frac{du^\alpha}{d^\alpha u}\frac{d^\alpha u}{dx^\alpha},$$

and then using (3.6) to calculate $d^\alpha u / du^\alpha$ yields the result.

3.2. Integration with Respect to (dx)$^\alpha$

The integral with respect to $(dx)^\alpha$ is defined as the solution of the fractional differential equation

$$dy = f(x)(dx)^\alpha, \quad x \geq 0, \quad y(0) = 0, \quad 0 < \alpha < 1, \quad (3.10)$$

which is provided by the following result:

Lemma 3.1 Let *f(x)* denote a continuous function, then the solution of the equation (3.9) is defined by the equality [21,22,23]

$$y = \int_0^x f(\xi)(d\xi)^\alpha = \alpha \int_0^x (x-\xi)^{\alpha-1} f(\xi)d\xi \quad , \quad 0 < \alpha \leq 1,$$

and more generally, one has the equality

$$\int_a^x f(\xi)(d\xi)^\alpha = \alpha \int_a^x (x-\xi)^{\alpha-1} f(\xi - a)d\xi \quad , \quad 0 < \alpha \leq 1 . \blacksquare \quad (3.11)$$

The *fractional integration by part* formula

$$\int_a^b u^{(\alpha)}(x)v(x)(dx)^\alpha = \alpha![u(x)v(x)]_a^b - \int_a^b u(x)v^{(\alpha)}(x)(dx)^\alpha \quad (3.12)$$

can be obtained easily by using (3.3).

Change of variable. Consider the variable transformation $y = g(x)$ in which $g(x)$ is a non-decreasing differentiable function then one has the equality

$$\int f(y)(dy)^{\alpha} = \int f(g(x))(g'(x))^{\alpha}(dx)^{\alpha}, \quad 0<\alpha<1,$$

and when $g(x)$ has a positive fractional derivative of order $\beta, 0<\alpha,\beta<1$, one has

$$\int f(y)(dy)^{\alpha} = \Gamma^{-\alpha}(1+\beta)\int f(g(x))(g^{(\beta)}(x))^{\alpha}(dx)^{\alpha\beta}, \quad 0<\alpha,\beta<1..$$ Some examples

(i) On making $f(x)=x^{\gamma}$ in (3.11) one obtains

$$\int_0^x \xi^{\gamma}(d\xi)^{\alpha} = \frac{\Gamma(\alpha+1)\Gamma(\gamma+1)}{\Gamma(\alpha+\gamma+1)}x^{\alpha+\gamma}, \quad 0<\alpha\le 1 \tag{3.13}$$

(ii) Assume now that *f(x)* is the Dirac delta generalized function $\delta(x)$, then one has

$$\int_0^x \delta(\xi)(d\xi)^{\alpha} = \alpha x^{\alpha-1}, \quad 0<\alpha\le 1. \qquad (3.14).$$

Application to the fractional derivative of the Dirac delta function

On using the equation (3.3) on the one hand, and extending the definition of $\delta'(x)$ on the other hand, we shall define the fractional derivative of the Dirac delta function by the equality

$$\int \delta^{(\alpha)}(\xi)f(\xi)(d\xi)^{\alpha} = -\int \delta(\xi)f^{(\alpha)}(\xi)(d\xi)^{\alpha}, \quad 0<\alpha\le 1 \tag{3.15}$$

and the equation (3.11), direct yields

$$\int \delta^{(\alpha)}(\xi)f(\xi)(d\xi)^{\alpha} = -\alpha x^{\alpha-1}f^{(\alpha)}(0), \quad 0<\alpha\le 1 \tag{3.16}$$

3.3. Fractional Derivative of Integrals with Respect to (Dx)^A

The relation between fractional integral and fractional derivative reads

$$\frac{d^{\alpha}}{dx^{\alpha}}\int_0^x f(\xi)(d\xi)^{\alpha} = \Gamma(1+\alpha)f(x) = \alpha!\,f(x). \tag{3.17}$$

$$\frac{d^{\alpha}}{dx^{\alpha}}\int_0^{u(x)} f(\xi)(d\xi)^{\alpha} = \alpha!\,f(u(x))(u'(x))^{\alpha}. \tag{3.18}$$

The proof of (3.17) results from the combination of the equalities $y^{(\alpha)}(x)=f(x)$ and $d^{\alpha}y=\alpha!dy$ which yields the useful formula

$$y = \frac{1}{\alpha!}\int_0^x y^{(\alpha)}(\xi)(d\xi)^\alpha . \tag{3.19}$$

Our task now is to exhibit practical significance for this fractional calculus in systems modeling, and to this end, we shall begin with velocity in coarse-grained space.

4. Fractional Derivative and Velocity in Coarse-Grained Space

4.1. The Basic Modeling Assumption

For fixing the thought, let us assume that we are dealing with a mechanical point with the mass m moving in a one-dimensional coarse-grained space defined by the space co-ordinate $x(t)$, where t denotes the time. On a modeling standpoint, we shall assume that in a coarse-grained space, the generic point is not infinitely thin but rather has a thickness. So, if dx and $(dx)_c$ refer to the sizes of the thin points in the standard space and in the coarse-grained space respectively, then we should have $dx < (dx)_c$.

Modeling coarse-graining phenomenon in space. In order to describe the coarse-graining phenomenon on an analytical point of view, we shall assume that the differential increment in such a space is not dx, but rather is $(dx)^\alpha$, where α, $0 < \alpha < 1$, is a real-valued parameter which characterizes the grade of the phenomenon.■ By this way, for small positive dx, we shall have $(dx)^\alpha > dx$. For negative dx, this modeling could be somewhat troublesome as far as $(dx)^\alpha$ could be complex-valued depending upon the value of α, and one way to circumvent this pitfall is to restrict ourselves to the values $\alpha = m/n$, $m = 2k$, $n = 2q+1$, where k and q are positive integers. We so have a large definition for α which encompasses a large class of real systems. At first glance, we could have to select a modeling among $(dx)^\alpha$ and $\alpha(dx)$ for the coarse-graining phenomenon, but an additional underlying condition is that $(dx)_c / dx$ should increase indefinitely as dx tends to zero, therefore the model via $(dx)^\alpha$. Or again $x(t)$ so obtained by this way would be a differentiable function, whilst according to the meaning of the quotient $(dx)_c / dt$, it should not.

Coarse-grained space and fractal space. It appears that α is exactly the fractal dimension of the space in which x is running. Indeed, in the plane, let us consider a curve line of length L with a covering defined with disjoint balls of diameter ε. Let $N(\varepsilon)$ denote the number of balls so involved. The fractal (Hausdorf) dimension f of the curve is defined as the limit

$$f := \lim_{\varepsilon \downarrow 0} -\left(\ln N(\varepsilon)/\ln \varepsilon\right). \tag{4.1}$$

Usually one has the equality $N = L/\varepsilon$ which provides the dimension 1, but in the case of a fractal curve, one will have $N = L/\varepsilon^\alpha$ therefore

$$f \ := \ \lim_{\varepsilon \downarrow 0} \ -\left(\ln \varepsilon^{\alpha} / \ln \varepsilon\right). \tag{4.2}$$

*Definition 4.*1 In this framework, the velocity on the right $\dot{x}_{\alpha}$ (t) of a particle will be defined by the expression

$$\dot{x}_{\alpha}(t) \ := \ \frac{(dx)^{\alpha}}{dt}, \quad dx > 0, \quad 0 < \alpha < 1. \tag{4.3}$$

In the special case when $\alpha = m/n$, $m = 2k$, $n = 2q+1$, k and q integers, this definition holds for positive and negative dx, and then defines the fractional velocity on the left and on the right.■

Coarse-graining in both space and time In the reference [18], we suggested to consider the velocity

$$v_{\beta}(t) \ := \ x^{(\beta)}(t) \ = \ \frac{d^{\beta}x}{dt^{\beta}} \ = \ \beta! \frac{dx}{(dt)^{\beta}}, \tag{4.4}$$

again with the condition $0 < \beta < 1$. Here one has $(dt)^{\beta} > dt$, in other words in this modeling assume that it is time which involves the coarse-grained phenomenon.

When we deal with a dynamics in which both space and time are coarse-grained, then in quite a natural way, we are led to introduce the velocity

$$\dot{x}_{\alpha,\beta}(t) \ := \ (dx)^{\alpha} / (dt)^{\beta}, \quad 0 < \alpha < \beta < 1 \tag{4.5}$$

$$= \ \left((dx)^{\alpha/\beta} / dt\right)^{\beta}$$
$$= \ \left(\dot{x}_{\alpha/\beta}(t)\right)^{\beta}. \tag{4.6}$$

In the sequel, we shall show how the velocity in coarse-grained space can be expressed in terms of fractional derivative with respect to time.

4.2. Mathematical Expressions of the Fractal Velocity

Preliminary result

Lemma 4.1 Given a function $y = f(x)$ and its inverse (function) $x = g(y)$, their fractional derivatives of order α, $0 < \alpha < 1$ satisfy he conditions

$$y^{(\alpha)}(x)x^{(\alpha)}(y) \ = \ ((1-\alpha)!)^{-2}(xy)^{1-\alpha}. \blacksquare \tag{4.7}$$

Proof. One has the equality

$$y^{(\alpha)}(x)x^{(\alpha)}(y) = \left(\frac{d^{\alpha}y}{dx^{\alpha}}\right)\left(\frac{d^{\alpha}x}{dy^{\alpha}}\right) = \left(\frac{d^{\alpha}y}{dy^{\alpha}}\right)\left(\frac{d^{\alpha}x}{dx^{\alpha}}\right),$$

and we take account of (3.2) which provides the relation (on setting $\gamma = 1$)

$$(1-\alpha)!d^{\alpha}x = x^{1-\alpha}(dx)^{\alpha},$$

for both x and y, to obtain the result.■

Application to the velocity $\dot{x}_{\alpha}(t)$

Assume that $x(t)$ is a function of time t, and that the latter is not continuously evolving, but is rather discontinuous, in such a manner that one can write $d^{\alpha}t = \alpha!dt$. In these conditions, the equation (4.3) yields the expression

$$\dot{x}_{\alpha}(t) = \frac{\alpha!(dx)^{\alpha}}{\alpha!dt} = \alpha!\frac{(dx)^{\alpha}}{d^{\alpha}t} = \alpha!\left(t^{(\alpha)}(x)\right)^{-1}, \tag{4.8}$$

which defines $\dot{x}_{\alpha}(t)$ in terms of the fractional derivative of time with respect to space. And substituting (4.7) into (4.8) yields the sought result

$$\begin{aligned} \dot{x}_{\alpha}(t) &= \alpha!\left((1-\alpha)!\right)^{2}(xt)^{\alpha-1}x^{(\alpha)}(t) \\ &=: \rho(\alpha)(xt)^{\alpha-1}x^{(\alpha)}(t). \end{aligned} \tag{4.9}$$

In the following, we shall show how modelling subjectivity by using observation with informational invariance, introduces the Hurst exponent in quite a natural way.

5. Introduction to Subjectivity and Self-Similarity in Observation

Observation with informational invariance

In an approach to define a quantitative theory of (human) subjectivity, we considered the problem in the setting of Shannon information theory [41], and we assumed that we do not observe one variable only, but rather a pair of variables in such a manner that the amount of information they so involved remains constant; what we referred to as the principle of observation with informational invariance. In the more general framework, this basic pair of observed variables is drawn from linguistics, and it is basically the pair (syntax, semantics) or in a like manner (symbol, meaning) of natural languages. By using very simple and straightforward arguments, we have shown that observed variables and real variables are related by the so-called Lorentz transformation of the special relativity. Various applications have been proposed and mainly we derived a concept of observed probability in the form $p_{obs} = ap^{b}$, where p is a probability and a, b denote two constants [16,17]. For further reading on information and meaning of information, see for instance [16,17].

Zooming and self-similarity

Recently, more exactly after Mandelbrot, the notion of self-similarity and of self-similar function, which was firstly a mathematical curiosity began pervasive in natural science. What is the matter? Loosely it is as follows. Consider a function or more generally a pattern $x(t)$ which is indexed by the time t, and assume that we observe it at the time at, $a > 0$. Then we shall say that $x(t)$ is a self-similar function when the pattern $x(at)$ can be obtained from $x(t)$ by a simple zooming effect, that is to say when we may write $x(at) = C(a)x(t)$, $t \in \Re$ The fact that the constant C depends upon a merely pictures the property that the magnitude of the zooming should depend upon the value of a. We can go a step farther. On taking the derivative of the equation $x(at) = C(a)x(t)$ with respect to a, we obtain the equality

$$\frac{dx(at)}{d(at)} = \frac{C(a)}{a}\dot{x}(t) = \frac{C'(a)}{t}x(t),$$

which provides

$$t\dot{x}(t)x^{-1}(t) = \lambda,$$

therefore

$$x(t) = Kt^{\lambda},$$

where K and λ denote two constants.

Shortly, the zooming effect in pattern modeling refers directly to the function t^{λ}.

Can observation with informational invariance create self-similarity?

Assume that we are observing the magnitude of the increment dx, $dx > 0$, with some subjectivity. On a practical standpoint, we shall say that the dx so observed is more or less infinitely small. Everything happens as if we had the equality

$$observed\ \ dx = (dx)^{\alpha}$$

where α denotes a real-valued parameter which describes the (grade of) subjectivity of the observation. We may have $(dx)^{\alpha} < dx$ when $\alpha > 1$, that is to say the observed dx is smaller than the actual dx, or on the contrary $(dx)^{\alpha} > dx$ when $\alpha < 1$ which pictures the fact that the observed dx is larger than the actual dx. But with $(dx)^{\alpha}$, we are in the depth of fractional calculus and self-similarity! And in quite a natural way this gives rise to the question as to whether observation with informational invariance could create fractal?

6. Learning Observation with Informational Invariance

6.1. Observation in the Presence of Syntax and Semantics

General setting of the modelling

The main goal of the following model is to provide an approach to introducing subjectivity in observation processes, and to this end, we shall proceed as follows: statement of the problem. Let (R) denote an observer (human being or physical device), who is observing a source of information (S). We assume that the observable (S) is defined in a pair of spaces Ω and Ω', where Ω denotes a space of symbols, that is to say the syntactic space, whilst Ω' is the space of meanings referred to as the semantic space. Let α and a denote the generic elements of Ω and Ω' respectively. We shall refer to the pair (α, a) as a lexem, that is to say a symbol taken with a given meaning. On another standpoint, a can be thought of as representing an object of our real world, while α is its image in the considered space of representation.

Coupling effects in observation processes

We assume that (R) is observing an object defined by the pair (α, a). (R) is unable to observe α and a separately (or one at a time), but rather he observes them simultaneously, and his purpose of course is to determine their exact respective values. In this framework, (R) uses α to identify a, and conversely he uses a to determine a, and as a result we so creates coupling effects between α and a. In addition, we shall assume that there is a complete symmetry between α and a, on all points of view, that is to say theoretical or applied.

Definition of the mathematical background

We shall restrict ourselves to the case when both Ω and Ω' reduce to $\Re$. Clearly both α and a are continuous real-valued variables, and we assume that it is possible to define continuous uncertainties in the information theoretic sense of this term, or likewise continuous informational Shannon entropies for α and a respectively.

6.2. Equations of Learning Observation Processes

We consider an observation which is characterized by the following properties.

(Axiom A1) The main purpose of the observer *(R)* is to define the exact values of α and a respectively, but the observation results into two observed values α' and a' which are linear combinations of α and a, in the form

$$\alpha' = g_{11}\alpha + g_{12}a, \tag{6.1.a}$$

$$a' = g_{21}\alpha + g_{22}a. \tag{6.1.b}$$

(Axiom A2) Principle of observation with informational invariance. The observation process works in such a manner that it neither creates information nor destroys the amount of information involved in the pair (α, a)

(Axiom A3) The variables α and a are co-varying in the sense that both of them increase or both of them decrease.

(Axiom A4) It is assumed that the amount of information involved in the pair (α, a) is defined by the Shannon entropy [41] of continuous random variables.

6.3. Derivation of the Equations of the Learning Observation

One has the following

Proposition 6.1 Assume that the pair $(\alpha, a) \in \Re \times \Re$ of variables is observed by an observer (R) according to an observation process with informational invariance as defined in sub-section 6.2; and let (α', α') denote the pair so measured by the observer; then there exists $\omega \in \Re$ such that the following equations hold:

$$\alpha' = \alpha \cosh \omega + a \sinh \omega , \tag{6.2.a}$$

$$a' = \alpha \sinh \omega + a \cosh \omega . \blacksquare \tag{6.2.b}$$

Indication on the proof.

(Step 1) Preliminary background. We first bear in mind that if $X \in \Re$ and $Y \in \Re$ denote two random variables with the joint probability density $p(x, y)$; then the Shannon informational entropy $H(X,Y)$ of the pair is defined by the expression [41].

$$H(X,Y) := -\int_{\Re} p(x,y) \ln p(x,y) dxdy$$

An interesting property of this entropy is the following one. If one makes the transformation

$$(X',Y')^T = A(X,Y)^T ,$$

where A is a 2×2 matrix

$$A := \begin{pmatrix} g_{11} & g_{12} \\ g_{21} & g_{22} \end{pmatrix},$$

with the determinant $|A|$, and the superscript T denotes the transpose, then one has the equality

$$H(X',Y') = H(X,Y) + \ln|A| .$$

(Step 2) According to the informational invariance, the above transformation (6.2) should be a unitary transformation, and then one should have the condition

$$|g_{11}g_{22} - g_{12}g_{21}| = 1 ,$$

which provides the quartet

$$g_{11} = \cosh\omega, \quad g_{12} = \sinh\omega, \quad g_{21} = \sinh\omega, \quad g_{22} = \cosh\omega, \tag{6.3}$$

or the quartet

$$g_{11} = \cos\theta, \quad g_{12} = \sin\theta, \quad g_{21} = -\sin\theta, \quad g_{22} = \cos\theta, \tag{6.4}$$

where ω and θ denote two real-valued parameters. According to the third axiom, we shall select the first transformation (6.3). This being the case as it is well known, on setting

$$u \; := \; \sinh\omega / \cosh\omega,$$

and

$$\rho(u) \; := \; (1-u^2)^{-1/2}, \tag{6.5}$$

one has as well the equations

$$\alpha' \; = \; \rho(u)(\alpha + ua), \tag{6.6.a}$$

$$a' \; = \; \rho(u)(u\alpha + a), \tag{6.6.b}$$

which are the well known Lorentz-Poincaré's equations.

For further details see [16,17].

Remarks of importance

(First remark) The point of importance is that, by using very simple and straightforward information theoretic arguments only, we once more come across the Lorentz-Poincaré's equations, irrespective to any physical framework.

(Second remark) Another very direct way to derive the equation of this model is to select the observed y of an observable x in the form $y = Gx$, where G is a scalar-valued gain coefficient which can be written in the form $G = \exp(\omega)$ so that we have

$$\begin{aligned} y \; &= \; xe^{\omega} \\ &= \; x\cosh\omega + x\sinh\omega, \end{aligned}$$

in such a manner that the coefficients $\cosh\omega$ and $\sinh\omega$ so appear in quite a natural way.

Remark that one cannot claim that it is a model selected for convenience only. Indeed, one can write y either in the form $y = x + \varepsilon$ where ε denotes an error term, or in the form $y = Gx$. One is used to refer to the first model, but here we rather work with the second one. The term of learning process has been coined to picture the fact that in this case x and y vary in the same way, whilst in the dislearning process, they vary in opposite ways. Sometimes also we use the terms of Minkowskian (learning) and Euclidean (dislearning) observation processes.

7. Dislearning Observation with Informational Invariance

7.1. Derivation of the Transformation Equations

We now assume that the observation process is defined by the following axioms.

(Axiom B1) Identical to *(Axiom A1)* in subsection (6.2)

(Axiom B2) Identical to *(Axiom A2)* in subsection (6.2)

(Axiom B3) The variables α and a vary in opposite way: one of them increases when the other one decreases

(Axiom B4) Identical to *(Axiom A4)* in subsection (6.2)

The Minkowskian observation and the Euclidean observation are different only by the axioms *A3* and *B3*. *A3* can be thought of as defining a learning process, whilst on the contrary *B3* refers to a dislearning one.

Proposition 7.1 Assume that the observation of the pair (α, a) is performed via an observation process which satisfies the axioms B1 to B4. Then there exists a real-valued parameter θ which provides the observation result (α', a') in the form

$$\alpha' = \alpha \cos\theta + a \sin\theta, \tag{7.1.a}$$

$$a' = -\alpha \sin\theta + a \cos\theta . \blacksquare \tag{7.1.b}$$

Indication on the proof. We refer to the proof of the proposition 6.1 above. According to (B3), we shall select the quartet (6.4).

Here, we shall set

$$v := tg\theta, \qquad \eta(v) = (1 + v^2)^{-1/2},$$

to write

$$\alpha' = \eta(v)(\alpha + va), \tag{7.2.a}$$

$$a' = \eta(u)(-v\alpha + a). \tag{7.2.b}$$

For more details see [16,17]

7.2. On the Selection of the Suitable Mode of Observation

Selecting the type of observation for a given problem is a matter of modelling, and depends heavily upon the axioms A3 and B3. When modeling the system, we shall have to clarify whether the two variables vary in the same way or in opposite ways.

Assume that we are observing an observable x. The basic popular observation model is to assume that the result of the observation provides a measure y in the form

$$y = x + e = x(1 + ex^{-1}), \qquad (7.3)$$

where e in most cases, is considered as an error term. But there is another model, quite significant as well, and which consists of writing the equality

$$y = Gx = x + (G-1)x, \qquad (7.4)$$

in which G denotes a gain coefficient, and of course, these two models are equivalent as far as we assume that e and G may be x-dependent.

As it is evident, if we assume that there is a coupling effect between x and y, then the learning model holds for $G > 0$, in which case x and y vary in the same way, whilst the dislearning model applies whenever $G < 0$, that is to say when x and y vary in opposite ways.

Regarding the equation (7.4), if y is the estimate of x, it is likely that they are co-varying, in which case the learning model is quite relevant, for the pair (x, y).

Let us assume now that it is the pair (x, e) which is observed. Here again we are entitled to expect that the larger x is, the larger e is, in such a manner that the learning observation would still hold. But one may have the opposite, that is to say e may decrease when x increases. This might happen for instance when the measurement device is more accurate for large magnitudes of the observable than for small ones.

8. Observation of Information via Informational Invariance.

8.1. Definition of the Local Observation Process Parameters

We consider an observable system *(S)* defined by the state $x \in \Re$ at the physical universal time *t*, say *x(t)*, and we restrict ourselves to the case $x > 0$. This system is locally observed by an external observer *(R)* as follows.

Definition 8.1. A local observation of the system (S) in the position (x_0, t_0) is an observation which is restricted to the vicinity of (x_0, t_0), irrespective of any trajectory $x(t)$ which may drive (x, t).■

The practical need of this definition can be explained as follows. Assume that *(R)* is observing the position of a physical particle with the velocity v, then we know that $x = x_0 + v(t - t_0)$.Now, assume that the observation process starts at (x, t) and that v is unknown to *(R)*. Then the latter will try to estimate $x(t+dt)$ by using an observation process which is assumed to involve informational invariance. After a long duration of observation, *(R)* will guess the equation $x = x_0 + v(t - t_0)$ and the relative observation will have no more meaning.

As a consequence of Definition 8.1, and this is of paramount importance, if *(R)* is observing a process defined by the equation $dx = f(x)dt$, the observation model will apply to the pair (dx, dt) only. This being the case, we make the following assumptions.

(Axiom H1) We assume that *(S)* is driven by its proper internal time τ which may be identical with or different from the universal time, and that it is only x and τ which are significant to a human external observer *(R)*. Clearly (x, τ) is substituted for (x, t) in the observation process. In other word it is (x, τ) which is observed by *(R)* instead of (x, t).

(Axiom H2) We simplify the model in assuming that $\tau = rt$, what amounts to say that the observed observable is now defined by (x, r) (instead of (x, τ)).

(Axiom H3) As far as subjectivity is involved in the observation process of (x, r) by *(R)*, we shall assume that it is only the significances of x and r which are of interest to *(R)*, and we shall assume that they are respectively defined by the quantities $\ln x$ and $\ln r$.

Further remarks and comments on these assumptions.

It is by-now taken for granted that any system is driven by its own internal proper time. In biology for instance this has been claimed a long time ago, and moreover very often one even states that biological time, for instance, is a random time. Consider now the perception of time by a human being. It is well known that the term "recently" has not the same meaning for a young people or for an elder. For a young boy this means "a few days ago" whilst for an elder its significance may be "one year ago". We are used to say that time goes slowly with elders. The proper time of an elder is $\tau = rt$ with $r < 1$.

The last assumption claims that we do not take interest in the value, the magnitude of $x(t)$ itself, but rather in the amount of information this value may have to the observer. And to arrive at $\ln x$, we used the following result. If X is a random variable with the entropy $H(X)$ and C denotes a real-valued constant, one then has the basic relation

$$H(CX) = H(X) + \ln|C|,$$

therefore we conclude that $\ln|C|$ could be thought of as measuring the amount of information involved in X (see for instance [16]).

Another way to think of the axiom *(H3)* is to state that we take for granted that the scale of subjectivity is logarithmic (instead of linear).

8.2. Functional Equations Generated by Learning Observation of Information

We consider the special case of the learning observation and we assume that the variables under consideration are not x and r, but rather are the pair of information densities $(\ln x, \ln r)$, in which case we have the observation equations

$$(\ln x)' = \rho(u)\ln x + u\rho(u)\ln r\,, \tag{8.1.a}$$

$$(\ln r)' = u\rho(u)\ln x + \rho(u)\ln r\,. \tag{8.1.b}$$

If we define the new variables x' and r' by the equation

$$\ln(x') := (\ln x)'$$

and

$$\ln(r') := (\ln r)',$$

we obtain the equalities

$$x' = r^{u\rho(u)} x^{\rho(u)} \tag{8.2.a}$$

and

$$r' = r^{\rho(u)} x^{u\rho(u)}, \tag{8.2.b}$$

where x' should be understood as $x'(t) := x(r't)$, and we then arrive at the functional equation

$$x\left(x^{u\rho}(t) r^{\rho} t\right) = r^{u\rho} x^{\rho}(t)\,. \tag{8.3}$$

In the next section, we shall see how we can exhibit the practical meaning of (8.3) in terms of self-similar processes. Shortly, we shall show that subjectivity, or more exactly observation with subjectivity, creates self-similarity.

9. Self-Similarity via Observation with Informational Invariance

9.1. Formal Definition of Self-Similar Processes

Main (non-random) definition

A self-similar process with the customary index parameter H (Hurst parameter) is a process $x(t)$ which satisfies the property

$$x(rt) = r^{H} x(t)\,, \tag{9.1}$$

which requires that

$$x(0) = 0\,. \tag{9.2}$$

Strictly speaking, the equation (9.1) should be understood in law in a probabilistic framework: $x(rt)$ and $r^H x(t)$ have the same probability density.

The zooming effect of this definition will be clearer if we re-write (9.1) in the form

$$x(rt_2) - x(rt_1) = r^H \left(x(t_2) - x(t_1)\right).$$

If we make the substitutions $r \leftarrow t$ and $t \leftarrow 1$ into (9.1) we obtain the equation

$$x(t) = t^H x(1). \tag{9.3}$$

The corresponding differential form reads

$$dx = x(1)(dt)^H, \tag{9.4}$$

and this is fully supported by the fractional calculus formula [21-23]

$$\int_0^t (du)^H = t^H, \quad 0 < H < 1$$

which holds in the framework of fractional calculus via modified Riemann-Liouville derivative.

9.2. Characterization of Self-Similarity via Informational Invariance

Derivation of approximate self-similarity properties

First of all, according to (9.1) one has necessarily $x(0) = 0$. This being the case, we make the substitutions $r \leftarrow t$ and $t \leftarrow 1$ into (8.3) to obtain the equality

$$x\left(x^{u\rho}(1) t^\rho\right) = t^{u\rho} x^\rho(1). \tag{9.5}$$

If we consider the special case when

$$x(1) = 1,$$

then (9.5) turns to be

$$x\left(t^\rho\right) = \left(t^\rho\right)^u x^\rho(1),$$

in other words the function $x(s)$, $x(1) = 1$, $s := t^\rho$, is self-similar with the Hurst parameter u.

Remark that for a self-similar function $x(t)$, $x(1)$ may have any value.

Next, one makes the change of variable (what amounts to change the scale of time)

$$\tau \; := \; x^{u\rho}(1)t^{\rho} \,, \tag{9.6}$$

which provides

$$t^{u\rho} \; = \; \tau^{u}\left(x^{-u\rho}(1)\right)^{u}$$

and in inserting into (9.5) we obtain the equation

$$x(\tau) \; = \; \tau^{u}\left(x(1)\right)^{\rho-\rho u^{2}},$$

which we re-write as

$$x(\tau) \; = \; \tau^{u}\left(x(1)\right)^{\sqrt{1-u^{2}}}, \quad 0<u<1. \tag{9.7}$$

and which could be considered as characterizing observation with informational invariance.

Setting $\tau=1$ into (9.7) yields

$$\left(x(1)\right)^{1-\sqrt{1-u^{2}}} \; = \; 1, \tag{9.8}$$

and we eventually obtain

$$x(\tau) \; = \; \tau^{u}, \quad 0<u<1, \tag{9.9}$$

or again, by virtue of (9.6)

$$x(t^{\rho}) \; = \; t^{u\rho(u)}, \quad 0<u<1. \tag{9.10}$$

Application to the special case when u is small

(i) Assume now that the magnitude of u is small as compared to the unity. One then has

$$u\rho \;\cong\; u, \qquad \rho \;\cong\; 1, \qquad \tau \;\cong\; x^{u}(1)t$$

and (9.7) yields

$$x(\tau) \; = \; \tau^{u}\, x(1), \quad 0<u<1. \tag{9.11}$$

This equation (9.11) is exactly the equation (9.1) which defines self-similar functions, and we so arrive at the following conclusion.

Relation between subjectivity and self-similarity. Assume that an observer *(R)* observes locally a system *(S)* defined by the vector $(x,t)\in\Re^{2}$, to obtain (dx,dt), via a Minkowskian (or learning) observation process with informational invariance defined by the parameter u. Then, for small values of u, there exists a scaling τ of time, which expresses the result of this observation in the form of a self-similar process $dx \propto (d\tau)^u$ of which the Hurst parameter is exactly the parameter u of the observation.

Shortly, observation with informational invariance creates self-similarity, that is to say fractals.

(ii) Another way to exhibit this self-similarity property is to start directly from (8.3) as follows. Assume that u is small so that we can write

$$\rho \cong 1 \quad \textit{and} \quad \rho u \cong u,$$

and on inserting into (8.3), we obtain the equality

$$x\left(x^u rt\right) \cong r^u x(t).$$

On setting

$$a := x^u r,$$

we have as well

$$x(at) \cong a^u x^{-u^2} x(t)$$

or merely (since u^2 is small with respect to u)

$$x(at) \cong a^u x(t).$$

10. Fractional Noises as a Result of Subjectivity

Self-similarity and probability distributions

Let us examine now how one can randomize the self-similarity condition (9.11). If we set the problem in a probabilistic framework and assume that $x(t)$ is a stochastic process, then we shall randomize (9.11) in the form

$$x(rt) \overset{D}{=} r^H x(t), \tag{10.1}$$

where the equality holds in probability distribution: clearly the left side and the right one have the same probability distribution. On making the substitutions $r \leftarrow t$ and $t \leftarrow 1$ we obtain the equality

$$x(t) \overset{D}{=} t^H x(1), \tag{10.2}$$

which is the starting point to define fractional Brownian motion. Indeed, if we assume that $x(1)$ is a Gaussian random variable and that $x(t)$ has stationary increments, then $x(t)$ turns to be exactly a fractional Brownian motion of order H.

Case when x(1) is defined as the value of a random variable

The most straightforward way to randomize $x(t)$ in the equation (10.2) is to assume that $x(1)$ is a random variable with prescribed probability distribution.

So, if we denote by $p(x(1))$ the probability density of $x(1)$, then according to the equation $x(t) = t^H x(1)$, we shall have the probability density $q(x(t))$ in the form

$$q(x(t)) = \frac{1}{t^H} p\left(\frac{x(1)}{t^H}\right).$$

Case when x(t) is defined as the value of a white noise w(t)

An alternative is to assume that $x(1)$ is a white noise $w(t)$ with the prescribed probability density $p(w)$ and to write

$$dx = w(t)(dt)^H. \tag{10.3}$$

According to the section related to observation with informational invariance, if the self-similarity is created by subjectivity, then we can substitute u for H into (10.3) to obtain

$$dx = w(t)(dt)^u. \tag{10.4}$$

We so obtained an alternative modeling for fractional Brownian motion which we have suggested an used to solve various problems [19,21,22] (remark that this model is different from the approach commonly accepted!) and which is based on fractional calculus via the so-called modified Riemann-Liouville derivative [21,23]. Remark that when $H = 1/2$ we obtain the notation introduced by Maruyama to defined Brownian motion, in such a manner that in our framework, Brownian motion can be given a meaning in terms of subjectivity. The incremental equation (10.4) can be thought of as a generalization of the Maruyama notation $db = w(t)\sqrt{dt}$.

11. Subjective Observation and Systems Dynamics

Preliminary background

First of all, let us remark that given a dynamical system

$$dx = f(x,t)dt, \quad x \in \Re, \tag{11.1}$$

under very large smooth mathematical conditions, it can be written in the form

$$G(x,t) = 0, \tag{11.2}$$

where $G(x,t)$ is the solution of the partial differential equation

$$G_x(x,t)f(x,t) + G_t(x,t) = 0. \tag{11.3}$$

Conversely, assume that the state $x(t) \in \Re$ of a dynamical system is defined by the equation (11.2) where $G(x,t)$ is a differentiable function. One then has the equality

$$G_x(x,t)dx + G_t(x,t)dt = 0, \tag{11.4}$$

therefore we obtain the dynamical equation of $x(t)$ in the standard form which reads

$$dx = -\frac{G_t(x,t)}{G_x(x,t)}dt, \tag{11.5}$$

$$=: f(x,t)dt.$$

Dynamical systems and subjective observation

Assume now that there is some uncertainty on the definition of $G(x,t)$ and that the latter is defined by an external observer (R) via an observation process with informational invariance. In order to picture this situation, we shall assume that $G(x,t)$ turns to be a function $G(x,\theta,t)$ where $\theta \in \Re$ denotes a hidden parameter which is considered as a structural parameter, and we shall so have the condition

$$G(x,\theta,t) = 0. \tag{11.6}$$

Differentiating (11.6) yields

$$G_x(x,\theta,t)dx + G_t(x,\theta,t)dt + G_\theta(x,\theta,t)d\theta = 0. \tag{11.7}$$

If we assume now that $d\theta$ is observed via an observation with information invariance, then according to the above sections, we shall have

$$d\theta = w(t)(dt)^\alpha, \quad 0 < \alpha < 1, \tag{11.8}$$

and substituting into (11.7) will yield

$$G_x dx + G_t dt + G_\theta w(t)(dt)^\alpha = 0. \tag{11.9}$$

Therefore the model, in the standard form,

$$dx = f(x,t)dt + g(x,t)w(t)(dt)^\alpha, \quad 0 < \alpha < 1, \tag{11.10}$$

with $f = -G_t / G_x$ and $g = -G_\theta / G_x$.

Further remarks and comments

Strictly speaking this equation is different from the fractional stochastic equation

$$dx = f(x,t)dt + g(x,t)\, db_H(t)$$

where $b_H(t)$ is the classical fractional Brownian motion, that is to say a Gaussian self-similar process with stationary increments and covariance

$$E\{b_H(t_2)b(t_1)\} = (\sigma^2/2)\left(t_1^{2H} + t_2^{2H} - |t_2 - t_1|^{2H}\right). \quad (11.11)$$

As an alternative we have suggested to, and we have used the model (11.8) which is mainly based upon the equality [21]

$$\int_0^t w(\tau)(d\tau)^H = H\int_0^t (t-\tau)^{H-1} w(\tau)d\tau, \quad 0 < H < 1 \quad .$$

Example 11.1

We consider the case when the parameter θ is hidden via $x+\theta$ in such a manner that $G(x,\theta,t)$ turns to be $G(x+\theta,t)$. On referring to (11.7), one has the equality $G_\theta = G_x$ therefore the equation

$$G_x(x+\theta)\, dx + G_\theta(x+\theta)\, d\theta + G_t(x+\theta)\, dt = 0,$$

which provides

$$dx = -\frac{G_t(x+\theta,t)}{G_x(x+\theta,t)}dt - w(t)(dt)^\alpha .$$

This being the case, $x+\theta$ is nothing else but the observed x to the observer (R) in such a manner that formally we can make the substitution $(x+\theta) \leftarrow x$ and write

$$dx = -\frac{G_t(x,t)}{G_x(x,t)}dt - w(t)(dt)^\alpha , \quad (11.12)$$

$$= f(x,t)dt - w(t)(dt)^\alpha . \quad (11.13)$$

Example 11.2

We now consider the case when θ is hidden in the form $G(\theta x, t)$. We then have the equality

$$G_v\theta\, dx + G_v x d\theta + G_t dt = 0$$

where G_v denotes the partial derivative with respect to $v := x\theta$, and $d\theta$ takes on the value $w(t)(dt)^\alpha$. It follows that

$$dx = -\frac{G_t(x\theta,t)}{G_v(x\theta,t)\theta}dt - \frac{G_v(x\theta,t)}{G_v(x\theta,t)}\frac{x}{\theta}w(t)(dt)^\alpha .$$

Here we shall make the substitution $x\theta \leftarrow x$ to eventually obtain the model

$$dx = -\frac{G_t(x,t)}{G_v(x,t)}dt - xw(t)(dt)^\alpha$$

$$= f(x,t)dt - xw(t)(dt)^\alpha . \qquad (11.14)$$

Example 8.4

This example is drawn from elementary mathematical finance. Consider the dynamics defined by the function

$$G(x,t) = t - \mu^{-1} Lnx ;$$

then (11.14) direct yields the equation

$$dx = -\mu x dt - xw(t)(dt)^\alpha , \qquad (11.15)$$

which is the basic model of stock exchange dynamics in the Black-Scholes framework.

12. A General approach to Modeling Systems Dynamics

12.1. Derivation of the General Model

We consider a system which is defined by the state variable x and two times t and τ. t is an indexing time which can be considered as being identic with the standard physical time, and τ is the internal proper time of the system, for instance a biological time. We shall select the model in the form

$$dx = f(x,t)dt + g(x,\tau)d\tau . \qquad (12.1)$$

We further assume that the proper internal time τ is the result of an observation with informational invariance between t and τ, and as a result of the above derivation, one can select a model for τ in the form

$$d\tau = \tau(1)(dt)^u , \qquad (12.2)$$

and on substituting into (12.1), we obtain the equation

$$dx = f(x,t)dt + \tilde{g}(x,t)x(1)(dt)^u . \tag{12.3}$$

On selecting $x(1) = w(t)$, we obtain the useful model

$$dx = f(x,t)dt + \tilde{g}(x,t)w(t)(dt)^u . \tag{12.4}$$

12.2. Consistency with Fractional Series

Here appears a new concern regarding the consistency, the soundness of the right member of (12.3), since then, one can claim that this right-side term (of (12.3)) should be identical with the first terms of the Taylor's series of $x(t)$. Assume that $u = 1/n$ in (12.3), then according to (2.8), one would have the fractional Taylor's series

$$dx = \sum_{k=1}^{n} \frac{x^{(k/n)}(t)}{(k/n)!}(dt)^{k/n}, \tag{12.5}$$

which should then be identical with

$$dx = f(x,t)(dt)^{n/n} + \tilde{g}(x,t)x(1)dt^{1/n} . \tag{12.6}$$

When $n = 1/2$, the consistency is complete as far as (12.5) turns to be

$$dx = \frac{x^{(1/2)}(t)}{(1/2)!}\sqrt{dt} + \dot{x}(t)dt , \tag{12.7}$$

while (12.6) provides

$$dx = f(x,t)(dt) + \tilde{g}(x,t)x(1)\sqrt{dt} . \tag{12.8}$$

Otherwise, comparing (12.5) and (12.6) would yield

$$\dot{x}(t) = f(x,t), \quad k = n , \tag{12.9}$$

$$x^{(k/n)}(t) = 0, \quad 2 \le k \le n-1 , \tag{12.10}$$

$$x^{(1/n)}(t) = (1/n)!\tilde{g}(x,t)x(1) . \tag{12.11}$$

But with our definition of modified Riemann-Liouville derivative, (12.9) contradict (12.10) as far as $x^{(k\alpha)}(t) = 0$ implies that $x^{(k\alpha+q\alpha)}(t) = 0$, $q \ge 1$. It follows that at first glance, it would be more convenient to select a stochastic differential equation model in the form

$$dx = \sum_{k=1}^{n} f_k(x,t)w_k(t)(dt)^{k/n} \tag{12.12}$$

but this remains to be investigated more carefully.

13. Human Factors in Modelling Systems Dynamics

13.1. Fractals in Human Systems

By the terms of human systems, we mean a system which involves human factors either as internal parameters or as external controllers. Basically, such systems are driven by subjectivity either in the dynamics of their internal structure, or in the making-decision process of the controllers.

Our claim is that such systems necessarily do involve fractals, as far as the subjectivity so involved is the result of an observation process via informational invariance. So, let us consider a system of which the absolute or the subjectivity-free dynamics is

$$dx = f(x,t)dt \,. \tag{13.1}$$

If the time is a driving time for the system, and if moreover it is subject to subjectivity, then the model

$$dx = f(x,t)(dt)^{\alpha}, \quad 0 < \alpha < 1, \tag{13.2}$$

would be more convenient. When on the contrary, it is the state x which is affected by the subjectivity, we should rather consider the equation

$$(dx)^{\alpha} = f(x,t)dt, \quad 0 < \alpha < 1, \tag{13.3}$$

or

$$dt = f^{-1}(x,t)(dx)^{\alpha}. \tag{13.4}$$

As an example, let us consider the short term interest rate $r(t)$ in finance of which the dynamics is usually selected in the form

$$dr = f(r,t)dt + g(r,t)w(t)db(t)\,, \tag{13.5}$$

where $b(t)$ is a Brownian motion (or Wiener-Levy process). With the Maruyama's notation one can write

$$db = w(t)\sqrt{dt} \, ,$$

therefore the equation

$$dr = f(r,t)dt + g(r,t)w(t)\sqrt{dt} \, . \tag{13.6}$$

If we further assume that the volatility is generated by subjectivity, then we are led to consider a more general model in the form

$$dr = f(r,t)dt + g(r,t)w(t)(dt)^{\alpha}, \quad 0 < \alpha < 1 \, . \tag{13.7}$$

in other words, the value $\alpha = 1/2$ would be considered as of special interest.

But here we must take care and, as pointed out above, in order to be fully consistent with fractional Taylor's series, it might be wise to rather select the model

$$df(x) = \sum_{k=0}^{\infty} \frac{f^{(k\alpha)}(x)}{(k\alpha)!}(dx)^{k\alpha} \, , \tag{13.8}$$

in such a manner that a suitable generalization of (13.6) which is fully consistent with (10.8) should read

$$dr = \sum_{k=1}^{n} f_k(x,t)w_k(t)(dt)^{k/n} \, , \tag{13.9}$$

where $w_k(t), k = 1,\dots,n$ is a sequence of Gaussian white noises of which the variances are $(\sigma_k)^2$. In order to conclude this first part on the motivation for fractal modelling, we shall once more consider the matter with a pedagogical point of view, by considering some prior desiderata which could be a prerequisite for a definition of fractional derivative. We beg the reader to forget for a moment all his preceding lecture.

13.2. Preliminary Desiderata for Fractional Derivative and Self-Similarity

Definition 13.1 A function $x(t)$ is said to have a fractional derivative $x^{(\alpha)}(t)$ of order α, $0 < \alpha < 1$ whenever the limit ($\Delta x(t) := x(t) - x(0)$, the symbol := means that the left-side is defined by the right-one)\

$$x^{(\alpha)}(t) \propto \lim_{\Delta t \to 0} \frac{\Delta x(t)}{(\Delta t)^{\alpha}} \, , \tag{13.10}$$

exists and is finite.■

Shortly, this amounts to write the equality

$$dx = b\,(dt)^{\alpha} \, . \tag{13.11}$$

As a direct result of this definition, when $x(t)$ is a constant, one has that $dx(t) = 0$, and *thus the fractional derivative of a constant should be zero*. In a like manner, when $x(t)$ is differentiable, one has the differential $dx = \dot{x}(t)dt$, and as a result its fractional derivative is zero. Clearly, on the practical standpoint, the definition 13.1 deals with non-differentiable functions only.

Definition 13.2. A function $x(t)$ is said to be self-similar with the Hurst parameter $\alpha > 0$ whenever there exists a positive constant a such that

$$x(at) \quad \propto \quad a^{\alpha} x(t) . \blacksquare \tag{13.12}$$

As a result of (13.12), one has the equivalence

$$x(t) \quad \propto \quad t^{\alpha} x(1) . \tag{13.13}$$

Some remarks.

(i) According to (13.11), for small t one has

$x(t) - x(0) \quad \cong \quad bt^{\alpha} ,$

therefore

$$x(at) - x(0) \quad \cong \quad a^{\alpha}\left(x(t) - x(0)\right),$$

in other words, if $x(t)$ is α-th differentiable, then $x(t) - x(0)$ is self-similar, with α as Hurst parameter, or again when $x(t)$ is locally self-similar at $t = 0$.

(ii) The converse is straightforward. If $x(t)$ is self-similar, then $x(t) = 0$, and according to (13.12) one has

$$x(dt) \quad = \quad dx \quad \propto \quad x(1)(dt)^{\alpha} ,$$

in such a manner that $x(t)$ is locally α-th differentiable at zero.

(iii) According to (13.10), the fractional derivative of a constant should be zero.

(iv) Assume that $x(t) = f(t) + g(t)$ where $f(t)$ is differentiable and $g(t)$ is α-th differentiable, then one has the equality

$$\Delta x \quad = \quad b_0 (\Delta t)^{\alpha} + b_1 \Delta t ,$$

therefore one still obtains the limit condition expressed by (13.10). It follows that, exactly like a derivative defines a function up to an arbitrary additive constant, a function of which only the fractional derivative is known, is defined up to an additive differentiable function, inclulding constant functions.More generally, if one has the equality

$$\Delta x = b_0(\Delta t)^{\alpha} + b_1(\Delta t)^{\beta}, \quad \alpha < \beta$$

then $x(t)$ is α-th differentiable at zero.

(v) And the last remark of importance is related to the value of $x(0)$. In (13.13) one has $x(0) = 0$, but in (13.12) $x(0)$ may have any value.

These simple remarks show how fractional derivative and self-similarity are deeply related, at such a point that, in a first approach, we could have restricted ourselves to the class of self-similar functions only.

13.3. Desiderata for Fractional Derivative and Mittag-Leffler Function

The Mittag-Leffler function $E_{\alpha}(y), y = t^{\alpha}$, is defined by the series

$$E_{\alpha}(t^{\alpha}) = \sum_{k=0}^{\infty} \frac{t^{\alpha k}}{(\alpha k)!}, \tag{13.14}$$

with the notation $(\alpha k)! := \Gamma(1+\alpha k)$ where $\Gamma(.)$ is the Euler function. Exactly like the exponential $e(t)$ is the solution of the differential equation

$$Dy(t) = y(t), \quad y(0) = 1,$$

we would like to define a fractional derivative D^{α} which would have $E_{\alpha}(t^{\alpha})$ as eigenfunction, that is to say which would provide the equality

$$D^{\alpha} E_{\alpha}(t^{\alpha}) = E_{\alpha}(t^{\alpha}). \tag{13.15}$$

By this way, we would be in a position to expand a fractional calculus parallel to the classical one, by substituting $E_{\alpha}(\lambda t^{\alpha})$ for $e(\lambda t)$ almost everywhere. On substituting (13.14) into (13.15), we then arrive at the following set of desiderata for a *suitable* definition of fractional derivative.

$$D^{\alpha}(K) = 0, \quad K\ constant, \tag{13.16}$$

$$D^{\alpha} x^{\alpha k} = \frac{(\alpha k)!}{(\alpha k - \alpha)!} x^{\alpha(k-1)}. \tag{13.17}$$

And well obviously this fractional derivative should provide the customary derivative when $\alpha = 1$. Remark that the condition (13.16) is not required for convenience only, but on the contrary is quite consistent with the remarks above.

Second Part: Stochastic Differential Equation of Fractional Order

14. Mean-Square Stochastic Calculus and White Noise

14.1. Derivation of the Maruyama's Notation

The basic tools (milestones) of the Itô's stochastic calculus are firstly, the Brownian motion $b(t)$ (with zero mean and the variance σ^2) and, secondly, the convergence in the mean-square sense. $b(t)$ is considered as being the elemental process by the means of which one can define (or describe) any stochastic process via the Levy-Itô-Doob's modelling

$$dx = f(x,t)dt + g(x,t)db(t), \tag{14.1}$$

The interesting feature of the mean-square convergence is summarized in the following property.

Proposition 14.1. Let $h(t,\omega)$ denote a random function of time (ω is the random event involved in the definition of h) with $\left\langle |h(t,\omega|^2 \right\rangle < \infty$ for all t. Assume that $h(t,\omega)$ is mean-square continuous; and assume further that $h(t,\omega)$ is independent of the increment $db(t)$, $t \le t_j \le a$; then the following equality holds, that is

$$\int_0^a h(t,\omega)(db)^2 = \sigma^2 \int_0^a h(t,\omega)dt, \tag{14.2}$$

for every a .■

Loosely speaking one has the equivalence

$$(db)^2 \equiv \sigma^2 dt. \tag{14.3}$$

In our attempt to consider white noise as the basic elemental process, we can go a step further, and state the following

Lemma 14.1. Conditions of proposition 14.1. But we now assume that $h(t,\omega)$ is independent of $w(t_j)$, $t \le t_j \le a$, where $w(t)$ is the white noise associated with $b(t)$. Then, in the mean-square sense, one has also the equality

$$\int_0^a h(t,\omega)(db)^2 = \int_0^a h(t,\omega)w^2(t)dt = \sigma^2 \int_0^a h(t,\omega)dt \text{ .■} \tag{14.4}$$

Proof. We have to prove that

$$\left\langle |\Delta^2| \right\rangle := \left\langle \left| \sum_{i=0}^{n-1} h_i (w_i^2 - \sigma^2)\Delta t_i \right|^2 \right\rangle,$$

tends to zero in the mean square sense when the span Δt_i tends to zero. One has successively

$$\begin{aligned}
\langle|\Delta^2|\rangle &:= \left\langle \sum_{i=0}^{n-1}\sum_{j=0}^{n-1} h_i h_j (w_i^2-\sigma^2)(w_j^2-\sigma^2)\Delta t_i \Delta t_j \right\rangle \\
&= \left\langle \sum_{i=0}^{n-1} |h_i|^2 (w_i^2-\sigma^2)^2 (\Delta t_i)^2 \right\rangle \\
&= \sum_{i=0}^{n-1} \langle h_i^2\rangle \langle (w_i^2-\sigma^2)^2\rangle (\Delta t_i)^2 .
\end{aligned} \tag{14.5}$$

But

$$\langle (w_i^2-\sigma^2)^2\rangle = \sigma^4, \tag{14.6}$$

therefore one then has

$$\begin{aligned}
\langle|\Delta^2|\rangle &:= 2\sigma^4 \sum_{i=0}^{n-1} \langle h_i^2\rangle (\Delta t_i)^2 \\
&\leq 2\sigma^4 \rho \sum_{i=0}^{n-1} \langle h_i^2\rangle (\Delta t_i)^2,
\end{aligned} \tag{14.7}$$

where ρ denotes $\rho := \max_i \Delta t_i$, therefore the convergence to zero as $\rho \downarrow 0$.■

Identification principle. Loosely speaking formally, this result provides the identification

$$(db)^2 = w^2(t)dt$$

which is exactly the Maruyama's notation $db = w(t)\sqrt{dt}$ [32].

14.2. White Noise and Basic Itô's Calculus Formulae

In this framework we can expand a definition of integral which is parallel to Itô's one and which reads

$$\int_0^a h(t,\omega)w(t)\sqrt{dt} = \lim_{\rho\downarrow 0} \sum_{i=0}^{n-1} h_i(\omega) w_i \sqrt{t_{i+1}-t_i} ,$$

with standard notation, that is to say $0 = t_0 < t_1 < ... < t_n = a$, $0 < a \leq T$ and $\rho = \max_i (t_{i+1} - t_i)$. We shall further assume that $h(t_i,\omega) \equiv h_i(\omega)$ is independent of w_i, and that

$$\int_0^T \langle h^2(t,\omega)\rangle dt < \infty .$$

With these assumptions one has the equalities

$$\left\langle \int_0^a h(t,\omega)w(t)\sqrt{dt} \right\rangle = 0$$

and

$$\left\langle \int_0^a h_1(t,\omega)w(t)\sqrt{dt} \int_0^a h_2(t,\omega)w(t)\sqrt{dt} \right\rangle = \sigma^2 \int_0^a \langle h_1 h_2 \rangle dt .$$

In the next section, we shall generalize the Maruyamas notation, and by using fractional calculus, we shall exhibit the main properties of the process so obtained.

15. Maruyama's Notation of Order *A* and Fractional Calculus

Very often now in the physical and the engineering literature, one comes across stochastic differential equations written in the form

$$dx = f(x,t)dt + g(x,t)w(t)\sqrt{dt} , \tag{15.1}$$

where $w(t)$ is the Gaussian white noise, instead of the usual one (14.1), which reads

$$dx = f(x,t)dt + g(x,t)db(t), \tag{15.2}$$

where $b(t)$ is the standard Brownian motion, what amounts to set *formally* (this is the so-called Maruyama notation)

$$db = w(t)\sqrt{dt} . \tag{15.3}$$

Let us go a step farther and let us define the process $\tilde{b}(t,\alpha)$ ($b(t,\alpha)$ is already selected to denote the fractional Brownian motion)

$$d\tilde{b}(t,\alpha) = w(t)(dt)^{\alpha}, \quad 0<\alpha<1, \tag{15.4}$$

or, what amounts to the same,

$$\tilde{b}(t,\alpha) = \tilde{b}(0,\alpha) + \int_0 w(\tau)(d\tau)^{\alpha} .$$

If we further assume that $\tilde{b}(0,\alpha)=0$ with the unit probability then we are led to introduce the

Definition 15.1 The *coarse-grained Brownian motion* $\tilde{b}(t,\alpha)$ of order α, $0<\alpha<1$, is defined by the expression

$$\tilde{b}(t,\alpha) = \int_0^t w(\tau)(d\tau)^\alpha . \tag{15.5}$$

$$= \alpha \int_0^t (t-\tau)^{\alpha-1} w(\tau)d\tau . \blacksquare \tag{15.6}$$

Remarks and comments.

(i) We do not use the term of fractional Brownian motion because this expression is already copyrighted for the process (C denotes a constant) [6,29,30,31].

$$b(t,\alpha) = C\int_0^t (t-\tau)^{\alpha-(1/2)} w(\tau)d\tau$$

$$= C\left(\alpha+\frac{1}{2}\right)^{-1} \int_0^t w(\tau)(d\tau)^{\alpha+(1/2)} . \tag{15.7}$$

(ii) The term $(dt)^\alpha$, $0<\alpha<1$, can be considered as picturing a time coarse-graining effect, therefore the word of coarse-grained Brownian motion, herein used to refer to this process $\tilde{b}(t,\alpha)$.

(iii) Remark the difference between $b(t,\alpha)$ and $\tilde{b}(t,\alpha)$. $b(t,1/2)$ is an integral with respect to dt, while $\tilde{b}(t,1/2)$ is an integral with respect to $(dt)^{1/2}$.

(iv) According to fractional calculus (via modified Riemann Liouville derivative) one has the fractional derivative

$$\tilde{b}^{(\alpha)}(t,\alpha) = \alpha! w(t) . \tag{15.8}$$

Making the change of variable $\tau = tu$ in (15.6) yields

$$\tilde{b}(t,\alpha) = t^\alpha \tilde{b}(1,\alpha) \ in\ law$$

whereby we obtain the equalities

$$\left\langle \tilde{b}(t,\alpha) \right\rangle = 0, \tag{15.9}$$

and

$$\left\langle \tilde{b}^2(t,\alpha) \right\rangle = t^{2\alpha} \left\langle \tilde{b}^2(1,\alpha) \right\rangle . \tag{15.10}$$

Moreover, a simple calculation provides

$$\left\langle \tilde{b}^2(1,\alpha) \right\rangle = \sigma^2 . \tag{15.11}$$

(v) An alternative to obtain (15.10) is as follows.
(Step 1) The equation (15.4) provides

$$\left\langle (d\tilde{b})^2 \right\rangle = \sigma^2 (dt)^{2\alpha} , \tag{15.12}$$

therefore

$$\left\langle \tilde{b}^2(t,\alpha) \right\rangle = \sigma^2 \int_0^t (d\tau)^{2\alpha} . \tag{15.13}$$

(Step 2) Assume that $0 < \alpha \le 1/2$, then the lemma 3.1, equation (3.11) direct yields

$$\left\langle \tilde{b}^2(t,\alpha) \right\rangle = \sigma^2 t^{2\alpha} . \tag{15.14}$$

(Step 3) Assume now that $1/2 < \alpha \le 1$. We then have to calculate y starting from the equation

$$dy = \sigma^2 (dt)^{2\alpha} ,$$

which provides

$$(dy)^{1/2} = \sigma (dt)^{\alpha} .$$

We then have

$$\int_0^y (du)^{1/2} = \sigma \int_0^t (d\tau)^{\alpha} ,$$

and (3.11) yields

$$\sqrt{y} = \sigma t^{\alpha} ,$$

therefore (15.11).

(vi) As expected when $\alpha = 1/2$ one recovers results related to the (standard) fractional Brownian motion.

(vii) he following result can be easily obtained

Lemma 15.1 Conditions of Lemma (14.1)

$$\int_0^a h(t,\omega)(d\tilde{b})^2 \quad = \quad \int_0^a h(t,\omega)w^2(t)(dt)^{2\alpha} \quad = \quad \sigma^2 \int_0^a h(t,\omega)(dt)^{2\alpha} \,. \qquad (15.15)$$

16. Maruyama's Notation of Order A and White Noise Calculus

Lemma 16.1 Let $f:\mathfrak{R} \to \mathfrak{R}, x \to f(x)$ denote a continuously infinitely differentiable function, and let $u:\mathfrak{R} \to \mathfrak{R}, t \to u(t)$ denote another function which has only fractional derivative of order $k\alpha$ where k denote a positive integer, and α is such that $0 < \alpha < 1$. Let K denote the smaller integer such that $K\alpha \geq 1$. Then the following expansion of fractional order holds, that is

$$f(u(t+h)) \quad = \quad f(u(t)) + \sum_{k=1}^{K} \frac{f_u^{(k)}(u)}{(\alpha!)^k k!} \left(u^{(\alpha)}(t)\right)^k h^{k\alpha} \,, \qquad (16.1)$$

or in the differential form

$$df(u(t)) \quad = \quad \sum_{k=1}^{K} \frac{f_u^{(k)}(u)}{(\alpha!)^k k!} \left(u^{(\alpha)}(t)\right)^k (dt)^{k\alpha} \,.\blacksquare \qquad (16.2)$$

Proof. According to the classical Taylor's expansion, we are entitled to write

$$f(u+\Delta u) \quad = \quad f(u(t)) + \sum_{k=1}^{\infty} \frac{f_u^{(k)}(u)}{k!} (\Delta u)^k \,, \qquad (16.3)$$

This being the case, since $u(t)$ is not differentiable, we have to manipulate it with fractional Taylors series, and applying (2.8) yields

$$\Delta u \quad \cong \quad (\alpha!)^{-1} u^{(\alpha)}(t) h^{\alpha} \,, \qquad (16.4)$$

and on substituting into (16.3) we obtain the result.■

A rigorous proof can be obtained as follows. Firstly one checks that the lemma applies to polynomial functions. Then one uses the Weierstrass theorem to consider functions which can be approximated by polynomials.

As expected when $\alpha = 1/2$, one recovers known results.

Illustrative example. Consider the stochastic process $\tilde{b}(t,\alpha)$ defined by the equation (15.4). One has the equality $\tilde{b}^{(\alpha)}(t,\alpha) \quad = \quad \alpha! w(t)$, and substituting into (16.2) directly yields the series

$$df(\tilde{b}) \quad = \quad \sum_{k=1}^{K} \frac{f_u^{(k)}(\tilde{b})}{k!} w^k(t)(dt)^{k\alpha} \,. \qquad (16.5)$$

which exhibits the striking role of the powers of Gaussian white noise, see for instance [1].

Assume that $f(\tilde{b})$, with $\alpha = 1/2$, is the exponential function $\exp(\tilde{b})$; then (16.5) provides the expression

$$de^{\tilde{b}} = e^{\tilde{b}}\left(w(t)\sqrt{dt} + 2^{-1}w^2(t)dt\right),$$

which is well known in the Ito's stochastic calculus

Lemma 16.2 Condition of Lemma 1. Assume that $f(u)$ is not differentiable with respect to *u*, but only βth-differentiable, with $0 < \beta < 1$. Let K denote the smaller integer such $K\alpha\beta > 1$. Then one has

$$f(u(t+h)) = f(u(t)) + \sum_{k=1}^{K} \frac{f_u^{(k\beta)}(u)}{(\alpha!)^k (k\beta)!}\left(u^{(\alpha)}(t)\right)^k h^{k\alpha\beta}, \tag{16.6}$$

or in the differential form

$$df(u(t)) = \sum_{k=1}^{K} \frac{f_u^{(k\beta)}(u)}{(\alpha!)^k (k\beta)!}\left(u^{(\alpha)}(t)\right)^k (dt)^{k\alpha\beta} .\blacksquare \tag{16.7}$$

Proof. The proof is similar to the proof of the above lemma 16.1, but it starts with the fractional Taylor's series

$$f(u+\Delta u) = f(u) + \sum_{k=1}^{\infty} \frac{f_u^{(k\beta)}(u)}{(k\beta)!}(\Delta u)^{k\beta} .\blacksquare$$

In the special case when $u(t)$ is the coarse-grained Brownian motion $\tilde{b}(t,\alpha)$, (16.7) yields

$$df\left(\tilde{b}(t,\alpha)\right) = \sum_{k=1}^{\infty} \frac{f_u^{(k\beta)}(\tilde{b})}{(k\beta)!} w^k(t)(dt)^{k\alpha\beta} . \tag{16.8}$$

Illustrative example. Assume that $f(\tilde{b})$ is the Mittag-Leffler function $E_\beta(\tilde{b})$, then one has

$$dE_\beta\left(\tilde{b}\right) = E_\beta\left(\tilde{b}\right)\sum_{k=1}^{K} \frac{w^k(t)}{(k\beta)!}(dt)^{k\alpha\beta} .$$

17. Itô's Lemmas of Fractional Order

17.1. Stochastic Dynamics and Maruyama's Notation

With the Maruyama's notation, we are led to consider the equation

$$dx(t) = f(x,t)dt + g(x,t)w(t)(dt)^{\alpha}, \quad 0 < \alpha \le 1, \tag{17.1}$$

which is the counterpart of the classical equation (see for instance [6,8,10,14,15,42]

$$dx(t) = f(x,t)dt + g(x,t)db(t,\alpha). \tag{17.2}$$

Analogously with the classical theory [15,42], one will define the solution of (17.1) by the recursion

$$x_{n+1}(t) = x(0) + \int_0^t f(x_n,\tau)d\tau + \int_0^t g(x_n,\tau)w(\tau)(d\tau)^{\alpha}$$

As an illustrative example of what happens with Itô's lemma, assume that $\varphi(x,t)$ is a function expandable in Taylor's series, and let us consider the equation

$$dx(t) = f(x,t)dt + g(x,t)w(t)(dt)^{1/3}.$$

One has successively

$$(dx)^2 = w^2 g^2 (dt)^{2/3},$$

$$(dx)^3 = w^3 g^3 dt,$$

and substituting into the Taylor's series

$$d\varphi(x,t) = \varphi_x dx + 2^{-1}\varphi_{xx}(dx)^2 + 6^{-1}\varphi_{xxx}(dx)^3,$$

yields

$$d\varphi = \varphi_x gw(dt)^{1/3} + 2^{-1}\varphi_{xx} g^2 w^2 (dt)^{2/3} + \left(\varphi_x f + 6^{-1}\varphi_{xxx} g^3 w^3\right)dt,$$

therefore the conditional mean variation

$$\langle d\varphi \rangle_x = 2^{-1}\varphi_{xx}\sigma^2 g^2 (dt)^{2/3} + \varphi_x f dt.$$

17.2. Background on Itô's Lemma

Lemma 17.1. (Itô's lemma). Let $x(t)$ be the solution of the Itô's stochastic differential equation

$$dx(t) = f(x,t)dt + g(x,t)db(t)\,, \tag{17.3}$$

where $\{b(t), t > t_0\}$ is a Brownian motion process with $E\{(db(t))^2\} = \sigma^2(t)dt$. Let $\varphi(x,t)$ be a scalar-valued real function, continuously differentiable in t and having continuous second partial derivatives with respect to x. Then the stochastic differential $d\varphi$ of φ is

$$d\varphi = \varphi_t dt + \varphi_x dx + (1/2)\sigma^2(t)\varphi_{xx}dt\,. \blacksquare \tag{17.4}$$

Indication on the proof. The proof is based on the fact that in the mean-square sense, one has the equality

$$\int_a^b \varphi(x,t)(db)^2 = \int_a^b \varphi(x,t)\sigma^2(t)dt\,.$$

This being the case, it is easy to show that the lemma applies to polynomials with respect to x. It is then sufficient to applies the Weierstrass theorem of approximation of functions to get the result.

Formal derivation. A formal derivation which can be soundly supported by a rigorous proof reads as follows. One rewrites the equation (17.1) in the form

$$dx(t) = f(x,t)dt + g(x,t)w(t)\sqrt{dt}$$

which provides the mean-square equality

$$(dx)^2 = g^2(x,t)w^2(t)dt = g^2(x,t)\sigma^2(t)dt$$

while one has $(dx)^k = o\left((dt)^{3/2}\right)$ when $k \geq 3$. It is then sufficient to substitute dx in the Taylor's series of $\varphi(x,t)$ to get the result.

17.3. Itô's Lemma and Maruyama's Notation

Lemma 17.2. Let $x(t)$ be the solution of the stochastic differential equation

$$dx = f(x,t)dt + g(x,t)w(t)(dt)^{1/2n}$$

where $n \in Z$, $n \geq 2$. Let $\varphi(x,t)$ be a scalar-valued real function, continuously differentiable in t and having continuous partial derivatives with respect to x, up to the order $2n$. Then the stochastic differential $d\varphi$ of φ is

$$d\varphi = \varphi_t dt + \sum_{k=1}^{2n-1} \frac{1}{k!}\varphi_x^{(k)}(dx)^k + \frac{\sigma^{2n}(t)}{n!2^n}\varphi_x^{(2n)}dt\,. \blacksquare \tag{17.5}$$

The proof proceeds as above and uses the property

$$E\{w^{2k}(t)\} = 1.3.5.7.....(2k-1)\sigma^{2k}(t) = \frac{(2k)!}{k!2^k}\sigma^{2k}(t).$$

Remark that if instead of $2n$, we have $2n+1$, then we shall not obtain the second dt-term in (17.5).

Lemma 17.3 Let $x(t)$ be the solution of the Itô's stochastic differential equation

$$dx(t) = f(x,t)dt + g(x,t)db(t), \tag{17.6}$$

where $\{b(t), t > t_0\}$ is a Brownian motion process with $E\{(db(t))^2\} = \sigma^2(t)dt$. Let $\varphi(x,t)$ be a scalar-valued real function, continuously differentiable in t, non-differentiable with respect to x, but with continuous fractional derivatives of order $k\alpha, 0 < \alpha < 1$, $\alpha = 1/2n$, k and n positive integers.Then the stochastic differential $d\varphi$ of φ is

$$d\varphi = \varphi_t dt + \sum_{k=1}^{2n-1}\frac{1}{(k\alpha)!}\varphi_x^{(k\alpha)}(dx)^{\alpha k} + \frac{\sigma^{2n}(t)}{n!2^n}\varphi_x^{(2n\alpha)}dt .\blacksquare \tag{17.7}$$

The proof is parallel to the proofs above.

18. Fokker-Planck Equation Fractal in Space and Time

18.1. Derivation of the Equation

In a previous paper [18] we dealt with the derivation of a class of FP-equations which are fractal in time only, and here we shall go a step farther, and we shall examine in which way we can meaningfully extend the approach in order to deal with dynamics which are fractal in both space and time. We have the following result.

Proposition 18.1. Assume that the stochastic process $x(t)$ under consideration satisfies the conditions

$$E\{(\Delta x)^\beta | x,t\} = f(x,t)(\Delta t)^\alpha, \quad 0 < \alpha, \beta < 1, \tag{18.1}$$

$$E\{(\Delta x)^\beta | x,t\} = g^2(x,t)\sigma^2(t)(\Delta t)^\alpha, \tag{18.2}$$

and is driven by the stochastic differential equation

$$(dx)^\beta = f(x,t)(dt)^\alpha + g(x,t)w(t)(dt)^{\alpha/2}, \tag{18.3}$$

where $w(t)$ is a Gaussian white noise with zero mean and the variance $\sigma^2(t)$. Then its probability density $p(x,t)$ is defined by the following fractional partial differential equation

$$\frac{\partial^\alpha p}{\partial t^\alpha} = -\frac{\alpha!}{\beta!}\frac{\partial^\beta (fp)}{\partial x^\beta} + \frac{\alpha!}{(2\beta)!}\frac{\partial^{2\beta}(g^2\sigma^2 p)}{\partial x^{2\beta}}. \blacksquare \tag{18.4}$$

Proof. Remark that the conditions (18.1) and (18.2) only are not quite equivalent to (18.3). Indeed (18.3) implies that

$$E\{(\Delta x)^{k\beta}|x,t\} = o((\Delta t)^{2\alpha}), \quad k \geq 3, \tag{18.5}$$

and in addition, it infers some special properties for the increment $\Delta x(t) := x(t+\tau) - x(t)$. Remark also that on taking account of the derivative

$$\frac{d^\beta x}{dx^\beta} = \frac{1!}{(1-\beta)!}x^{1-\beta},$$

one can rewrite (18.3) in the form

$$d^\beta x = ((1-\beta)!)^{-1} x^{1-\beta}\left(f(dt)^\alpha + gw(dt)^{\alpha/2}\right).$$

This being the case, the proof runs as follows.

(i) We start with the characteristic function

$$\phi_\beta(u,t) := \int_{\Re} E_\beta\left(-v^\beta x^\beta\right)p(x,t)(dx)^\beta, \tag{18.6}$$

we calculate its D_t^β-derivative in two different ways, and on equating the different results so obtained, we shall get the equation. Clearly, one has the equality

$$\partial_t^\alpha \phi_\beta(u,t) = \int_{\Re} E_\beta\left(-v^\beta x^\beta\right)\partial_t^\alpha p(x,t)(dx)^\beta, \tag{18.7}$$

$$= \frac{\left\langle d^\alpha E_\beta(-v^\beta x^\beta\right\rangle}{dt^\alpha}. \tag{18.8}$$

(ii) The fractional Taylor's series applied to $E_\beta\left(-v^\beta x^\beta\right)$ yields

$$dE_\beta(-v^\beta x^\beta) = \left(-\frac{v^\beta}{\beta!}(dx)^\beta + \frac{v^{2\beta}}{(2\beta)!}(dx)^{2\beta} + \ldots\right)E_\beta(-v^\beta x^\beta).$$

We multiply both sides by $\alpha!$, we remark that

$$\alpha! dE_\beta(-v^\beta x^\beta) = d^\alpha E_\beta(-v^\beta x^\beta),$$

and we take the mathematical expectation to obtain the equality.

$$\frac{\langle d^\alpha E_\beta(-v^\beta x^\beta\rangle}{dt^\alpha} = \alpha! \int_{-\infty}^{+\infty} \left(-\frac{v^\beta}{\beta!} f + \frac{v^{2\beta}}{(2\beta)!} g^2 \sigma^2 \right) E_\beta(.) p(x,t)(dx)^\beta \cdot \tag{18.9}$$

(iii) The right side above can be slightly re-written by using integrations by parts. Indeed, one has successively

$$-\int_{\Re} v^\beta f(x,t) E_\beta(-v^\beta x^\beta) p(dx)^\beta = -\int_{\Re} E_\beta(-v^\beta x^\beta) \partial_x^\beta (fp)(dx)^\beta$$

and

$$\int_{\Re} v^{2\beta} E_\beta(-v^\beta x^\beta) g^2 \sigma^2 p(dx)^\beta = \int_{\Re} E_\beta(-v^\beta x^\beta) \partial_x^{2\beta} \left(g^2 \sigma^2 p \right)(dx)^\beta$$

in such a manner that (18.9) can be re-written in the form

$$\partial_t^\alpha \phi = \int_{\Re} E_\beta(-v^\beta x^\beta) \left(-\frac{\alpha!}{\beta!} \partial_x^\beta (fp) + \frac{\alpha!}{(2\beta)!} \partial_x^{2\beta} (g^2 \sigma^2 p) \right) (dx)^\beta . \tag{18.10}$$

Comparing (18.7) and (18.10) yields the result. ■

18.2. Comments on Previous Results in the Literature

As we mentioned it in the introduction, there are a huge number of fractional diffusion equations which have been proposed recently to generalize the FP-equation, and one of them, for instance, is [2]

$$\partial_t^\alpha p(x,t) = -\mu (I - \Delta)^{\gamma/2} (-\Delta)^{\beta/2} p(x,t), \quad \mu > 0,\ t \in \Re,\ x \in \Re^n, \tag{18.11}$$

where $p(x,t) = 0$ for $t \le 0$, $\alpha \in [0,2]$, $\gamma \ge 0$, $\beta > 0$. Here, Δ is the n-dimensional Laplace operator, and the operators $-(I - \Delta)^{\gamma/2}$ and $(-\Delta)^{\beta/2}$ are "interpreted" (we quote the author) as inverses of the Bessel and Riesz potentials respectively. The fractional derivative in time is taken in the Caputo-Djrbashian sense.

(i) As we already mentioned it, we are reluctant to use the Djrbashian-Caputo derivative because clearly, it defines D^α by the relation $D^\alpha := D^{\alpha-1}(D^1)$, $0 < \alpha < 1$; in other words it explicitly assumes that the function under consideration is differentiable, whilst the very purpose of fractional calculus is to deal with non-differentiable functions. For instance, with the Caputo definition, defining the fractional derivative of $f(x) := x^{\alpha}$, $0 < \alpha < 1$ appears already to be a problem, since the derivative $\alpha x^{\alpha-1}$ is meaningless at $x = 0$.

(ii) In order to get more insight in the difference between (18.11) and (18.4), let us consider the simple case $\Delta = D_x^2$, in which case (18.11) turns to be

$$D_t^{\alpha} = (-1)^{\beta/2} \sum_{k=0}^{\infty} (-1)^k \binom{\gamma/2}{k} D_x^{\beta+2k},$$

therefore the following main remark. The model (18.11) is a generalization in the form

$$D_t^{\alpha} = \sum_{k=0}^{\infty} a_k D_x^{\beta+2k} = D_x^{\beta} f(D_x^2), \tag{18.12}$$

whilst (18.4) proposes the generalized model

$$D_t^{\alpha} = \sum_{k=0}^{\infty} b_k D_x^{k\beta} = g(D_x^{\beta}). \tag{18.13}$$

In (18.12), the "unit" of space derivative is D_x^2, whilst it is D_x^{β} in (18.13).

19. Application to Fractional Stochastic Differential Equations

19.1. A basic equation of mathematical finance

A stochastic differential equation which is useful in mathematical finance is

$$dx = \mu x dt + \sigma x w(t)\sqrt{dt}, \tag{19.1}$$

where x is a wealth price for instance, μ and σ are given real-valued parameters, and $w(t)$ is a Gaussian white noise with zero mean and unit variance, while t denotes time. This equation is the basic model which is involved in the derivation of the famous so-called Black-Scholes equation. Loosely speaking, in a non-random framework, the dynamical equation should be merely $dx = \mu x dt$, but the effects of various external factors which are not easy to be apprehended, are introduced by means of the additional term $\sigma w(t)dt$, which can be furthermore supported by arguments similar to those of statistical physics for instance. Another point of view is exhibited in writing the equation in the form

$$dx = \mu x\left(dt + \frac{\sigma}{\mu} w(t)\sqrt{dt}\right) = \mu x(d\tau)$$

where τ could be considered as the internal time of the system

A more general model can be considered in the form

$$dx = \mu(t)xdt + \sigma(t)xw(t)(dt)^{\alpha}, \qquad (19.2)$$

where α, $0 < \alpha < 1$, can be used for instance to take account of volatility in finance.

Let us look for a solution in the form

$$x(t) = u(t)y(t),$$

then, on substituting into (19.2), we obtain the equation

$$\frac{dy}{y} + \frac{du}{u} = \mu(t)dt + \sigma(t)w(t)(dt)^{\alpha}$$

which can be split in the two equations

$$du = \mu(t)udt, \qquad (19.3)$$

$$dy = \sigma(t)yw(t)(dt)^{\alpha}. \qquad (19.4)$$

The equation (19.3) does not give rise to any problem and its solution is

$$u(t) = u(0)\exp\left(\int_0^t u(\tau)d\tau\right), \qquad (19.5)$$

but the matter is quite different with (19.4) which requires a more detailed discussion.

19.2. Geometric Fractional Brownian Motion via Exponential

In the following, for pedagogical reasons, we shall more especially consider the equation

$$dy = yw(t)(dt)^{1/4} = yd\tilde{b}(t,1/4), \qquad (19.6)$$

where $w(t)$ is now a Gaussian white noise with zero mean and the constant standard deviation σ. The reader will generalize easily to $\alpha = 1/n$ for any positive n.

On the surface, it is attractive to write

$$\int_0^y \frac{dv}{v} = \int_0^t w(\tau)(d\tau)^{1/4} = \tilde{b}(t,1/4)$$

to obtain

$$y(t) = y(0)\exp\{\tilde{b}(t,1/4)\}, \tag{19.7}$$

but on doing so this would not be quite correct.

Indeed, according to the fractional Itô's lemma 17.2, we have

$$de^{\tilde{b}(t,1/4)} = e^{\tilde{b}}\left(w(dt)^{1/4} + \frac{1}{2}w^2\sqrt{dt} + \frac{1}{6}w^3(dt)^{3/4} + \frac{1}{24}w^4 dt\right),$$

that is to say, on replacing $w^4(t)$ by its expectation $E\{w^4\} = 3\sigma^4$,

$$de^{\tilde{b}(t,1/4)} = e^{\tilde{b}}\left(w(dt)^{1/4} + \frac{1}{2}w^2\sqrt{dt} + \frac{1}{6}w^3(dt)^{3/4} + \frac{1}{8}\sigma^4 dt\right), \tag{19.8}$$

which contradicts (19.6).

In order to take account of this remark, we shall rather look for a solution of (19.6) in the form

$$y(t) = \exp\left\{\int_0^t \left(w(d\tau)^{1/4} + \lambda w^2\sqrt{d\tau} + \mu w^3(dt)^{3/4} + \eta\sigma^4 dt\right)\right\}, \tag{19.9}$$

$$=: \exp\{U(\tilde{b})\},$$

where λ, μ and η are three constant parameters to be determined. To this end, we shall proceed as follows. One has successively

$$dU = w(dt)^{1/4} + \lambda w^2\sqrt{dt} + \mu w^3(dt)^{3/4} + \eta\sigma^4 dt,$$

$$(dU)^2 = w^2\sqrt{dt} + 2\lambda w^3(dt)^{3/4} + (\lambda^2 + 2\mu)w^4 dt,$$

$$(dU)^3 = w^3(dt)^{3/4} + 3\lambda w^4 dt,$$

$$(dU)^4 = w^4 dt.$$

We then obtain

$$dy = y\left(dU + \frac{1}{2}(dU)^2 + \frac{1}{6}(dU)^3 + \frac{1}{24}(dU)^4\right),$$

or more explicitly,

$$dy = y\left(w(dt)^{1/4} + \left(\lambda + \frac{1}{2} \right) w^2 (dt)^{2/4} + \left(\mu + \lambda + \frac{1}{6} \right) w^3 (dt)^{3/4} + \left(\eta\sigma^4 + \frac{\lambda^2 + 2\mu}{2} w^4 + \frac{\lambda}{6} w^4 + \frac{2\lambda}{6} w^4 + \frac{1}{24} w^4 \right) dt \right). \quad (19.10)$$

Here again, we replace $w^4(t)$ by its expectation $E\{w^4\} = 3\sigma^4$, and we eventually obtain

$$dy = y\left(w(dt)^{1/4} + \left(\lambda + \frac{1}{2} \right) w^2 (dt)^{2/4} + \left(\mu + \lambda + \frac{1}{6} \right) w^3 (dt)^{3/4} + \left(\eta + \frac{3}{2}\left(\lambda^2 + 2\mu\right) + \frac{9}{6}\lambda + \frac{3}{24} \right) \sigma^4 dt \right).$$

We now select λ, μ and η so as to have the equalities

$$\lambda + (1/2) = 0,$$

$$\mu + \lambda = -(1/6),$$

$$\eta + (3/2)(\lambda^2 + 2\mu) + (9/6)\lambda + (3/24) = 0$$

which provides

$$\lambda = -\frac{1}{2}, \quad \mu = \frac{1}{3}, \quad \eta = -\frac{3}{4},$$

therefore the solution

$$y(t) = \exp\left\{ \int_0^t \left(w(d\tau)^{1/4} - \frac{1}{2} w^2 \sqrt{d\tau} + \frac{1}{3} w^3 (dt)^{3/4} - \frac{3}{4}\sigma^4 t \right) \right\} \quad (19.11)$$

19.3. Geometric Fractional Brownian Motion via Mittag-Leffler Function

19.3.1. Introduction of the Mittag-Leffler Function

In the following, we shall show how fractional Taylor's series provides a new look on the definition of geometric fractional Brownian motion, and to this end, we once more consider the equation (19.6) which we recall below for convenience

$$dy = yw(t)(dt)^{\alpha} = yd\tilde{b}(t,\alpha), \quad 0 < \alpha < 1, \quad (19.12)$$

where $w(t)$ is a Gaussian white noise with zero mean and the constant standard deviation. $w(t)$ is nowhere differentiable, it is the same for $y(t)$, and as a result, the basic equation (3.1), which relates $d^\alpha y$ and dy, holds here. In other words, we can meaningfully multiply both sides of (19.12) to obtain the new equation

$$d^\alpha y = \alpha! y w(t)(dt)^\alpha , \tag{19.13}$$

therefore

$$\int_0^y \frac{d^\alpha v}{v} = \alpha! \int_0^t w(\tau)(d\tau)^\alpha ,$$

and the preliminary approximation

$$\begin{aligned} y(t) &= y_0 E_\alpha\left(\alpha! \int_0^t w(\tau)(d\tau)^\alpha\right) \\ &= y_0 E_\alpha\left(\alpha! \tilde{b}(t,\alpha)\right). \end{aligned} \tag{19.14}$$

In order to test the soundness of (19.14) , we shall seek an estimate $\hat{y}$ of y in (19.14) on using a formal framework. To this end, we proceed as follows.

(Step 1) We use the approximation

$$\begin{aligned} \int_0^t w(\tau)(d\tau)^\alpha &= w(\hat{t}) \int_0^t (d\tau)^\alpha \\ &= w(\hat{t}) t^\alpha , \end{aligned}$$

with $0 < \hat{t} < t$, to obtain the estimate

$$\hat{y}(t) = y_0 E_\alpha\left(\alpha! w(\hat{t}) t^\alpha\right). \tag{19.15}$$

(Step 2) Strictly speaking $\hat{t}$ is a function $\hat{t}(t)$ of t, but we shall assume that it is slowly varying in such a manner that we can write the approximate fractional derivative

$$D^\alpha D^\alpha \ldots D^\alpha \hat{y}(t) = D^{k\alpha} \hat{y}(t) = \left(\alpha! w(\hat{t})\right)^k \hat{y}(t) .$$

(Step 3) With this result, we have the approximate fractional Taylor's series

$$d\hat{y}(t) = \hat{y}(t) \sum_{k=0}^{\infty} \frac{\left(\alpha! w(\hat{t})\right)^k}{(\alpha k)!} (dt)^{\alpha k} . \tag{19.16}$$

(Step 4) As an illustrative example, assume that $\alpha = 1/4$

$$d\hat{y}(t) = \hat{y}(t)\left(w(\hat{t})(dt)^{1/4} + \frac{(\alpha!)^2}{(2\alpha)!} w^2(\hat{t})(dt)^{2/4} + \frac{(\alpha!)^3}{(3\alpha)!} w^3(\hat{t})(dt)^{3/4} + \frac{(\alpha!)^4}{(4\alpha)!} w(\hat{t})dt \right). \tag{19.17}$$

This equation (19.17) is exactly the counterpart of (19.10) in such a manner that the search for solution will work in exactly the same way.

19.3.2. Mittag-Leffler Function and Exponential. A Comparison

Background on Itô's calculus and exponential function

As we know it, the wrong (incomplete) solution of the equation

$$dy = ydb(t), \tag{19.18}$$

where $b(t)$ denotes the standard Brownian motion $\left(\langle b^2 \rangle = \sigma^2\right)$ is

$$y_e(t) = e^{b(t)}. \tag{19.19}$$

And in effect, according to the Itô's lemma, we have the equality

$$\begin{aligned} dy_e(t) &= e^{b(t)}\left(db + (1/2)(db)^2\right) \\ &= e^{b(t)}\left(db + (1/2)\sigma^2 dt\right), \end{aligned}$$

which is not consistent with (19.18). Therefore the solution

$$y(t) = \exp\left(b(t) - (\sigma^2/2)t\right). \tag{19.20}$$

White noise calculus and Mittag-Leffler function

We re-write (19.18) in the form

$$dy = yw(t)\sqrt{dt}, \tag{19.21}$$

which provides, on multiplying both sides by (1/2)!,

$$\frac{d^{1/2}y}{y} = \left(\frac{1}{2}\right)! w(t)\sqrt{dt}. \tag{19.22}$$

If we consider (19.22) as a (deterministic) non-random equation, then its solution will be selected in the form

$$y_f(t) = E_{1/2}\left((1/2)!\int_0^t w(\tau)\sqrt{d\tau}\right); \tag{19.23}$$

but if we come back to the stochastic framework, then the fractional Taylor's series will yield

$$\begin{aligned} dy_f &= y_f\left(w\sqrt{dt} + ((1/2)!)^2 w^2 dt\right) \\ &= y_f\left(w\sqrt{dt} + ((1/2)!)^2 \sigma^2 dt\right). \end{aligned} \tag{19.24}$$

We are then led to look for a solution in the form

$$y = E_{1/2}\left((1/2)!\int_0^t w(\tau)\sqrt{d\tau} - \gamma\sigma^2 t\right), \tag{19.25}$$

where γ is a constant to be determined. The fractional Taylor's series yields

$$dy = y\left(\frac{(1/2)!w}{(1/2)!}\sqrt{dt} - \gamma\frac{\sigma^2}{(1/2)!}dt\right) + y\frac{((1/2)!)^2}{1!}\sigma^2 dt.$$

It is then sufficient to select γ so as to have

$$\frac{\gamma\sigma^2}{(1/2)!} = \left(\frac{1}{2}!\right)^2 \sigma^2,$$

to obtain the solution

$$y = E_{1/2}\left(\left(\frac{1}{2}!\right)\int_0^t w(\tau)\sqrt{d\tau} - \left(\frac{1}{2}!\right)^3 \sigma^2 t\right), \tag{19.26}$$

which is the counterpart of (19.20).

Third Part: Fractional Differential Geometry

20. Fractional Derivative of Implicit Functions

20.1. Background on the Fractional Derivative of Compounded Functions

Introductory remarks A curved line which is continuous everywhere but nowhere differentiable, exhibits random-like features in the sense that it cannot be duplicated, and moreover has an infinite length. This is a well known result. Nevertheless, on looking more closely, with the present fractional calculus, we have all the main elements which could help us to extend the classical framework of differential geometry to fractal differential geometry, and in the following, we shall outline how this can be done.

The reader may wonder why we are dealing with fractional differential geometry in a text related to Brownian motion, and the reason to that is the following. Geodesics which are continuous everywhere but nowhere differentiable begin to be of some interest in theoretical physics, and the white noise is exactly an archetype model for such functions.

This being the case, we come back again to the equation (3.4) which we shall clarified by the following statement

Proposition 20.1 Let be given two $\Re \to \Re$ functions $f: u \to f(u)$ and $u: x \to u(x)$, then, depending upon the mathematical assumptions which are made about the differentiability of $f(.)$ and $u(.)$ respectively, one has the following derivative chain rules

$$\frac{d^\alpha f(u(x))}{dx^\alpha} = \frac{d^\alpha f(u)}{du^\alpha}\left(\frac{du}{dx}\right)^\alpha, \tag{20.1}$$

$$f_x^{(\alpha)}\big(u(x)\big) = (1-\alpha)! u^{\alpha-1} f_u^{(\alpha)}\big(u\big) u_x^{(\alpha)}(x). \tag{20.2}$$

$f(u)$ is αth-differentiable and $u(x)$ is differentiable in (20.1); $f(u(x))$ and $u(x)$ are αth-differentiable in (5.2).■

Proof. First one obtains (20.1) in writing

$$\frac{d^\alpha f(u(x))}{dx^\alpha} = \frac{d^\alpha f(u)}{du^\alpha}\frac{(du)^\alpha}{(dx)^\alpha}. \tag{20.3}$$

Next, on substituting (3.6) into (20.3) yields

$$\frac{d^\alpha f}{dx^\alpha} = \frac{d^\alpha f(u)}{du^\alpha}(1-\alpha)! u^{\alpha-1}\frac{d^\alpha u}{dx^\alpha}. \tag{20.4}$$

These properties result from the following remark. If a function has a derivative then it has also an αth-derivative for any α, $0<\alpha<1$. But, on the opposite, a function may have an αth-derivative whilst its derivative is not bounded.

Example 20.1. Consider the function $f(x) = E_\alpha\big(\lambda \ln^\alpha x\big)$, $x>0$, where λ denotes a constant, then (20.1) direct yields

$$\frac{d^\alpha f(x)}{dx^\alpha} = \frac{d^\alpha E_\alpha\big(\lambda \ln^\alpha x\big)}{\big(d(\ln x)\big)^\alpha}\left(\frac{d\ln x}{dx}\right)^\alpha$$

$$= \lambda x^{-\alpha} E_\alpha\big(\lambda \ln^\alpha x\big).$$

Example 20.2. Inverse function. We consider the pair

$$y = f(x), \qquad x = g(y),$$

in other words $y = f(g(y))$. One has successively

$$y^{(\alpha)}(x)x^{(\alpha)}(y) = \frac{d^\alpha y}{dx^\alpha}\frac{d^\alpha x}{dy^\alpha} = \frac{d^\alpha y}{dy^\alpha}\frac{d^\alpha x}{dx^\alpha},$$

and on taking account of (3.6) we obtain the formula

$$y^{(\alpha)}(x)x^{(\alpha)}(y) = ((1-\alpha)!)^{-2}(xy)^{1-\alpha}. \tag{20.5}$$

As an application, let us consider the pair $y = E_\alpha(x^\alpha)$, $x^\alpha = Ln_\alpha y$. We already know that

$$y^{(\alpha)}(x) = E_\alpha(x^\alpha),$$

in such a manner that (20.5) yields

$$E_\alpha(x^\alpha)x^{(\alpha)}(y) = ((1-\alpha)!)^{-2}(xy)^{1-\alpha},$$

therefore

$$x^{(\alpha)}(y) = D^\alpha\left((Ln_\alpha y)^{1/\alpha}\right) = \frac{1}{((1-\alpha)!)^2}\frac{(xy)^{1-\alpha}}{E_\alpha(x^\alpha)},$$

that is to say

$$D^\alpha\left((Ln_\alpha y)^{1/\alpha}\right) = \frac{1}{((1-\alpha)!)^2}\frac{y^{1-\alpha}}{y}(Ln_\alpha y)^{(1-\alpha)/\alpha}$$

$$= \frac{1}{((1-\alpha)!)^2}\frac{(Ln_\alpha y)^{(1-\alpha)/\alpha}}{y^\alpha}. \tag{20.6}$$

20.2. Implicit Function

Proposition 5.2 Let $\Phi(x, y)$ denote a mapping from an open set $U \subset \Re^2$ to $\Re$. We assume that there exists at least an $x_0 \in U$ such that $\Phi(x_0, y_0) = 0$. If for every (x, y) in U, $\Phi(x, y)$ is αth-differentiable with respect to x and y respectively, with the additional condition that $\Phi_y^{(\alpha)}(x, y) \neq 0$, then, in a neighbourhood $V \subset U$ of (x_0, y_0), the equation $\Phi(x, y) = 0$ defines y as a function of x, of which the αth-derivative is

$$\frac{d^\alpha y}{dx^\alpha} = -\frac{y^{1-\alpha}}{\Gamma(2-\alpha)}\frac{\Phi_x^{(\alpha)}(x, y)}{\Phi_y^{(\alpha)}(x, y)}, \tag{20.7}$$

and the derivative, when it exists, is such that

$$(y'(x))^{\alpha} = -\frac{\Phi_x^{(\alpha)}(x,y)}{\Phi_y^{(\alpha)}(x,y)}. \blacksquare \tag{20.8}$$

Proof. Since $\Phi(x,y)=0$ is a constant its differential $d\Phi = d_x\Phi + d_y\Phi$ should be zero. According to the fractional Taylor's series via modified Riemann-Liouville derivative, one has

$$d_x\Phi = (\alpha!)^{-1}\Phi_x^{(\alpha)}(x,y)(dx)^{\alpha}$$

therefore the differential

$$d\Phi = (\alpha!)^{-1}\Phi_x^{(\alpha)}(dx)^{\alpha} + (\alpha!)^{-1}\Phi_y^{(\alpha)}(dy)^{\alpha}$$

which provides

$$\Phi_x^{(\alpha)}(dx)^{\alpha} + \Phi_y^{(\alpha)}(dy)^{\alpha} = 0. \tag{20.9}$$

According to (3.6), one can re-write (20.9) in the form

$$\Phi_x^{(\alpha)}(x,y)(dx)^{\alpha} + \Phi_y^{(\alpha)}(x,y)(1-\alpha)!\,y^{\alpha-1}d^{\alpha}y = 0.$$

and dividing by $(dx)^{\alpha}$ yields the result.

Example 20.3. Assume that

$$\Phi(x,y) := x^{\alpha} + y^{\alpha} - R^{\alpha}, \quad 0<\alpha<1.$$

A simple calculation yields

$$\Phi_x^{(\alpha)}(x,y) = \Phi_y^{(\alpha)}(x,y) = \alpha!$$

and on inserting into (20.8) we obtain

$$\begin{aligned} y^{(\alpha)}(x) &= -((1-\alpha)!)^{-1}y^{1-\alpha} \\ &= -((1-\alpha)!)^{-1}\left((R^{\alpha}-x^{\alpha}\right)^{(1-\alpha)/\alpha}. \end{aligned} \tag{20.10}$$

20.3. Further Extension to Multivariate Functions

The preceding result can be generalized in a straightforward way to multivariate functions, what can be illustrated on a specific example as follows.

We assume that we have two given functions $\Phi_1(x_1,x_2,x_3)$ and $\Phi_2(x_1,x_2,x_3)$ and on equating them to zero, we so define x_1 and x_2 as functions of x_3, the derivatives of which are obtained on equating to zero the differentials

$$\alpha! d\Phi_1 = (\partial_1^\alpha \Phi_1)(dx_1)^\alpha + (\partial_2^\alpha \Phi_1)(dx_2)^\alpha + (\partial_3^\alpha \Phi_1)(dx_3)^\alpha , \tag{20.11.a}$$

and

$$\alpha! d\Phi_2 = (\partial_1^\alpha \Phi_2)(dx_1)^\alpha + (\partial_2^\alpha \Phi_2)(dx_2)^\alpha + (\partial_3^\alpha \Phi_2)(dx_3)^\alpha . \tag{20.11.b}$$

We then so obtain the equations

$$(\partial_1^\alpha \Phi_1)\left(\frac{dx_1}{dx_3}\right)^\alpha + (\partial_2^\alpha \Phi_1)\left(\frac{dx_2}{dx_3}\right)^\alpha = -\partial_3^\alpha \Phi_1$$

$$(\partial_1^\alpha \Phi_2)\left(\frac{dx_1}{dx_3}\right)^\alpha + (\partial_2^\alpha \Phi_2)\left(\frac{dx_2}{dx_3}\right)^\alpha = -\partial_3^\alpha \Phi_2 .$$

Using (3.6) to re-write $(dx_1)^\alpha$ and $(dx_2)^\alpha$ yields the system

$$x_1^{\alpha-1}(\partial_1^\alpha \Phi_1)\frac{d^\alpha x_1}{(dx_3)^\alpha} + x_2^{\alpha-1}(\partial_2^\alpha \Phi_1)\frac{d^\alpha x_2}{(dx_3)^\alpha} = -\left((1-\alpha)!\right)^{-1}\partial_3^\alpha \Phi_1$$

$$x_1^{\alpha-1}(\partial_1^\alpha \Phi_2)\frac{d^\alpha x_1}{(dx_3)^\alpha} + x_2^{\alpha-1}(\partial_2^\alpha \Phi_2)\frac{d^\alpha x_2}{(dx_3)^\alpha} = -\left((1-\alpha)!\right)^{-1}\partial_3^\alpha \Phi_2 .$$

The fractional derivatives $D^\alpha\left(x_1(x_3)\right)$ and $D^\alpha\left(x_2(x_3)\right)$ appear to be suitably defined whenever one has the condition

$$(x_1 x_2)^{\alpha-1}\left(\partial_1^\alpha \Phi_1 \partial_2^\alpha \Phi_2 - \partial_1^\alpha \Phi_2 \partial_2^\alpha \Phi_1\right) \neq 0 .$$

Definition 20.1 The fractional Jacobian matrix of order α (or αth-Jacobian matrix) or tangent αth-mapping) $D^\alpha \Phi$ is the matrix

$$D^\alpha \Phi(x) = \left(a_{ij}(x)\right) := \left(\partial_j \Phi_i(x)\right) \tag{20.12}$$

and on introducing the vector

$$(dx^\alpha)^T := \left((dx_1)^\alpha, (dx_2)^\alpha, \ldots, (dx_n)^\alpha\right),$$

we have the differential

$$d\Phi(x) \quad = \quad (\alpha!)^{-1} D^{\alpha}\Phi(x)(dx^{\alpha})^{T} . \qquad (20.13).$$

In short, the fractional derivative of order α is substituted for the derivative in the classical definition of the Jacobian matrix.

20.4. Fractional Differentiable Manifold

Definition 5.2 Given an αth-differentiable vector mapping $\Phi^{T} := (\Phi_1, \Phi_2, ..., \Phi_{n-p})$ from $\Re^{n}$ to $\Re^{n-p}$, $0 \le p \le n$, we assume that there exists at least an x_0 such that $\Phi(x_0) = 0$ and that the Jacobian matrix of order α, $D^{\alpha}\Phi$, has full rank $(n-p)$ in a neighbourhood V of x_0. The set X defined by the implicit equation $\Phi(x) = 0$ is referred to as an *αth-differentiable manifold of dimension p.* Clearly,

$$X \quad := \quad \{x \in V | \Phi(x) = 0\} . \blacksquare$$

Introductory remark for the continuation A curve (or graph) which is continuous everywhere, but is not differentiable, has an infinite length (se for instance Nottale [34]). Loosely speaking, this can be understood as follows. Assume that $y = f(x)$ and that $f(x)$ is not differentiable, then we can write $dy = o(dx^{\alpha})$, $o < \alpha < 1$, where $o(.)$ denotes the Landau's symbol. If $s(x)$ denotes the curve length, we then have

$$\begin{aligned} ds^{2} &= dx^{2} + o(dx^{2\alpha}), \quad 0 < \alpha < 1 \\ &= o(dx^{2\alpha}), \end{aligned}$$

Therefore

$$s \quad = \quad \int o(dx^{\alpha}) . \qquad (20.14)$$

In quite a natural way, we are led to examine whether it could not be possible to generalize the definition of curve length in such a way that it applies to non-differentiable functions, and provides a finite value to the right side of (20.15). To this end, we shall use fractional calculus.

21. Arc Length for Non-Differentiable Curves

21.1. First Proposal for Non-Differentiable Curves Defined by Functions Y=F(X)

Definition 21.1 Let $f : \Re \to \Re, x \to f(x)$ denote a function which is non differentiable but is endowed with a fractional derivative of order α, $0 < \alpha < 1$. The arc length of the curve defined by the point $(x, y = f(x))$ from $x = 0$ to $x = x, 0 < x$,is defined by the expression

$$s_1(x,\alpha) = \frac{1}{\alpha!}\int_0^x \sqrt{\frac{1}{((1-\alpha)!)^2}x^{2(1-\alpha)} + \left(f^{(\alpha)}(x)\right)^2}\,(dx)^\alpha \tag{21.1}$$

where the integral is defined by (3.11).

An alternative definition is

$$s_{11}(x,\alpha) = \int_0^x \sqrt{\frac{1}{((1-\alpha)!)^2}x^{2(1-\alpha)} + \left(f^{(\alpha)}(x)\right)^2}\,(dx)^\alpha \tag{21.2}$$

that is to say

$$s_{11}(x) = \alpha!\, s(x) . \blacksquare$$

Motivation. This definition is supported by the following arguments
(Step 1) Recall that if one has the equality

$$y^{(\alpha)}(x) = g(x),$$

then it follows that

$$y(x) = \frac{1}{\alpha!}\int_0^x g(\xi)(d\xi)^\alpha ,$$

where the integral is taken in the sense of (3.11) that is to say

$$\int_0^x g(\xi)(d\xi)^\alpha = \alpha\int_0^x (x-\xi)^{\alpha-1} g(\xi)(d\xi)^\alpha .$$

(Step 2) Consider the case when $g(x) = x$. One has $g^{(\alpha)}(x) = \left((1-\alpha)!\right)^{-1} x^{1-\alpha}$, and according to the remark above, we can write

$$\frac{1}{\alpha!}\int \frac{1}{(1-\alpha)!} x^{1-\alpha}(dx)^\alpha = x = \int dx ,$$

whereby we obtain the equality

$$\frac{1}{\alpha!(1-\alpha)!} x^{1-\alpha}(dx)^\alpha = dx . \tag{21.3}$$

Remark that this equality (21.3) yields the equality

$$\alpha! dx = d^{\alpha} x = ((1-\alpha)!)^{-1} x^{1-\alpha} (dx)^{\alpha},$$

which is exactly the known equation

$$\frac{d^{\alpha} x}{dx^{\alpha}} = \frac{x^{1-\alpha}}{(1-\alpha)!}$$

(Step 3) This being the case, let us now consider the curve defined by the equation $y = f(x)$; the square of its element of arc length is still

$$(ds)^2 = (dx)^2 + (dy)^2. \tag{21.4}$$

Taking account of the equality (21.3) on the one hand, and of the equality (provided by the fractional Taylor's series)

$$dy = (\alpha!)^{-1} f^{(\alpha)}(x)(dx)^{\alpha}, \tag{21.5}$$

on the other hand, and substituting into (21.4) direct yields

$$(ds)^2 = \frac{1}{(\alpha!)^2}\left(\frac{1}{((1-\alpha)!)^2} x^{2(1-\alpha)} + \left(f^{(\alpha)}(x)\right)^2\right)(dx)^{2\alpha}. \tag{21.6}$$

(Step 4) The expression (21.2) is obtained by setting

$$(d^{\alpha} s)^2 = (d^{\alpha} x)^2 + (d^{\alpha} y)^2. \tag{21.7}$$

Notice that, as expected, when $\alpha = 1$, one recovers the classical formula.

21.2 Second proposal for curves defined by non-differentiable functions *y=f(x)*

Definition 21.2 Let $f : \Re \to \Re, x \to f(x)$ denote a function which is both differentiable and fractional differentiable of order α, $0 < \alpha < 1$. The arc length of the curve defined by the point $(x, y = f(x))$ from 0 to $x = x, 0 < x$, is defined by the expression

$$s_2(x) = \int_0^x \sqrt{\frac{x^{2(1-\alpha)}}{(\alpha!(1-\alpha)!)^2}\left(1 + \left((1-\alpha)! f^{(\alpha)}(x)\right)^{2/\alpha} f^{2(\alpha-1)/\alpha}(x)\right)}\,(dx)^{\alpha} \tag{21.8}$$

where the integral is defined by (3.11).■

Motivation This definition can be supported by the following considerations.

(Step 1) Assume that $f(x)$ is both αth-differentiable and differentiable. The expression (21.4) for $(ds)^2$ still holds, and in the same way (21.3) still applies, to yield

$$(dx)^2 = (\alpha!(1-\alpha)!)^{-2}x^{2(1-\alpha)}(dx)^{2\alpha}. \tag{21.9}$$

(Step 2) This being the case, one has the fractional derivative

$$\frac{d^\alpha f}{(df)^\alpha} = \frac{1}{(1-\alpha)!}f^{1-\alpha}, \tag{21.10}$$

whereby we obtain the equality (on dividing both sides by $(dx)^\alpha$)

$$\frac{d^\alpha f}{(dx)^\alpha} = \frac{1}{(1-\alpha)!}f^{1-\alpha}\left(\frac{df}{dx}\right)^\alpha,$$

therefore

$$f'(x) = \left((1-\alpha)!f^{(\alpha)}(x)\right)^{1/\alpha}f^{(\alpha-1)/\alpha}(x). \tag{21.11}$$

(Step 3) Inserting (21.9) and (21.11) into (21.4) we eventually obtain

$$(ds)^2 = (dx)^2 + (f'(x))^2(dx)^2$$

$$ds^2 = \frac{x^{2(1-\alpha)}}{(\alpha!(1-\alpha)!)^2}\left(1+\left((1-\alpha)!f^{(\alpha)}(x)\right)^{2/\alpha}f^{2(\alpha-1)/\alpha}(x)\right)(dx)^{2\alpha}. \tag{21.12}$$

In other words, (6.8) would be the expression of tthe classical definition in terms of fractional derivative.

21.3. Arc Length of Non-Differential Parametric Curves

Definition 21.3 Assuming now that the curve is defined in the parametric form $x = f(t)$ and $y = g(t)$, where $f(.)$ and $g(.)$ both are non-differentiable but with a fractional derivative of order α, then the arc length is provided by the expression

$$s_1(t,\alpha) = \frac{1}{\alpha!}\int_0^t\sqrt{\left(f^{(\alpha)}(\tau)\right)^2+\left(g^{(\alpha)}(\tau)\right)^2}\,(d\tau)^\alpha, \tag{21.13}$$

and

$$s_2(t,\alpha) = ((1+\alpha)!)^{\frac{1}{\alpha}}\int_0^t\sqrt{\left(f^{(\alpha)}(\tau)\right)^{\frac{2}{\alpha}}(f(t))^{\frac{2(\alpha-1)}{\alpha}}+\left(g^{(\alpha)}(\tau)\right)^{\frac{2}{\alpha}}(g(t))^{\frac{2(\alpha-1)}{\alpha}}}\,(d\tau)^\alpha \blacksquare. \tag{21.14}$$

Indeed, once more we start from (21.7), but here we have

$$d^{\alpha}x \;=\; f^{(\alpha)}(t)(dt)^{\alpha}$$

and

$$d^{\alpha}y \;=\; g^{(\alpha)}(t)(dt)^{\alpha}$$

21.4 Further remarks and comments

Remark 1. The characteristic property of the expression (21.1) is that it involves the term $(dx)^{\alpha}$, in other words, it is not linear with respect to dx.

Remark 2. We then have two candidates for a possible definition of arc length in terms of fractional derivatives, and the question of course is which one is the right one on a modeling standpoint.

As expected, both $s_1(x)$ and $s_2(x)$ yield the well known classical

$$s(x) \;=\; \int_0^x \sqrt{1+(f'(x))^2}\,dx$$

when $\alpha = 1$. The $s_1(x)$ formula refers to (21.5) which holds only for non-differentiable functions (when a function is differentiable, it does not have fractional Taylor series) in other words, *strictly speaking,* $s_1(x)$ *should apply to non-differentiable functions only.*

In contrast, the $s_2(x)$ formula applies to a function which is both differentiable and fractional differentiable. This means that, irrespective of the value of α, $s_2(x)$ and $s(x)$ yield exactly the same result that is to say the same length, what, at first glance, is not the case with with $s_1(x)$.

The conclusion is straightforward. If we want a fractional approach to differentiable curve, then the second model is quite relevant. In the opposite way, if we deal with curves which are non-differentiable then the first model would be compulsory. Nevertheless, we think that at the present stage, the issue is still opened.

Remark3. Dealing with curves which are continuous everywhere but nowhere differentiable involves many traps. For instance their lengths in the customary Euclidean sense following (21.4) is infinite. Or again, these functions cannot be replicated, in such a manner that they exhibit random-like features. It is clear that our modelling should be consistent with these properties which are strictly relevant to direct observation and this is a topic which remains to be clarified.

For instance, it is likely that the fractional derivative does introduce a smoothing effect which results in finite length.

Remark 4. Other suggestion for a possible definition. In the framework of the definition 21.1, at first glance, an alternative to define the arc length of the curve *y=f(x)* could be

$$\sigma(b,\alpha) \;:=\; \frac{1}{\alpha!}\int_0^b \left(\frac{x^{2(1-\alpha)}}{((1-\alpha)!)^2} + \left(f^{(\alpha)}(x)\right)^2 \right)^{1/2} dx \;\blacksquare \tag{21.15}$$

Motivation. This definition would be suggested by the following considerations.

(Step 1) The equation (21.3) still holds and reads

$$(d\sigma)^2 = \frac{1}{(\alpha!)^2}\left[\frac{x^{2(1-\alpha)}}{((1-\alpha)!)^2} + \left(f^{(\alpha)}(x)\right)^2\right](dx)^{2\alpha},$$

therefore we obtain the equality

$$(d\sigma)^{1/\alpha} = \frac{1}{(\alpha!)^{1/\alpha}}\left[\frac{x^{2(1-\alpha)}}{((1-\alpha)!)^2} + \left(f^{(\alpha)}(x)\right)^2\right]^{1/2\alpha} dx,$$

which clearly shows that $(d\sigma)^{1/\alpha}$ is linear with respect to *dx*.

(Step 2) This being the case, we define the arc length $\sigma(a,x)$ by the mean value

$$\sigma(x,\alpha) := \left(\int_0^x (d\sigma)^{1/\alpha}\right)^\alpha$$

to obtain the expression (21.15).

We shall see more about in the following.

21.5. Some Illustrative Examples

Example 21.1 Assume that $f^{(\alpha)}(x) = 0$, which amounts to say that *y(x)* is constant (with the modified Riemann-Liouville derivative!).

One has successively (see Equ. (4.20))

$$s_1(x,\alpha) = s_2(x,\alpha) = \frac{1}{\alpha!}\int_0^x \frac{\xi^{1-\alpha}}{(1-\alpha)!}(d\xi)^\alpha = \frac{1}{\alpha!}\int_0^x \frac{\partial^\alpha \xi}{\partial \xi^\alpha}(d\xi)^\alpha = x, \tag{21.16}$$

and

$$\sigma(x,\alpha) = \frac{1}{\alpha!}\int_0^x \frac{\xi^{1-\alpha}}{(1-\alpha)!}d\xi = \frac{1}{\alpha!(1-\alpha)!}\left[\frac{\xi^{2-\alpha}}{2-\alpha}\right]_0^x = \frac{x^{2-\alpha}}{\alpha!(1-\alpha)!(2-\alpha)}. \tag{21.17}$$

This calculus enlightens the difference between $s_1(x)$ and $\sigma(x)$. The first one looks like quite consistent with what we should be entitled to expect whilst the second results is rather surprising at first glance. Indeed when *y(x) = constant,* we come across an horizontal straight line and the length provided by (21.1) should be equal to x, in quite a natural way. But, at first glance, this is not sufficient to disqualify (21.15) as a possible definition, and there remains to get more insight in its exact meaning.

Example 21.2 Application to white noise. Assume that (here we switch from space *x* to time *t*)

$$dy(t) \quad = \quad w(t)(dt)^{\alpha}, \tag{21.18}$$

where $w(t)$ denote a Gaussian white noise with zero mean and the variance v^2. This amounts to say that

$$y(t) \quad = \quad C + \int_0^t w(\tau)(d\tau)^{\alpha},$$

with C denoting a constant. Formally, on multiplying both sides of (21.18) by $\alpha!$, this equation provides

$$y^{(\alpha)}(t) \quad = \quad \alpha! w(t).$$

This being the case, irrespective of any stochastic framework, with the point of view of engineering mathematics, we shall merely consider $w(t)$ as a *random function*, that is to say function of which the path is a random variables, and the formula (21.1) direct yields

$$s_1(t,\alpha) \quad = \quad \int_0^t \sqrt{\frac{\tau^{2(1-\alpha)}}{(\alpha!)^2((1-\alpha)!)^2} + w^2(\tau)}\ (d\tau)^{\alpha}.$$

That is to say

$$s_1(t,\alpha) \quad = \quad \alpha \int_0^t \sqrt{\frac{\tau^{2(1-\alpha)}}{(\alpha!)^2((1-\alpha)!)^2} + w^2(\tau)}\ (t-\tau)^{\alpha-1}\, d\tau.$$

As a special case, assume that $\alpha = 1/2$, which is more or less equivalent to a model of Brownian motion, we find (recall that $\Gamma(1/2) = \sqrt{\pi}$)

$$s_1(t,1/2) \quad = \quad \frac{4}{\pi} \int_0^t \sqrt{\tau + \frac{\pi^2}{16} w^2(\tau)}\ (d\tau)^{1/2}. \tag{21.19}$$

(i) Assume that the variance v^2 is zero so that one has

$$s_1(t,1/2) \quad = \quad \frac{4}{\pi} \int_0^t \tau^{1/2} (d\tau)^{1/2}.$$

Starting from the fractional derivative

$$\frac{d^{1/2}(t)}{dt^{1/2}} = \frac{1!}{(1-1/2)!}t^{1-1/2} = \frac{1}{(1/2)!}t^{1/2}$$

one obtains the equality

$$t^{1/2} = (1/2)!\, t^{(1/2)}$$

with $\Gamma(1/2) = \sqrt{\pi}$,whereby we have that (see (3.19))

$$s_1(t,1/2) = \frac{4}{\pi}\int_0^t \left(\frac{1}{2}\right)!\, \tau^{(1/2)}(d\tau)^{1/2} .$$

$$= \frac{4(1/2)!}{\pi}\int_0^t \frac{d^{1/2}\tau}{d\tau^{1/2}} d\tau^{1/2}$$

$$= \frac{4}{\pi}[(1/2)!]^2 \left([\tau]_9^t\right)$$

$$= t . \qquad (21.20)$$

which is quite realistic.

(ii) Assume now that v is small, then Taylor's expansion provides the expression

$$s_1(t,1/2) = t + \frac{4}{\pi}\sum_{k=1}^{\infty}\binom{1/2}{k}\left(\frac{\pi^2}{16}\right)^k \int_0^\tau \tau^{-k} w^{2k}(\tau)(d\tau)^{1/2} , \quad (21.21)$$

therefore the approximation

$$s_1(t,1/2) \cong t + \frac{\pi}{8}\int_0^t \tau^{-1} w^2(\tau)(d\tau)^{1/2} .$$

$$= t + \frac{\pi}{16}\int_0^t \frac{w^2(\tau)}{\tau\sqrt{t-\tau}} d\tau . \qquad (21.22)$$

It follows that $s_1(t,1/2) = \infty$, at least when $w(0) \neq 0$. We can now come back to the preceding third remark above. It is clear that the function $y(t)$ defined by (21.18) is nowhere differentiable. It is also continuous in the stochastic mean square sense. This being the case $w(\tau)$ in (21.22) is a Gaussian random variable with zero mean value and as a result $w(0) = 0$ a.s. (almost surely), and then $s_1(t,1/2) = \infty$ a.s.

Example 21.3. Self-similar functions. A self-similar function with the Hurst parameter H , $0 < H < 1$, is a function $y(t)$ which satisfies the condition (see Section 5)

$$y(at) = a^H y(t), \quad a > 0 , (21.23)$$

which provides

$$y(t) = t^H y(1) . \qquad (21.24)$$

Close to zero, one has the equality

$$dy = (dt)^H y(1),$$

and this suggests to consider more especially the fractional derivative of order H,

$$y^{(H)}(t) = H!y(1).$$

On inserting into (21.1), we obtain

$$s_1(t,H) = \frac{1}{H!}\int_0^t \sqrt{\frac{t^{2(1-H)}}{((1-H)!)^2} + (H!)^2 y^2(1)}\ (dt)^H . \tag{21.25}$$

For large values of t, one can drop the term $(H!)^2 y^2(1)$ to obtain

$$s_1(t,H) \cong \frac{1}{H!}\int_0^t \frac{t^{(1-H)}}{(1-H)!}(dt)^H ,$$

and (3.11) yields

$$s_1(t,H) \cong t . \tag{21.26}$$

This result is quite consistent with (21.20), example 6.2, as far as a Brownian motion is a self-similar random process.

22. Radius of Curvature for Non-Differentiable Curves

22.1. Angle of the Tangent Line with the Horizontal Axis

Curves defined by the equation y=f(x)

Exactly like for differentiable continuous paths, we shall define the angle θ of the tangent line of the curve at the point (x, y) with the horizontal axis by its tangent $tg\theta = dy/dx$. Here, in the fractional framework, one has (see (2.8))

$$dy = (\alpha!)^{-1} y^{(\alpha)}(x)(dx)^{\alpha},$$

and (see (21.3))

$$dx = (\alpha!(1-\alpha)!)^{-1} x^{1-\alpha}(dx)^{\alpha},$$

in such a manner that we eventually obtain

$$tg\theta = dy/dx = (1-\alpha)!x^{\alpha-1}y^{(\alpha)}(x), \quad 0<\alpha<1. \tag{22.1}$$

Parametric curves (x(t),y(t))

Here one has $dy = (\alpha!)^{-1} y^{(\alpha)}(t)(dt)^{\alpha}$ and $dx = (\alpha!)^{-1} x^{(\alpha)}(t)(dt)^{\alpha}$ therefore

$$tg\theta \quad = \quad \frac{y^{(\alpha)}(t)}{x^{(\alpha)}(t)}. \qquad (22.2)$$

Example 22.1 Assume that $y(x) = E_{\alpha}(x^{\alpha})$ is the Mittag-Leffler function. One has easily $y^{(\alpha)}(0) = 1$, therefore $(tg\theta)_{x=0} = \infty$.

Example 22.2 Consider now the curve defined by the equation

$$dy \quad = \quad w(t)(dt)^{\alpha}$$

where t is time, and $w(t)$ is a Gaussian white noise with zero mean and the variance σ^2. Then the equation (22.1) direct yields

$$tg\theta \quad = \quad (1-\alpha)!\alpha!\frac{w(t)}{t^{1-\alpha}},$$

and we conclude that θ tends to zero as time increases. Loosely speaking, this can be explained by the fact that one has $-3\sigma < w(t) < +3\sigma$ with the probability 0,997 and that qualitatively speaking, everything happens as if $w(t)$ were more or less constant. θ exhibits an oscillation around zero as time increases, but on the average it is zero.

23. Concluding Remarks

Summary of the basic idea

Strictly speaking a fractional Brownian motion is a self-similar Gaussian process with the stationary increment

$$b(t,\alpha) - b(\tau,\alpha) \quad = \quad |t-\tau|^{\alpha} b(1,\alpha) \text{ in law,} \qquad (23.1)$$

or, what amounts to the same with the covariance

$$\langle b(t,\alpha), b(\tau,\alpha)\rangle \quad = \quad 2^{-1}\left(\left|t^{2\alpha}\right| + \left|\tau^{2\alpha}\right| - |t-\tau|^{2\alpha}\right) Var\, b(1,\alpha).$$

Here, we have merely replaced (23.1) by the increment differential equation (or the Maruyama's notation)

$$d\tilde{b}(t,\alpha) \quad = \quad w(t)(dt)^{\alpha} \qquad (23.2)$$

and by using fractional calculus via modified Riemann Liouville derivative combined with integral with respect to $(dt)^{\alpha}$, we have shown that, when $\alpha = 2^{-1}$, the equation (23.2) allows us to recover the framework of mean-square stochastic calculus. In other words, starting from Gaussian white noise only, we can expand a white noise calculus quite similar to stochastic calculus.

Motivation for fractional Brownian motion

Re-writing (23.2) provides the equation (which is meaningful in information theory)

$$\ln\left|d\tilde{b}(t)\right| = \alpha \ln(dt) + \ln\left|w(t)\right|$$

which has been observed in many real practical phenomena, like for instance, mathematical finance. The usefulness of this stochastic process should be thought of as being the result of mutual interactions between self-similarity, observation with informational invariance and coarse-graining phenomena.

Motivation for fractional calculus

The equation (23.2) is exactly the first two terms of the fractional Taylor's series, and then, in quite a natural way, we are led to set the problem in the framework of fractional calculus.

We so arrived at a unified approach combining fractional calculus, self-similarity, white noise and stochastic white noise calculus.

The formal equation (23.2) can be ascribed a meaningful significance by using fractional calculus involving fractional increment, and can be considered as defining a new fractional process referred to *as coarse-grained Brownian motion.*

REFERENCES

[1] Accardi, L. and Boukas, A.; White noise calculus and stochastic calculus, in *Stochastic Analysis:Classical and Quantum*, pp 260-300, World Scientific, 2005.

[2] Anh, V.V.. and Leonenko, N.. N.; Scaling laws for fractional diffusion-wave equations with singular initial data, *Statistics and Probability Letters*, 48 (2000), pp 239-252.

[3] Anh, V.V. and Leonenko, N. N.; Spectral theory of renormalized fractional random fields, *Teor. Imovirnost. Matem. Statyst.*, 66 (2002), pp 3-14.

[4] Bakai, E; Fractional Fokker-Planck equatin, solutions and applications, *Physical review E*, 63 (2001), pp 1-17.

[5] Caputo, M.; Linear model of dissipation whose Q is almost frequency dependent II, *Geophys. J.R. Ast. Soc.* 1967 (13), pp 529-539.

[6] Decreusefond, L. and Ustunel, A.S.; Stochastic analysis of the fractional Brownian motion, *Potential Anal.*, 1999(10), pp 177-214.

[7] Djrbashian, M.M. and Nersesian, A.B.; Fractional derivative and the Cauchy problem for differential equations of fractional order (in Russian), *Izv. Acad. Nauk Armjanskoi SSR*, 1968(3), No 1, pp 3-29.

[8] Duncan, T.E., Hu, Y., and Pasik-Duncan, B.; Stochastic calculus for fractional Brownian motion, I. Theory, *SIAM J. Control Optim.* 38 (2000), pp 582-612.

[9] El-Sayed, A.; Fractional order diffusion-wave equation, *Int. J. Theor. Phys.*, 1996 (35), pp 311-322.

[10] Gikhman, I. I. and Skorohod, A.W., *Stochastic Differential Equations*, Springer-Verlag, Berlin, 1972.

[11] Hanyga, A.; Multidimensional solutions of time-fractional diffusion-wave equations, *Proc. R Soc. London, A*, 458 (2002), pp 933-957.

[12] Hida, T.; Analysis of Brownian Functionals, *Carleton Mathematical Lecture Notes 13* (1975), 2nd ed.., 1978.

[13] Hida, T.; Selected papers (Accardi Ed), World Scientific (2001).

[14] Hu, Y. and Øksendal, B.; Fractional white noise calculus and applications to finance, *Infinite Dim. Anal. Quantum Probab. Related Topics* 6 (2003), 1-32.

[15] Itô, K.; On stochastic differential equations, *Mem. Amer. Soc.*, 1951 (4).

[16] Jumarie, G.; *Relative Information. Theories and Applications*, Springer Verlag, Berlin, 1990.

[17] Jumarie, G.; *Maximum Entropy, Information Without Probability and Complex Fractals*, Kluwer, Dordrecht, 2000.

[18] Jumarie, G.; Further results on Fokker-Planck equation of fractional order, *Chaos, Solitons. Fractals*, 2001 (12), pp 1873-1886.

[19] Jumarie, G.; Stock exchange dynamics involving both Gaussian and Poissonian white noises: approximation solution via a symbolic stochastic calculus, *Insurance: Mathematics and Economics*, 2002 (31), pp 179-189.

[20] Jumarie, G.; Fractional Brownian motions via random walk in the complex plane and via fractional derivative. Comparison and further results on their Fokker-Planck equations, *Chaos, Solitons and Fractals,* 2004 (4), pp 907-925.

[21] Jumarie, G.; On the representation of fractional Brownian motion as an integral with respect to $(dt)^{\alpha}$, *Applied Mathematics Letters*, 2005(18), 739-748.

[22] Jumarie, G.; On the solution of the stochastic differential equation of exponential growth driven by fractional Brownian motion, *Applied Mathematics Letters*, 2005(18), 817-826.

[23] Jumarie, G. (Guy.J); Modified Riemann-Liouville derivative and fractional Taylor series on non-differential functions. Further results, *Computer and Mathematics with Applications*, 2006 (51), Nos 9-10, pp 1367-1376.

[24] Kober, H.; On fractional integrals and derivatives, *Quart. J. Math. Oxford*, 1940 (11), pp 193-215.

[25] Letnivov, A.V.; Theory of differentiation of fractional order, *Math. Sb.*, 1868(3), pp 1-7

[26] Kolwankar, K.M.; Gangal A.D.; Holder exponents of irregural signals and local fractional derivatives, *Pramana J. Phys*, 1997 (48), pp 49-68.

[27] Kolwankar K.M., Gangal, A.D.; Local fractional Fokker-Planck equation, Phys. Rev. Lett., 1998 (80), pp 214-217.

[28] Liouville, J.; Sur le calcul des differentielles à indices quelconques(in french), *J. Ecole Polytechnique*, 1832(13), p 71-162.

[29] Mandelbrot, B.B. and van Ness, J.W.; Fractional Brownian motions, fractional noises and applications, *SIAM Rev.*, 1968(10), pp 422-437.

[30] Mandelbrot, B.B. and Cioczek-Georges, R.; A class of micropulses and antipersistent fractional Brownian motions, *Stochastic Processes and their Applications*, 1995(60), pp 1-18.

[31] Mandelbrot, B.B. and Cioczek-Georges, R.; Alternative micropulses and fractional Brownian motion, *Stochastic Processes and their Applications*, 1996(64), pp 143-152.

[32] Maruyama, G.; Continuous Markov processes and stochastic equation, *Rend. Circolo Mat. Palermo*, 1955 (2), No. 4, pp 48-90.

[33] Miller K.S. and Ross, B.; *An Introduction to the Fractional Calculus and Fractional Differential Equations*, Wiley, New York, 1993.

[34] Nottale, L. Scale relativity. A fractal matrix for organization in nature, *Electronic Journal of Theoretical Physics*, 2007, Vol 4, No 16(III), pp 15-102.

[35] Oldham, K.B. and Spanier, J.; *The Fractional Calculus. Theory and Application of Differentiation and Integration of Arbitrary Order*, Academic Press, New York, 1974.

[36] Osler, T.; Difference of fractional order, *Mathematics of Computation*, 1974 (28), pp 185-202.

[37] 37. Osler, T; Leibnitz rules for fractional derivatives used to generalize formulas of Walker and . Cauchy*Bull. Inst. Politechn Iasi* (NS), 1975 (21), No 25, p. 12.

[38] 38. Osler, T. J., Taylor's series generalized for fractional derivatives and applications, *SIAM. J. Mathematical Analysis*, 1971 (2) , No 1, pp 37-47.

[39] Podlubny, I.; *Fractional Differential Equations*, Academic Press, San Diego, 1999.

[40] Ross B.; *Fractional Calculus and its Applications*, Lecture Notes in Mathematics, Vol 457, Springer, Berlin, 1974.

[41] Shannon, C. E.; A mathematical theory of communication, *Bell System Technical Journal*, 1948 (27), pp 379-423, 623-656.

[42] Stratonovich, R.L.; A new form of representing stochastic integrals and equations, *J. SIAM. Control*, 1966 (4), pp 362-371.

[43] Wyss, W.; The fractional Black-Scholes equation, *Fract. Calc. Appl. Anal.*, 3, No 1 (2000), pp 51-61.

In: Brownian Motion: Theory, Modelling and Applications ISBN: 978-1-61209-537-0
Editors: R.C. Earnshaw and E.M. Riley

Chapter 2

APPLICATIONS OF MARKOV CHAINS AND BROWNIAN MOTION MODELS IN INSECT ECOLOGY

Petros T. Damos [1]***, Alexandros Rigas*** [2]
and Matilda Savopoulou-Soultani [1]

[1] Aristotle University of Thessaloniki, Faculty of Agriculture, Laboratory of Applied Zoology and Parasitology, Greece

[2] Democritus University of Thrace, Department of Electrical and Computer Engineering, Xanthi, Greece

ABSTRACT

This chapter discusses some applications of the fundamental theory of probability laws and stochastic processes such as the Brownian motion in an attempt to describe the time evolution of certain developmental events during the life cycle of an insect. In describing several ecological phenomena, deterministic and stochastic models are widely applied in which the theory of certain processes play an important role. One such process is the Wiener process which was introduced by N. Wiener (1923) as a mathematical model for the random movement of colloidal particles first observed by a British Botanist, R. Brown (1828), but has now applications in several fields of science. Markov processes are considered as one of the most important and fundamental concepts in today's probability theory having several applications in ecology and other fields. Here we are interested in the time evolution of certain developmental events in the life cycle of an insect and therefore most values are treated as discrete states in a Markov process. Fundamental theoretical results are first outlined and the mathematical approach that was applied is presented. The work proceeds by expressing the transition probabilities of stationary Markov chains by means of matrices and probability laws. The method is further illustrated by using real data sets applied on periodic phenomena in the biology of an insect such as diapause induction, diapause termination as well as insect seasonality in temperate climates. The applied stochastic models are able to give part of the realization of the above physiological processes. The time evolution of the developmental events is generated as a scaling limit of a random walk with stationary independent increments while, in the discussion of the examples based on original and simulated data, improved estimates are computed. The presented approach provides a relatively flexible method for handling discrete-valued biological data and modeling their time evolution. In particular, it appears quite useful for detecting abrupt or steady changes in the structure and the time evolution of the process as a result of seasonal and long term variations of environmental factors such as temperature and photoperiod. Although the models are implemented under the Markov assumptions for stationary independent increments, the work also discusses the presence of non stationary processes and memory effects, and how they can be assessed, both by change of deviance as well as graphically, via periodogram plots of residuals. In fact, the presented Brownian motion arises as a limit of many discrete stochastic processes (displacements of the particles) in much the same way that Gaussian random variables appear as a limit of other random variables through the central limit theorem. In this context the time evolution of several biological random variables can follow certain probability laws. Finally, fundamental for a potential field application of such models, is also a biological interpretation, which is explained by clearly understanding causality and association to several external factors.

1. Introduction

1.1. Stochasticity in Biological Phenomena

Stochasticity is evident in all biological processes at a molecular or macroscopic level including, among others, cellular processes, population growths, disease epidemics, and birth death processes. In most cases the above processes are influenced by internal (inherent - genetic) or external (environmental) factors. Much of the underlying mathematical and computational methods are well described for molecular processes (e. g. gene transcription, biochemical kinetics, heredity), but only few are emphasized on ecological events of poikilothermic (cold - blood) organisms and even fewer are translated to biological practice in an attempt to imply 'laws' to processes in the field.

Whenever practical results are derived from first defined principles the corn stone of the mathematical model is a stochastic kernel involving random variables in order to estimate probability of potential outcomes.

Hence, the biological system is treated as a dynamic random process having certain probability on its physical states and spontaneous evolves in time under certain external conditions. A random process is called Markovian if the probability distribution of any future occurrence depends on the history of the process up to the present time only via the present situation. In general, the system is assigned a set of transition probabilities per unit time for the process to go from a current state to another, while the form of these transitions depends upon the process itself as well as upon external factors and should actually reflect the nature of the casual interactions among system and environment. However, it should be made clear that biological phenomena are very complex and because of this complexity probabilistic models are interested in providing a holistic functional frame rather than to define a detailed functional and componential description of the system. Hence, the interest is on the behavior of the biological process itself and how to extrapolate practical inference under certain conditions. Poikilothermic organisms have life histories characterized by a progression through several developmental phases. In this chapter efforts are made to present two mathematical approaches in order to model important developmental events of a poikilothermic organism during its life cycle (i.e. insects). Both approaches enabling fundamental assumptions in the theory of Markov Chains in an attempt to apply well known diffusion processes, such as the Wiener of Brownian motion process on biological data but under a mathematical framework.

Moreover, they are based on common assumptions such as stationarity and time independency. That simple means that the future of the process does only depend on the present status and not the past. However, since most ecological phenomena are very complicated slight deviation of the above assumption can be observed and therefore efforts are also made to detect such kind of deviations. Throughout the article we challenged to unify, as possible, physical and mathematical theory with original experimental results derived by prior work in an attempt to make the chapter self containing and comprehensive. Therefore we restricted the introduction to a brief draft in some of the fundamental physical and mathematical concepts of stochastic processes that have applications in modern population ecology. For brevity we give in most cases only explicit equations, as abstracted by references and progressing by the formulation of general aspects and techniques with illustration on original and simulated data. Finally we summarize and discuss some of the applications and relevant results by a population - ecological standpoint.

1.2. Insect Seasonality and Important Life Cycle Adaptations

Temperate terrestrial environments are characterized by periodically lethal periods of time to poikilothermic organisms living in them due to extremes in temperature, illumination, moisture and host availability. To cope with unfavourable conditions during their life cycle, insects have therefore evolved several adaptation aspects including among others dispersal, habitat selection and physiological modifications such as cold resistance and diapause induction. However, such mechanisms must be synchronized with the environment to be successful. In most cases sensitivity is perceived by the insect due to direct environmental signals such as temperature and photoperiod [7] [33]. Seasonal alteration of these factors are a cause of significant developmental rate modifications which triggering diapause induction and termination and affecting the seasonal life cycle patterns, including multiple alternatives in one species and several types of temporal and spatial variations in development and phenology [6] [12] [15] [31] [45] [55] [68]. Insect phenology is the seasonal occurrence of stages, including the temporal pattern of adult emergence and can characterized by events such as the onset of emergence, its duration, the synchrony of individuals, any skew from normality, and several variations including irregularities and modalities [15] [30] [32] [45]. A typical life cycle of an Insect is presented in Figure 1 and describes the phenological patterns of a species with typical emergence and stages transition (i.e adult, egg, larval, and pupal stage) that takes place over a growth season in a temperate climate. In a typical life cycle of a moth, adult phenology evolves during the growth season, in which more than one generation are observable, and later it induces diapause and overwinters at a particular stage of development. Generally, many insects cease their development and initiate diapause at species-specific stages during overwintering [7] [32]. Diapause is a genetically determined state of suppressed development, the expression of which may be controlled by environmental factors. In most cases diapause acts as an important adaptive mechanism for dormancy during periods of unfavourable environmental conditions [21].

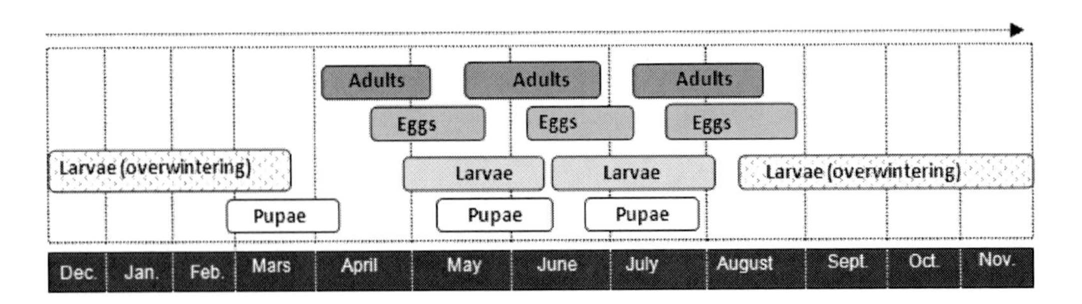

Figure 1. Typical life cycle of an holometabolous insect (i.e. moth) during season in a temperate climate.

However diapause must be distinguished from simple thermal quiescence. Thermal quisence ends when favourable conditions recur, while diapause does not cease immediately in response to favourable conditions. Hence, diapause begins long before the onset of unfavourable conditions and may not be terminated until long after the disappearance of such conditions [7] [32] [33].

The insect model that we choose for modelling most procedures is the peach twig borer *Anarsia lineatella* Zeller (Lepidoptera: Gelechiidae). Apart of its significant importance as pest of stone fruits worldwide, it displays a typical life cycle of a moth in temperate climates. In northern Greece, *A. lineatlla* has 3 or usually 4 generations per year according to prevailing temperatures [29]. The overwintering larvae that terminated diapause and the 1st generation larvae feed primarily on buds and tender twigs, while the larvae of the summer generations feed mainly on fruits, causing significant injury to the yield. By the end of summer and early autumn, larvae forming hibernaculae and overwinter as 2nd or 3rd instars [27] [28] [29].

The termination of diapause in insects is thought to be controlled and related by the status of one or more environmental factors, including photoperiod, humidity, and temperature [30] [78] [79]. However, the termination of diapause requires, or is at least accelerated by exposure to cold for a fixed minimum duration [28] [30] [78] [79]. Accordingly, the diapause state may be of very long duration and eventually comes to completion 'spontaneously', or its completion may be hastened by exposure of the insects for several weeks of low winterlike temperatures. In the field diapause development for many insect species, is completed by midwinter, but postdiapause development and growth do not occur until rising temperatures in spring permit resumption of growth and morphogenesis [78] [79].

1.3. Diffusion Models and Brownian Motion in Physical and Biological Sciences

From a mathematical point of view the definition of diffusion is a very simple statement, since diffusion is a Markov process in continuous time. However, in most cases not all diffusions can be easily estimated statistically. Therefore, the diffusion process may be represented as the solution to a stochastic differential equation if one closely approximates the motion from each point by a Brownian motion that may not drift to some direction. Furthermore, it is noteworthy, also for review reasons, to define diffusion according to the microscopic theory and macroscopic physical's theory because of its wide analogy with applications in some fields of ecology.

By definition diffusion is the random migration of molecules or small particles arising from motion due to thermal energy. The average instantaneous velocity over time over an ensemble of similar particles of mass m and velocity u_x on a one dimensional x axis is:

$$\langle u_x^{\ 2} \rangle = kT / m \ , \qquad (1),$$

and the root-mean-square velocity :

$$\langle u_x^{\ 2} \rangle^{1/2} = (kT / m)^{1/2} \ , \qquad (2),$$

where k is Boltzmann's constant and T is the absolute temperature[11].

In general, the one dimensional diffusion spreading from the particle starts at time t=0, at position x=0 and executes a random walk in which each particle steps to the right or to the left once every τ time unit, moving at velocity $\pm u_x$ a distance $\delta = \pm u_x$ having probability ½ and each successive step is statistically independent. Hence the position of the particle after the nth step differs from its position after the (n-1)th step by $\pm\delta$:

$$x_i(n) = x_i(n-1) \pm \delta \ , \qquad (3).$$

The mean displacement $<x(n)>$ of a particle, or of any measurable variable that exhibits a random walk is:

$$< x(n) >= \frac{1}{N} \sum_{i=1}^{N} x_i(n) \ , \qquad (4),$$

or

$$< x(n) >= \frac{1}{N} \sum_{i=1}^{N} x_i(n-1) =< x(n-1) > \ , \qquad (5).$$

The relative root mean square displacement is:

$$< x^2(n) >= \frac{1}{N} \sum_{i=1}^{N} x_i^{\ 2}(n) , \qquad (6),$$

or

$$< x^2(n) >=< x^2(n-1) > +\delta^2 \qquad (7)$$

for a particle that executes n steps in a time period $t = n\tau$ and has a diffusion coefficient:

$$D = \delta^2 / 2\tau, \quad (8).$$

Since the particles all starts at the origin (mean position zero) following a Brownian motion the mean position remains zero and the spreading (the mean displacement) of the particles does not change from time to time. However, the standard deviations (root mean square widths) of the distribution increase with the square root of the time. In order to find the physical law behind Brownian movement [16], Einstein concludes that the probability $f(x,t)$ to find a particle at a given time t in a distance x from its origin, follows a Gaussian distribution which is the solution of the diffusion equation [34] [35]:

$$f(x,t) = \frac{1}{\sqrt{4\pi Dt}} \exp(-\frac{x^2}{4Dt}) \quad (9).$$

The above equation (9) is interpreted as a normal probability density function in space, evolving in time and is a one dimensional solution of the differential equation:

$$D\frac{\partial^2 f(x,t)}{\partial x^2} = \frac{\partial}{\partial t} f(x,t) \quad (10).$$

For more details refer to [23]. The diffusion coefficient D in equation (9) is related to the square deviation $<x^2>$ as follows [11]:

$$<x^2> = 2\,D\,t \quad (11),$$

in two dimensions the square of the distance from the origin to the point (x,y) is $z^2 = x^2+y^2$ and therefore

$$<z^2> = 4D\,t \quad (12),$$

while in three dimensions:

$$<z^2> = 6D\,t \quad (13).$$

Equation (9) provides fundamental information about the statistics of randomly moving particles[1] in one dimension. Under the same context several biological processes exhibiting diffusive-like behavior can be modeled at one or higher dimensions [1] [17] [50].

In Figure 2, there is illustrated for instance a hypothetical two dimensional Brownian movement of a beetle. The most probable distance R which the randomly moving organism travels from the starting point after N steps is:

$$R = \sqrt{[\sum_{i=1}^{N} s_i^{\,2}]}, \quad (14)$$

[1] Einstein work on Brownian motion encouraged Jean Perrin to pursue his experimental work for confirming the kinetic theory and showing the existence of atoms [70].

where the length of each step is s_i. Starting point for the movement is the origin of a two dimensional coordinate system and successive movements derived after the projections of s on x and y axes that define the new coordinates.

In an attempt to understand and predict ecological processes of an organism occurring at landscape, continental or even global scale, numerous field ecologists have also followed stochastic approaches. Probably the most cited area in literature is that of modelling animal dispersal, ecological invasions and the emergence of spatial patterns. A classical, and noteworthy application, is formed under the assumption that insect's time dependent dispersal is a Brownian random movement [42] [50] [61]. Thus, when animals are released at a central point in the field (mark-recapture studies in homogenous habitat) their dispersion patterns can be described according to a diffusion model in a two-dimensional environment:

$$\frac{\partial u(x,y,t)}{dt} = D\left(\frac{\partial^2 u}{\partial x^2} + \frac{\partial^2 u}{\partial y^2}\right) \quad (15),$$

where $u\ (x,y,t)$ is the density of the organism at spatial coordinates x, y, and time t while D is the diffusion rate (distance^2/time) and their resulting distribution is Gaussian with mean square displacement of $4Dt$ [67] [80] [86]. Equation (15) is in fact same as the diffusion model (Ficks second equation), according to the macroscopic theory of diffusion [11] in two dimensions and describes the distribution of the concentration (C) in a given place over time t having diffusion coefficient D:

$$\frac{dC}{dt} = D\left(\frac{\partial^2 C}{\partial x^2} + \frac{\partial^2 C}{\partial y^2}\right) \quad (16),$$

which in three dimensions is:

$$\frac{dC}{dt} = D\nabla^2 C \quad (17),$$

where ∇ is the three dimensional Laplacian.

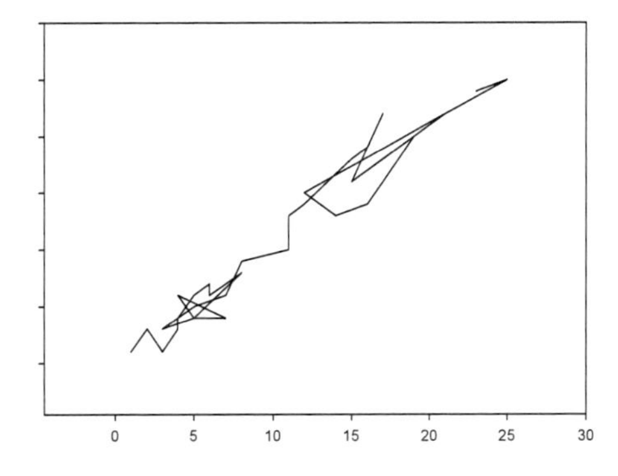

Figure 2. A two dimensional random-walk of a hypothetical insect movement (i.e beetle). Movement starts at time 0 and ends after infinite number of time steps.

Equation (16) states that the rate of change in concentration with time (at x and y) is proportional to the curvature of the concentration function, if the slope is constant and the concentration is stationary.

For more detailed information on the concept of diffusion models in modeling spatial dynamics of animal and insects the reader should refer to the references cited [22] [42] [50] [76] [81][86] [88]. Wider application on fields of biology, ecology, medicine and physics are given in [1] [2] [9] [17] [22] [43] [44] [50] [61] [72] [77].

2. Theoretical and Mathematical Considerations in Modelling Insect Populations

2.1. The Markov Chain Models

To address the challenge of modelling representative developmental events during the life cycle of an insect (i.e. diapause induction and diapause termination) we intend to use a theoretical approach based on Markov state models (MSMs) [27]. Such a model describes the kinetics of a system as Markovian between a set of states characterized by history independency [60] [62].

It is noteworthy to refer that in most of the ecological studies in which MSM's are applied they are concentrated in modeling plant communities and solving problems such as ecosystems vegetation dynamics, forest successions, recovery and restoration of natural habitats [3] [4] [5] [46] [47] [51] [52] [54] [56] [57] [58] [83][84] [87]. Nevertheless, in this chapter we are interested in insect population dynamics in which the building blocks of the MSMs, are transition probabilities, which are determined from original or simulation data collected at a certain time intervals.

Let $\{X_t, t \in T\}$ be a random process on a probability space (Ω, F, P), where T is a set of real or discrete numbers and X_t denotes a random variable defined at the index t. Thus, the random process defined for instance values of particular developmental events on the individuals of a population, will be finite (or infinite) sequences $\{X_t\}_{t\geq 1}$ or $\{X_1, X_2, \ldots X_n\}$ of random variables taking values in a discrete (countable or finite) set S, the elements of which will be called states. The index t is thought as a time index and thereby X_t represents the state of physiological or ecological processes measured or observed at the respective time.

Usually, a process (or a system) evolves in time from one state to another, and is therefore characterized by its own dynamics.

For example, the state of a physiological system can change due to the effect of an external stimulus. These state modifications are called transitions. In this chapter we consider certain developmental events in insect ecology as random process that change in time. The random observation variable (i.e. developmental time) is confined to a discrete set of states $[X_1, X_2, \ldots, X_n]$ in which transitions allowed only between nearest neighboring states. Thus, if X_n denotes the state of the process at time n, then the sequence $X_1, X_2, \ldots, X_n$ describes a sample path (or realization) of the process in the state space, from the beginning of the observation, up to a fixed time n [48].

The behavior of the process is random and therefore a certain probability to a sample path of the procces should be associated defined by

$$\mathrm{P}[X_1 = x_1, X_2 = x_2, \ldots, X_n = x_n] \qquad (18),$$

where $x_1, \ldots, x_n$ denote states which belong to a state space S. From the probability theory follows that these probabilities can be expressed in terms of the conditional probabilities:

$$\mathrm{P}[X_{n+1} = x_{n+1} | X_n = x_n, \ldots, X_1 = x_1] \qquad (19),$$

for all natural numbers n and for states $X_1, X_2, \ldots, X_{n+1}$ which belong to the state space S. The computation of all these conditional probabilities makes the study of a real phenomenon very complicated. The statement that the process is Markovian, is equivalent to saying that only the current state counts for its future evolution. This simplifies the computation of the conditional probabilities greatly, since the following property holds:

$$\mathrm{P}[X_{n+1} = x_{n+1} | X_n = x_n, \ldots, X_1 = x_1) = \mathrm{P}[X_{n+1} = x_{n+1} | X_n = x_n] \qquad (20)$$

This is known as Markov property. In this case the sequence of the random variables is a Markov chain [23][48]. We further define the one step transition probability from state i to state j at instant n as follows:

$$p_{ij} = \mathrm{P}[X_n = j | X_{n-1} = i] \qquad (21).$$

If the Markov chain satisfies the homogeneity property, this implies that one step transition probability p_{ij} whill be independent of n (i.e. the instant when the transition actually occurs). In relation to the one step transition probabilities, it is convenient to define a stochastic or probability transition matrix (M) and thus specify the table of probabilities associated with transitions from any state to any other state:

$$M = \{p_{ij}\} = \begin{pmatrix} p_{11}\ p_{21} \cdots\cdots p_{1n} \\ p_{21}\ p_{22} \cdots\cdots p_{2n} \\ . \\ . \\ p_{n1}\ p_{n2} \cdots\cdots p_{nn} \end{pmatrix} \qquad (17),$$

M is homogenous and has by definition the following properties:

$$p_{ij} \geq 0, \forall i, j \in S \qquad (18)$$

$$\sum_{j \in S} p_{ij} = 1, \forall i \in S \qquad (19).$$

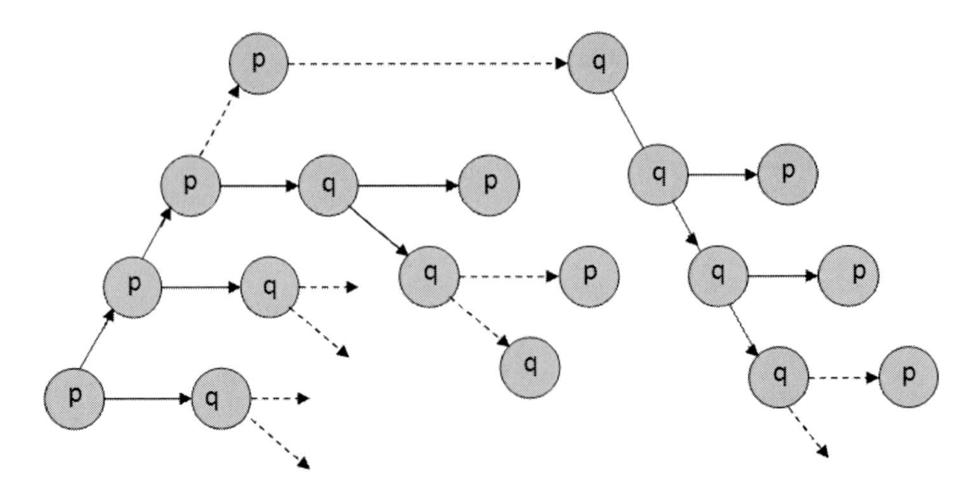

Figure 3. Oversimplified model flow, of a two state probability transition matrix, for an insect being on a diapause state (q) or on a non diapause state (p), p=1-q.

A convenient visual representation of a Markov chain model can be made according to a general two state transitions diagram as given in Figure 3. where a particular case of modeled induction of diapause and termination is presented and is characterized by irreversibility. This means one discrete flow in which a) all individuals progressively entering diapause and b) vice versa all individuals that have entered diapause do not break it until the onset of favorable conditions (see also 4.2).

For a given Markov chain with transition probability matrix M and state space S={0,1,2,…,k} the initial probability of state i is given by:

$$\rho_i^{(0)} = \mathrm{P}(X_0 = i), i \in S \tag{20),}$$

It can be shown [23] that:

$$\rho_i^{(n)} = \rho^{(n-1)} M \tag{21}$$

Hence

$$\rho_i^{(n)} = \rho_i^{(n-1)} M = \ldots = \rho_i^{(0)} M^{(0)} \tag{22}$$

and if there is a steady state as $n \to \infty$, then the stationary distribution is governed by the evolution of successive powers of the transition matrix. If the parameter space T is continuous then the transition probabilities from state i to state j at time t are defined:

$$p_{ij}(t) = \mathrm{P}(X_t = j | X_{t-1} = i), \quad \forall i, j \in S \text{ and } t \geq 0 \tag{23}$$

We can now proceed to define the homogeneity property, probability transition matrix and limiting behavior of the Markov chain as in the case where T is discrete [23] [48].

2. 2. The Brownian Motion Model or the Wiener Process

One fundamental question in applied stochastic modeling of insect ecology is, the degree of which alterations of environmental factors, namely temperature (since seasonal photoperiod alterations are standard for a respective region), can cause modification on the possible outcome of the process (i.e. irreversibility in an one direction model flow). Thus in solving such kind of ecological problems it is of interest to define how the probability evolves as time increases under certain conditions and how it deviates from universality. In other words to present a *probability law* which describes the time evolution of these kind of processes (for instance, a main concern in modern population genetics is to explain the allele frequencies in natural populations). The Brownian motion or the Wiener process is a limiting probability process of a continuous random walk having Markov properties and has been widely applied in different fields of science as predictive models [13]. The Brownian motion enables two of the most fundamental concepts in the theory of stochastic processes: the Markov property and the Martingale property. Actually, the Brownian motion is a canonical example of both a Markov process and a Martingale and by means of stochastic integration and random time changes all continuous path martingales and Markov processes can be represented in terms of Brownian motion [13] [73]. In order to describe the time evolution of the probability density functions (pdfs) of certain biological variables, having Markov properties, as a one dimensional diffusion process we applied a simplified Brownian motion model. Based on a standard Gaussian process on an interval I=[0, ∞) we can define the standard Brownian motion (or a Wiener W) process if:

i. W_o=0
ii. $\forall k \in N$ and $\forall t_1 < \ldots < t_k$ in I: $W_{t_k} - W_{t_{k-1}}, \ldots, W_{t_2} - W_{t_1}$ are independent (stationary increments)
iii. $\forall t, s \in I$ with $t < s$, $W_S - W_t$ has $N(0, s-t)$ normal distribution with zero mean and variance equal to *s-t*
iv. $t \mapsto W_t$ is almost surely continuous on I

The respective probability of a given realization can be computed:

$$\mathrm{P}\{W_t \in [x, x+dx]\} = \frac{1}{\sqrt{2\pi t}} \exp\left(-\frac{x^2}{2t}\right) dx \qquad (24),$$

$$P\{W_t < a\} = N\left(\frac{a}{\sqrt{t}}\right) \qquad (25),$$

where $$N(y) = \int_{-\infty}^{y} \frac{1}{\sqrt{2\pi}} \exp(-x^2) dx \qquad (26),$$

is the cumulative function of a standard Gaussian distribution, while for all $t, s \in I$:

Cov $(W_t, W_s) = s \wedge t$ or $E[W_t W_s] = \min(t, s)$ and $\lim_{t \to \infty} \frac{W_t}{t} = 0$ consist the law of large numbers for Brownian motion. Since the time evolution of a pdf can vary in time one is interested on a formalism that describes these deviations. Hence if X_t is the state variable (i.e. developmental time) on a probability space (*Ω, F, P*) that has standard Gaussian distribution $\forall X_t \in \Re^N$ having initial distribution $X_0 \approx f(x,0)$ and probability density function $f(x,t)$ and $\forall$ $W_t \in \Re^M$ (W_t is the Wiener process) then we denote the limiting process by $\{Z(t), t \geq 0\}$ and in the case that σ is a constant we have a Brownian motion:

$$dZ(t) = \sigma dW(t) \qquad (27),$$

with mean $E(dZ) = 0$ and variance $Var(dZ) = \sigma^2 dt$. Although Brownian motion, as theoretical presented, is a process that starts moving from the origin position, it is convenient to add a drift (μ) to the model and hence generalize the process that makes rapid movements also about a fixed linear trajectory (i.e. Brownian particle movement under the force of gravity). The above general definition is also important in modelling several stochastic processes in Biology in which the mean of the measurable variable is in most cases $\mu \geq 0$ [or $E(dZ) \geq 0$]. The added drift rate consists the deterministic velocity and the produced probability law is known as arithmetic Brownian motion[2] [71]:

$$Z(t) = \mu t + \sigma W(t) \qquad (28)$$

It is noteworthy to state that the above equation (28), can be transformed to a standard Brownian motion since the amount $[Z(t)-\mu t]/\sigma$ has a normal distribution with mean 0 and variance t. In the case of dealing with rates of change in the motion as $\Delta t \to 0$ differential calculus is involved and the arithmetic Brownian motion is given under its infinitesimal form:

$$dZ(t) = \mu(X_t, t)dt + \sigma(X_t, t)dW_t \qquad (29.1),$$

in which $\mu(x,t)$ is the drift coefficient or drift rate and $\sigma(x,t)$ is the velocity of diffusion coefficient, respectively. Properties of the solution can be determined by finding the mean value of properly chosen functions of the solutions. Although less accurate, for the finite and discrete number of time-steps t that correspons to the observation points of the variable, it is convenient to substituting directly the differential of time Δt and thereby simplify the calculations of a generalized Wiener process. Hence, if W is the standard Wiener process, then considering that values of ΔW for any two different short intervals of time Δt are independent and ε is a random drawing from a standardized normal distribution, $f(0,1)$ so that $\varepsilon\sqrt{\Delta t} = \Delta W$, or in other words, considering the random variable ε as the standardized

[2] In cases in which the variance is growing proportionally to the time interval, such as in real options problems, a log-normal diffusion process known as geometric Brownian motion is applied to simulate the process, or, a normal arithmetic Brownian motion is transformed upon the logarithm of the project value.

Gaussian noise, then the model can be generated for average drift and variance rates per unit of time as follows:

$$\Delta Z = \mu \cdot \Delta t + \sigma \cdot \varepsilon \sqrt{\Delta t} \quad (29.2).$$

Values of ΔZ have normal distribution with mean of $\mu\Delta t$ and variance of $\sigma^2\Delta t$ and the degree of uncertainty about the time evolution of the variable is proportional to the square root of the magnitude of the future observation time. Equation (29.2) thus contains all the essential components to model the dynamics of Brownian motion. From a general point of view, as in the case of particle movement in one spatial dimension x, for a process with drift $D_1(X,t)$ and diffusion D_2 (X,t), the time evolution of the pdf is described according to Kolmogorov differential forward equation (or Fokker-Planck equation), while a generalization of the Wiener process is presented according to Ito's stochastic differential equation:

$$\frac{\partial}{\partial t} f(x,t) = -\frac{\partial}{\partial x}[D_1(x,t) f(x,t)] + \frac{\partial^2}{\partial x^2}[D_2(x,t) f(x,t)] \quad (30).$$

For a more detailed mathematical description of diffusion models, and stochastic calculus involving differential solution operations, applied in solving diffusion problems in ecology, the reader should advised the references [1] [24] [40] [49].

2. 3. The Birth and Death Process

A special case of a Markov chain, with wide applications in modelling organism population dynamics, is also the one dimensional random walk model which in general has the following transition probabilities:

$$p_{ij} = \begin{cases} p & \\ 1-p, & \text{if} \\ 0 & \end{cases} \quad \begin{matrix} j = i+1; \\ j = i-1; \\ otherwise. \end{matrix} \quad (31),$$

the process X_t represents the position of a walker (i.e. one dimensional Brownian movement of a particle) moves to the right x_{t+1} with probability p and to the left x_{t-1} with probability $1-p$. The above definition is applied on a simple *birth and death* (BD) process in terms of transition rates (λ_i and μ_i). If we consider that the process X_t is a continuous-time Markov chain with states 0,1,... for which p_{ij}=0 whenever $|i-j|>1$, then it is called a *birth and death process*. Thus a birth and death process is a continuous time Markov chain with state space S={0,1,...} for which transitions from state i can only go to either state $i-1$ or state $i+1$. The general frame for population dynamics holds that each state corresponds to some population size at a specific time, and when the state increases by 1 we say that a birth occurs, while opposite a death occurs when it decreases by 1.

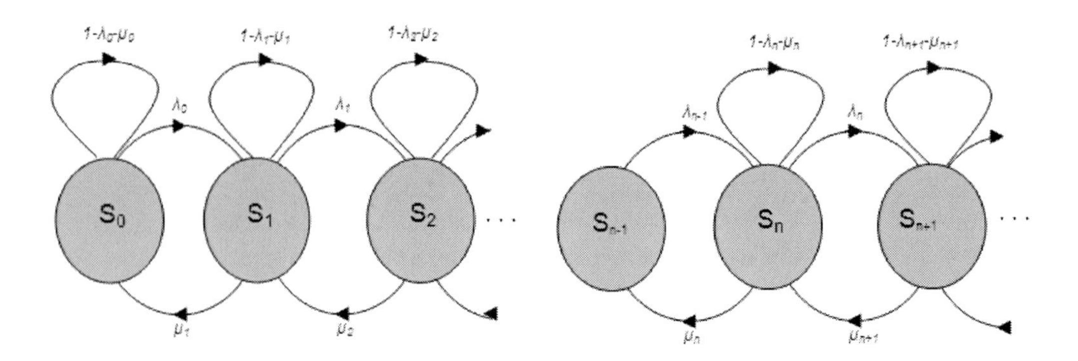

Figure 4. Simplified state transition diagram of a birth-death process. *X(t)=birth-death* on respective time intervals (*Δt*). Each state (S) is addressed to a probability *pi(t)*=P[*X(t)=i*] which equals the probability that the system is in state *i* at time *t*.

Let λ_i and μ_i given by: $\lambda_i = q_{i,i+1}$, $\mu_i = q_{i,i-1}$ and the values $\{\lambda_i, i \geq 0\}$ and $\{\mu_i, i \geq 1\}$ are birth and death rate respectively. If the process is in the state S_i at t, it can go into state S_{i+1} at time $t+\Delta t$ with probability $\lambda_i \Delta t$, or to state S_{i-1} with probability $\mu_i \Delta t$ (Figure 4). In this case the respective transition probabilities including also an additional possibility of the process to stay at the same state:

$$p_{ij} \cong \begin{cases} \lambda_i \Delta t \\ 1-(\lambda_i - a_i)\Delta t \\ \mu_i \Delta t \\ 0 \end{cases}, \quad \text{if} \quad \begin{array}{l} j = i+1; \\ j = i \\ j = i-1; \\ otherwise. \end{array} \tag{32}$$

and $u_i = \lambda_i + \mu_i$ is the rate at which the process leaves state S_i and p_{ij} is the probability that it goes to j, while $q_{ij} = u_i$ is the transition rate when the process makes a transition from state i to j [37]. The respective probabilities of a state transition diagram for very small time intervals (i.e. from t to t+Δt) are:

$$p_{ij}(t+\Delta t) = p_{i,j-1}(t)\lambda_{i-1}\Delta t + p_{i,j+1}(t)\mu_{j+1}\Delta t + [1-(\lambda_j + \mu_j)\Delta t]p_{ij}(t) \qquad (33).$$

If the time intervals are very short so that $\Delta t \to 0$, then we obtain:

$$p_{ij}'(t) = \lambda_{j-1} p_{i,j-1}(t) + \mu_{j+1} p_{i,j+1}(t) - (\lambda_i + \mu_j) p_{ij}(t) \tag{34}$$

The above relations hold for j=0 as well. In the case of a steady state (a limiting distribution exist) we have that $p_{ij}(t) \to \pi_j$ and $p_{ij}'(t) \to 0$ as $t \to \infty$. The equilibrium probabilities satisfy the following equation:

$$\lambda_{j-1}\pi_{j-1} + \mu_{j+1}\pi_{j+1} = (\lambda_j + \mu_j)\pi_j \tag{35}$$

For j=0 we get $\mu_1\pi_1 = \lambda_1\pi_0$, which leads to: $\lambda_{j-1}\pi_j = \mu_j\pi_j$, (j=1,2,…). Therefore

$$\pi_j = \frac{\lambda_o\lambda_1\ldots\lambda_{j-1}}{\mu_1\mu_2\ldots\mu_j}\pi_o = \rho_j\pi_0 \ , \tag{36}$$

Since $\sum_{j=0}^{\infty}\pi_j = 1$, this implies $\pi_0 = 1/\sum_{j=0}^{\infty}\rho_j$ provided that $\sum_{j=0}^{\infty}\rho_j < \infty$. Thus

$$\pi_j = \frac{\rho_j}{\sum_{j=0}^{\infty}\rho_j} \tag{37}.$$

For more details refer to [23] [73]. In the special case of a linear BD process we have $\lambda_j = j\lambda$ and $\mu_j = j\mu$, where λ and μ are constants. By substitution these values of λ_i and μ_i we find

$$p_{ij}(t) = \lambda(j-1)p_{i,j-1}(t) - (\lambda+\mu)jp_{ij}(t) + \mu(j+1)p_{i,j+1}(t) \tag{38}$$

If we define the probability generating function as follows:

$$G(t,s) = E(s^{x_t}|x_0 = i) = \sum_{j=0}^{\infty} p_{i,j}(t)s^j \ , \tag{39}$$

It can be shown that equation (34) takes the form:

$$\frac{\partial G}{\partial t} = (\lambda s - \mu)(s-1)\frac{\partial G}{\partial t} \tag{40}.$$

Solving equation (40) [23], with the initial condition $G(0,s) = s^i$, we find that

$$G(t,s) = [\frac{\mu(1-s)-(\mu-\lambda s)e^{-(\lambda-\mu)t}}{\lambda(1-s)-(\mu-\lambda s)e^{-(\lambda-\mu)t}}]^i, \ \lambda \neq \mu \tag{41}.$$

If we know $G(t,s)$, then we can obtain $p_{ij}(t)$ from (39). From an ecological perspective, such as in population modelling of insects and other animals, the transition rates

λ and μ may not be constant but a function of t [65]. This is a more interesting problem and can be solved by noticing that $G(t,s)$ satisfies the following partial differential equation:

$$\frac{\partial G}{\partial t} = [\lambda(t)s - \mu(t)](s-1)\frac{\partial G}{\partial t} \quad (42).$$

The solution of (42) is

$$G(t,s) = [1 + (\frac{\exp[r(t)]}{s-1} - \int_0^t \lambda(u)\exp[r(u)]du)^{-1}]^i \quad (43).$$

Where $x_0 = i$ and $r(t) = \int_0^t [\mu(u) - \lambda(u)]du$.

For more details refer to [41].

3. Assessing Statistical Properties of the Biological Data

3. 1. General Exploratory Data Analysis

When samples are taken from the population to estimate parameters such as the mean number of insects completing a specific developmental event (i.e. adult eclosion and emergence patterns during growth season), the estimate will almost certainly not equal to the true parameter. That means that for a sequential stochastic process several different estimates of the random variable parameters will be scattered about the true mean in some pattern which causes *Bias*[3] in precision of the sample estimates. Therefore the calculation of first and second moments, the frequency histograms and related frequency functions are probably a first source of exploratory data analysis. Considering the general case of data appearance during the stochastic process that follows a normal (or Gaussian) distribution, then the probability density function (pdf) of the random variable x, with parameters μ (first moment) and σ^2 (second moment) is given by:

$$f_X(x) = \frac{1}{\sqrt{2\pi\sigma^2}} e^{\frac{-(x-\mu)^2}{2\sigma^2}} \quad (44).$$

This is a bell shaped-curve, symmetric around the parameter μ, and its cumulative distribution function (cdf) is given by:

[3] Bias is any systematic deviation of the sample estimate from the true parameter and is defined as the size of difference between the expectation of a sample estimate and the population parameter being estimated.

$$F_X(x) = \int_{-\infty}^{x} \frac{1}{\sqrt{2\pi\sigma^2}} e^{\frac{-(y-\mu)^2}{2\sigma^2}} dy \qquad (45),$$

while $f_X(x) = \dfrac{dF_X(x)}{dx}$ (46)

(or $f(x) = \lim_{\Delta x \to 0} = \dfrac{P\{x \leq x \leq x + \Delta x\}}{\Delta x}$).

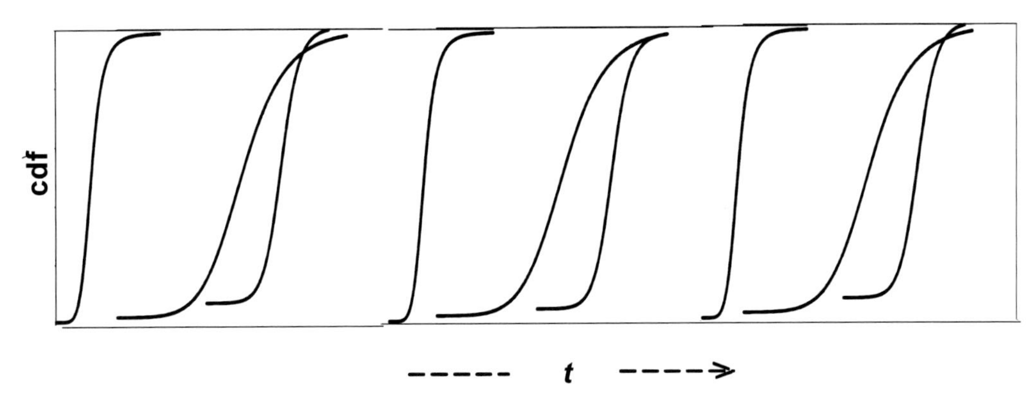

Figure 5. Time evolution of a typical cumulative density function (cdf) of moth emergence of three generations and during three successive cultivation seasons.

The function *f(x)* must be non negative and its area 1, while the cumulative distribution function must be continuous from the right and, as *x* increases from $-\infty$ to ∞, the function must increase monotonically from 0 to 1. Parameter estimation in the above functions can be obtained by applying several statistical procedures including maximum likelihood estimations, or ordinary least squares algorithms [6] [12]. Figure 5 illustrates the time evolution of a typical cumulative distribution function which corresponds to successive moth generations during the growth season.

3.2. Stationarity

By definition, a stochastic process (X_t) is stationary if its joint probability distribution does not change when shifted during the observations (either in time or space). The practical aftermath of this statement is that parameters of the distribution, such as the mean and variance, do not change in time. Hence, if $F_{(X_t)}(x)$ is the cdf, then the stochastic process X_t is stationary (for all *t=1,2,...*) if:

$$F_{(X_t)}(x) = F_{(X_{t+k})}(x) \qquad (47).$$

Apart from the presented mathematical formalism and by following practical ways in detecting stationarity[4] (i.e. seasonality on data appearance) is quite useful to apply some common graphical techniques in order to detect deviation from stationarity. The application of graphical techniques has the benefit of a quite easy application, even for statistically-untrained readers and grasps fundamental distributional information of the stochastic process which is displayed by the plots. By using these techniques typical values and general differences in the distribution of the values in the time series are easily detectable [18] [19] [20] [23] [25] [27] [82] [85]. Actually, a sequence plot of the observed variable X_t provides the first indication of seasonality [26]. Figure 6 for instance, illustrates the box plots of the number of male moths captured in pheromone traps during growth season. The primer plot of the respective time series provides as a first feel about the shifts in location and variation of the data [9] [39]. However, a box plot emphasize a few key percentiles and mark them on the vertical scale and there are some problems which can arise by the construction of box plots, such as the length of the whiskers, while the extreme values treatment are also somewhat arbitrarily determined. Therefore, Quantile – Quantile (Q-Q) and residual error plots are probably more suitable in detecting stationarity deviation from over the entire range of data values. Figure 7 illustrates the respective to the first flight (i.e. Figure 5) Q-Q plot and Figure 8. the respective residual error plot. These plots actually displaying the data trends all over the range. It is noteworthy that the applications of residual error plots are not only an excellent tool for conveying location and variation information in data sets but also particularly for detecting and illustrating location and variation changes between different groups of data. For the construction of normal Q-Q plot one plots the empirical quantiles against those of a standard normal distribution (or any kind of distribution in which we are interested to detect deviations) in which the plus signs (+) provide a straight reference line that is drawn by using the sample mean and standard deviation. For the detection of deviations in respect to the normal distribution, the vertical coordinate is the data value, and the horizontal coordinate is $\Phi^{-1}(p_i)$, where Φ^{-1} the inverse of the normal distribution and p_i is given by:

$$p_i = (r_i - \frac{3}{8}) / (n + \frac{1}{4}) \qquad (48)$$

and r_i is the rank of the ith value (when ordered from smallest to largest) and n the number of non missing values [20] [18] [74] [75] [89]. Thus, if the data are normally distributed with

[4] In the case that data of the process are to some degree not stationary it is convenient to transform them to a new time series which is stationary, by following the 'difference method' and thus create a new series with location and scale. If there is an observable trend it is suitable to fit some type of curve to the data and then model the residuals from that fit. In addition, for a non-constant variance, different kind of transformations can applied such as taking the logarithm or square root of the series and thus stabilizes the variance.

mean μ standard deviation σ where each observation has an identical weight[5] w, the points on the normal probability plot should lie approximately on a straight line.

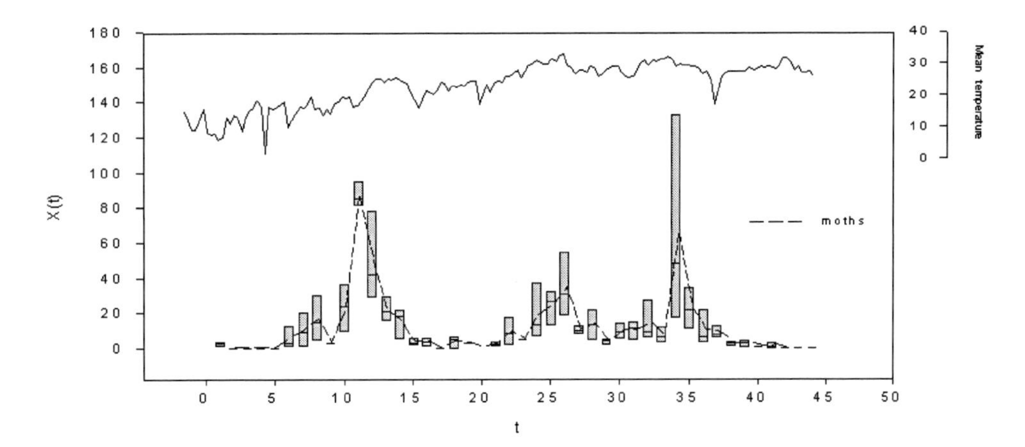

Figure 6. Time series (dashed line) and respective box-plot of adult population dynamics of a moth during a typical cultivation season in relation to mean daily temperatures (solid line in °C).

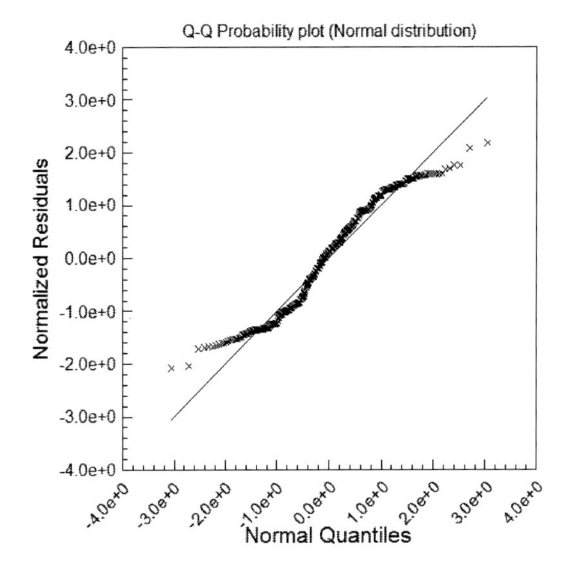

Figure 7. Normal quantile-quantile (Q-Q) plot in describing patterns of cumulative male moth data during a typical generation.

5 For a weighted normal probability, one first define the weighted probability as: $p_i^{(w)} = \sum_{j=1}^{i} w_i (1 - \frac{3}{8i}) / W (1 + \frac{1}{4n})$, where w_i is the weight associated with value x_i for the ith observation and $W = \sum_{i=1}^{n} w_i$ is the sum of weights and the above function is finally plotted against the normal quantile.

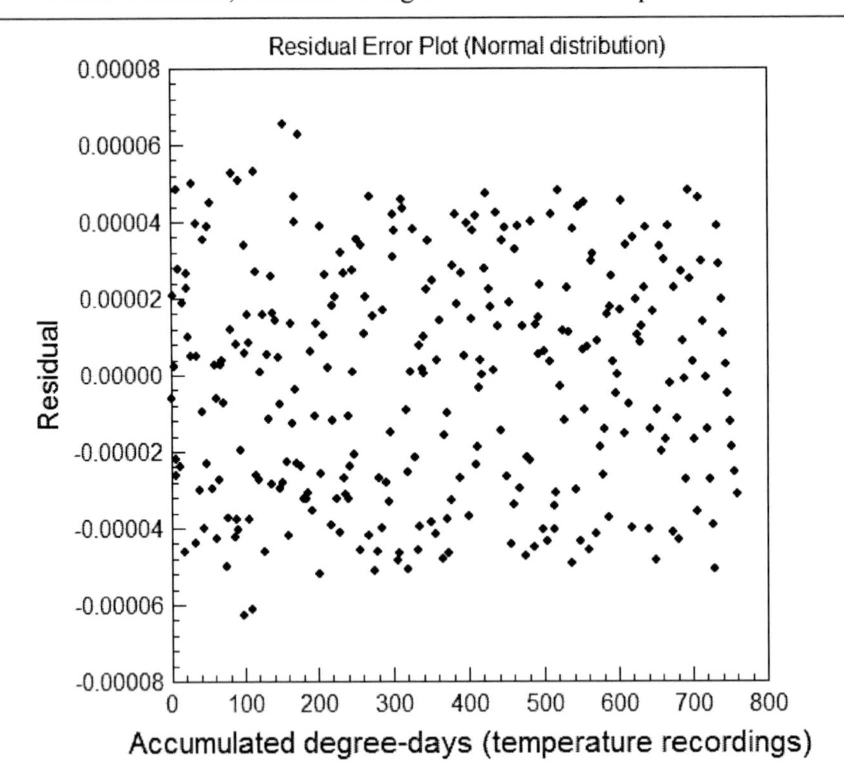

Figure 8. Residual error scatter plot in describing patterns of cumulative male moth data during a typical generation.

3.3. Memory Effects

One important assumption of stochastic process (i.e. on time series analysis and Markov Chain models) is the detection of memory effects. That is the degree in which an observation is related to an adjacent observation [10].

One practical way in detecting randomness on data analysis is by computing autocorrelations for data values at varying time lags. Ideally, if random, such autocorrelations should be near zero for any and all time-lag separations, while the validation of the correlation between random variables at two different points in space or time, can be made according to the correlation function.

The autocorrelation plots are prior prerequisite in application of Box – Jenkins models and defining the order of the model (i.e. Autoregressive Moving Average models, ARMA [14]).

The Correlation functions are a useful indicator of dependencies as a function of distance in time or space, and they can be used to assess the distance required between sample points for the values to be effectively uncorrelated and can form the basis of rules for interpolating values at points for which there are observations and by definition for random variables $X(s)$ and $X(t)$ at different points s and t of some space, the correlation function is:

$$C(s,t) = corr(X(s), X(t)) \tag{49}$$

A schematic illustration of a generated autocorrelation function applied in detecting the autocorrelation trends on data which corresponds to the seasonal emergence patterns of moth during three growth seasons are given in Figure 9. In this case seasonal moth appearance is treated as a time series.

In particular, for the construction of Markov chains with some degree of time dependency it is convenient to define successive probabilities related to some extend by a prior probability [63]. A special case is the *Memory Markov Chain* (MMC) which in fact is a modified first order Markov Chain with enhanced correlation. One other representative example is that of describing correlation of complex dynamic systems with long-range memory based on the concept of additive Markov chains [10] [62] [66].

Hence, the major assumption for the construction of an additive MMC holds that the probability of a random variable X_t to have a certain value under the condition that the values are dependent from the probabilities of more than one previous values, or:

$$P(X_{t+1} = x_{t+1} | X_t = x_t, X_{t-1} = x_{t-1}) \qquad (50).$$

In this case the probability that a random variable X_n has a certain value under the condition that the values of all previous variables are fixed depends on the values of m previous variables only, and the influence of previous variables on a generated one is additive [63]:

$$\Pr(X_n = x_n | X_{n-1} = x_{n-1}, X_{n-2} = x_{n-2}, \ldots, X_1 = x_1) = \sum_{r=1}^{m} f(x_n, x_{n-r}, r) \qquad (51).$$

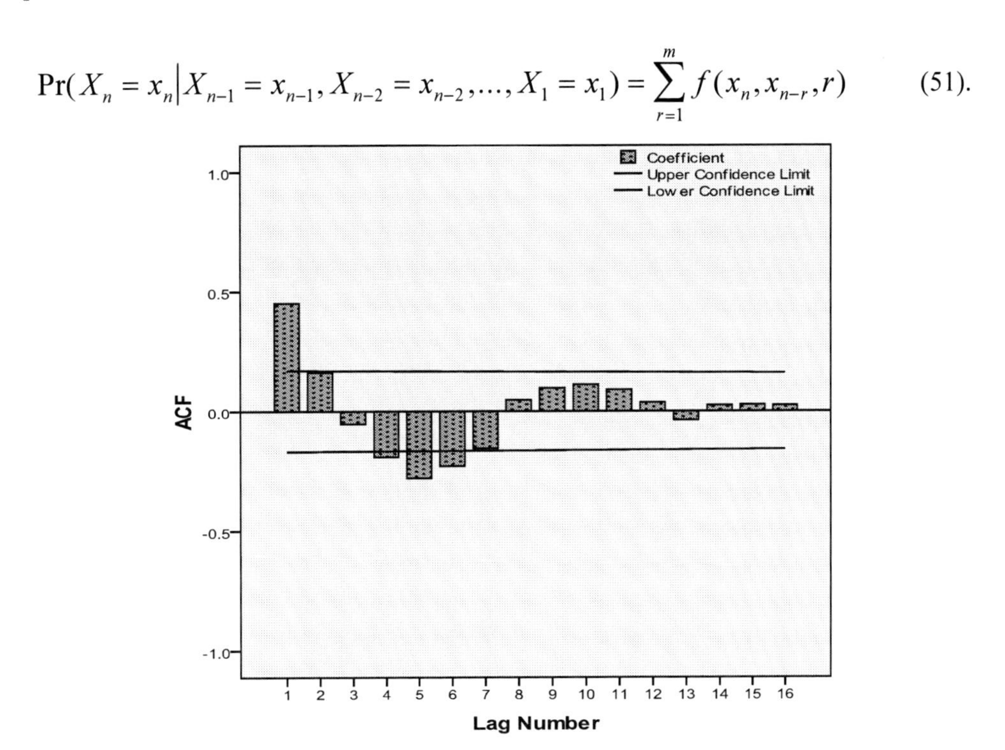

Figure 9. Autocorrelation plot concerning the data of moth generations, during successive growth seasons and modeled as a time series.

4. Modelling the Time Evolution of Insect Developmental Events

4.1. Insect Seasonality

In general, in most poikilothermic[6] organisms the speed of their metabolism, and related developmental events, varies with temperature. Hence the insect's 'physiological clock' run fast or slows depending on how warm or cold their habitat is. Most modelling procedures, either deterministic or stochastic, which are applied to simulated insect phenology during the growth season, are to a high degree related to accumulation of temperatures.

Generally, and for simplicity reasons, it is assumed that development is cumulative in a variable-temperature environment although other factors such as predation or individual variability also affecting population dynamics. Figure 10, for instance displays the empirical cumulative distribution functions of the summer generations of a moth (*A. lineatella*) as observed during 4 successive years in Mediterranean temperate climate (northern Greece), while Figure 11 plots the respective simulated cumulative density functions.

The cumulative probability distribution trajectories are generated for 12 discrete times ($t=1,...,12$) and the applied models are superimposed for comparison. In the above example the model describes the cumulative probability functions only for the adult stage (moths are apparent only during growth season) and is generated for successive growth periods (4 years).

In addition, the first and second moments of the empirical as well as the generated probability density function appears quite stable during all seasons. In fact each generation has a probability distribution $S(t)$ that changes over time (i.e. diffusion) as time intervals t increases. As in the case of general stochastic modelling of insect phenology, $S(t)$ and t are usually measured in degree-days and describe insects stage transition during time for all individuals of a cohort [15] [86].

The generated probability function in describing the seasonal moth phenology as a diffusion process is presented in Figure 12, while Figure 13 illustrates the respective to the diffusion process Brownian motion model, moth emergence and population dynamics temperature recordings correspond to a typical Mediterranean (Greece) temperate climate. Model is generated for 12 discrete time intervals corresponding 3 generations of 4 successive observation years. Statistical uncertainties are inherent in the generated Brownian dynamic simulation, while the trajectory step in the dynamics simulation appeared quite linear.

[6] They have body temperature directly tied to the temperature of the environment they live in, in contrast to homeothermic or warm-blood organisms which have evolved a homeostatic insight temperature regulation mechanism.

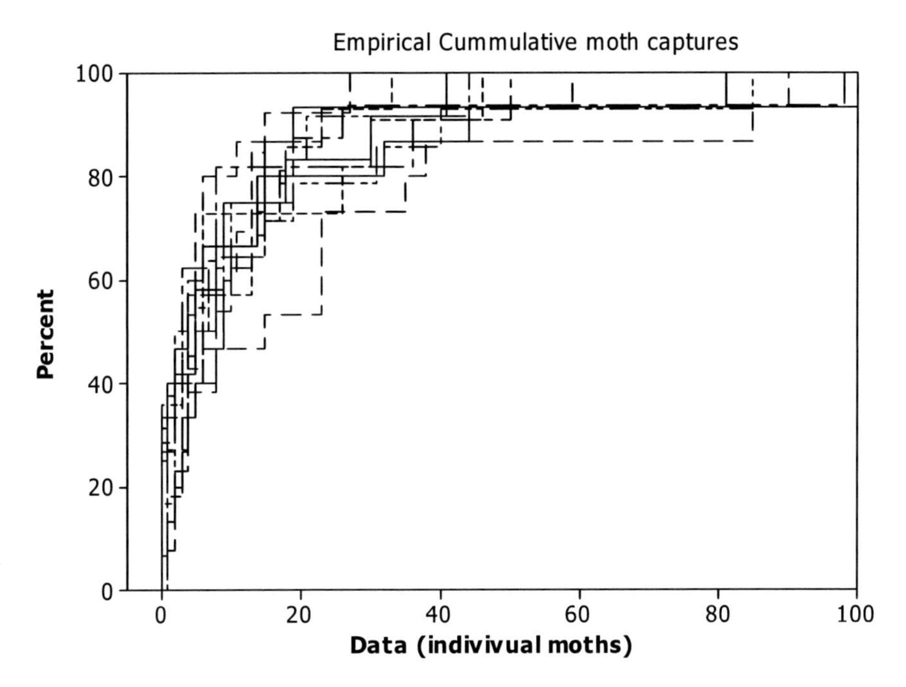

Figure 10. Empirical cumulative moth captures from data observed during successive growth seasons in a typical Mediterranean temperate climate (Greece).

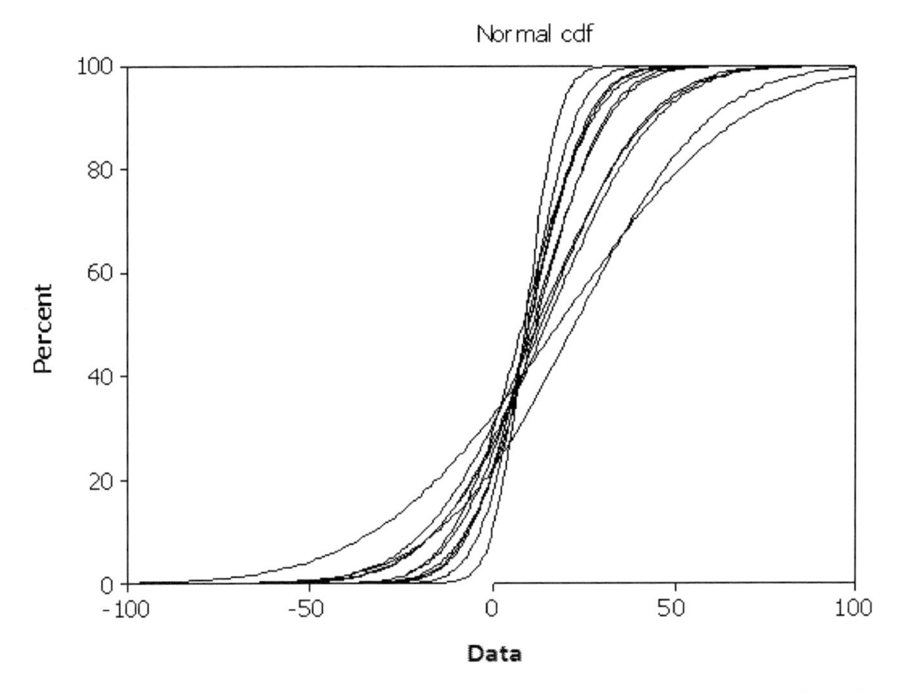

Figure 11. Comparative plots of simulated cumulative distribution functions (cdf), based on observed - empirical data of typical adult moth phenology during successive seasons in a temperate region.

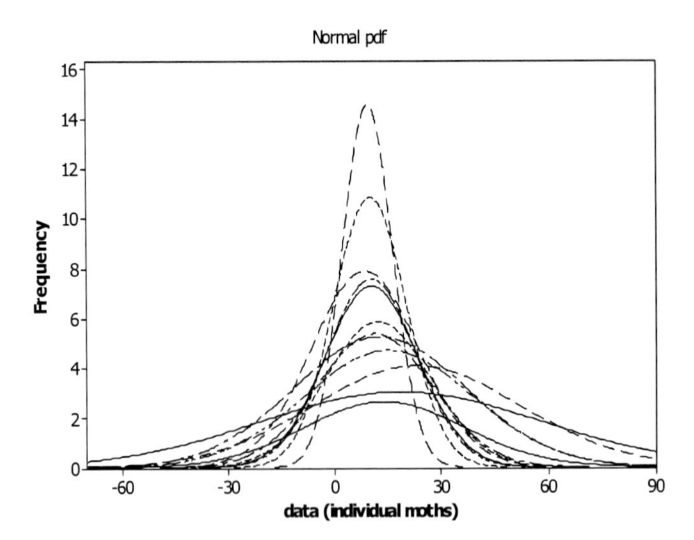

Figure 12. Generated normally probability distribution (pdf) of seasonal adult moth phenology regarded as a diffusion process or random Brownian movement. Simulated trajectories made at discrete time intervals (t=1,..,12).

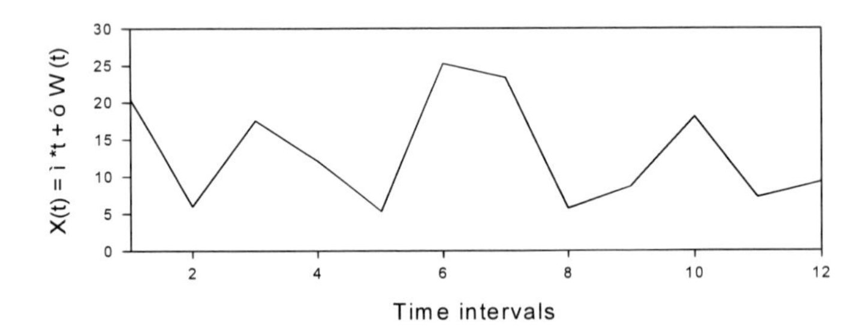

Figure 13. Observed emergence patterns of a moth, regarded as a sample function of a Brownian motion model or Wiener stochastic process. Set of points at which measurements were made, represent successive sampling time intervals during successive cultivation seasons.

4.2. Diapause Induction and Termination

By assuming that *A. lineatella* life cycle is a Gaussian process having continuous paths and stationary independent increments a theoretical stochastic description of the transition is made (see also 2.2 and related figure 3). That is, either to initiate diapause at the larval stage, mostly due to alterations on temperature and photoperiod, or to give an additional generation – state at the end of growth season. During winter the species have the possibilities either to terminate diapause or to 'hold' the diapause state for more time, weeks or days. Both outcomes are depending mostly upon critical environmental conditions (i.e. chilling temperatures and photoperiod) which the larvae are experienced during overwintering. In a more complex case (probably more closer to reality) the physiological state of the overwintering larvae could be divided to several diapause states with certain degree on its intensity (i.e. S=[*1,2,3,4,5*]) and the probability of diapause termination among the individuals flows through those states according to a respective homogenous transition matrix.

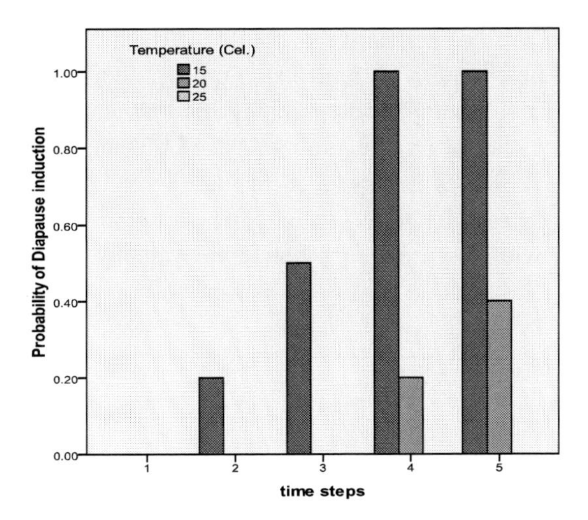

Figure 14. Temperature exposure effects for successive time steps on the probability of diapause induction on overwintering larva of *A. lineatella.*

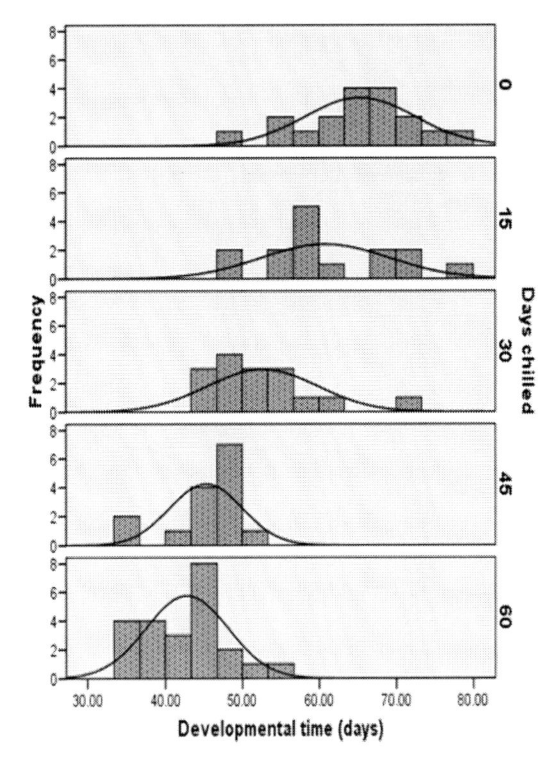

Figure 15. Observed frequencies and generated probability density functions (pdfs) on developmental time until insect pupation, for five fixed time values in which overwintering larvae were chilled at 4°C.

On the opposite case of diapause induction, on the end of growth season, the probability of larvae to enter diapause can be theoretically modelled by following the same approach and by applying a respective 2x2 transition matrix. Figure 14, for instance illustrates the effect of temperature on the probability of diapause induction in relation to successive time steps for larvae that where developed on a short day photocycle (8:16h L:D). The underling processes of diapause termination via the respective physiological states of different intensity has a probability distribution function $S(t)$ as a diffusion process that evolves in time as shown in Figure 15. The effect of two photoperiodic regimes and low temperatures experienced in the field during the winter season on the probability of diapause termination at controlled laboratory conditions is presented in Figure 16, while Figure 17, illustrates the effect of two photoperiodic regimes (short or long day) on the time evolution of diapause termination of an insect during a typical winter season on a temperate climate modelled as general Wiener process.

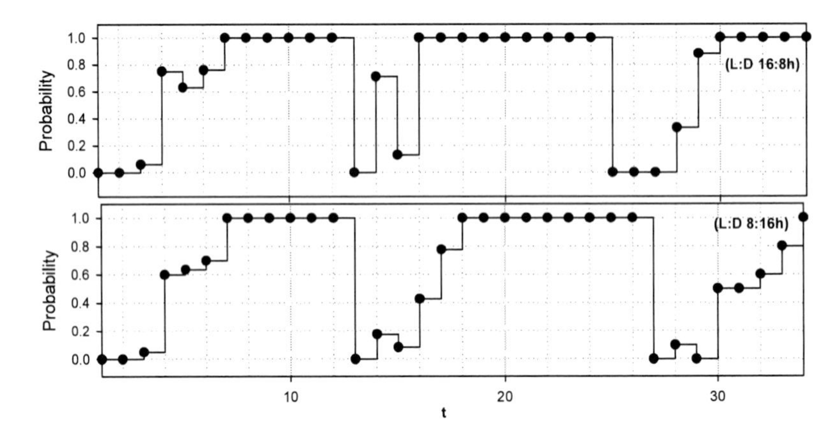

Figure 16. Effect of two different photoperiodic regimes (either long or short day) on the observed probability of diapause termination for different time intervals which corresponds to three successive winter seasons.

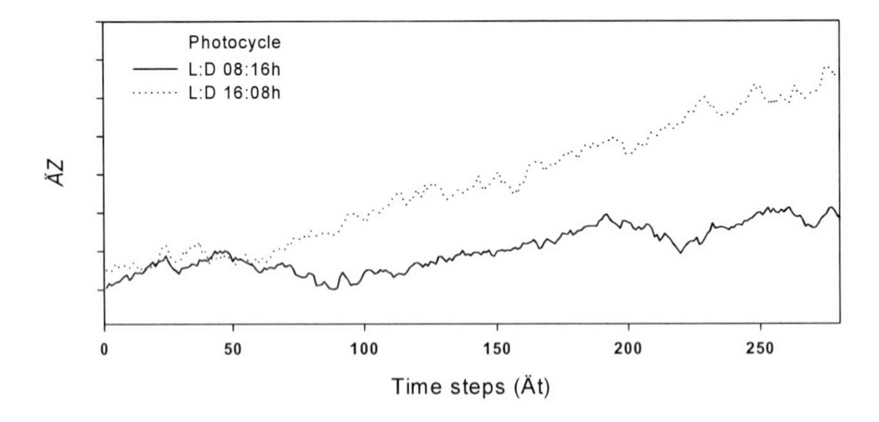

Figure 17. Diapause termination of overwintering larvae of a moth modelled as a Brownian motion, or a general Wiener process. Field collected larvae transferred and developed in the laboratory under two different photoperiodic life cycles (short or long day) at 20°C and 65±5% RH. Brownian movement is generated for 300 time steps and for two different (empirical) drift and variance rates.

4.3. Improved Estimates and Brownian Motion Monte Carlo Simulations

In modelling insect developmental events, parameter estimation is based on a particular number of data sets that are available during a restricted number of observation years and replications. One is therefore interested on applying optimization techniques in calculating improved estimates. For a given model for instance, which is selected for a finite number of time series and data sets, improved estimates of aspects of interest (i.e. confidence intervals for parameters such as the mean and variance), can easily follow if it is possible to obtain and replicate the data sets. In such a case the model could be fitted to each data set in turn, each time resulting in an alternative estimate of the parameters [65].

The idea behind the Monte Carlo methods is to use a class of computational algorithms to initiate and generate a sequence of pseudo-random variables (i.e. random number generator). Figure 18 is a Brownian Motion Monte Carlo (BMMC) simulation, for the mean moth population dynamics shifting from generation to generation during a sequence of growth seasons.

The models were generated for 10 hypothetical trials (replications) during 100 discrete time intervals and for empirical drift and velocity rates as estimated by field observations. This time approximates 33 insect generations which corresponding to 12 successive observation years. As in the case of particle movement and Markov chain models the observed random variable, which corresponds to insect population dynamics, has the property that previous paths are all irrelevant to the present position and therefore the prediction of future positions should take into account this uncertainty which is incorporated by the non deterministic term of the Brownian motion model as simulated.

The ability, to predict long terms seasonal flight patterns, by performing a deterministic computation, using inputs of pseudorandom generators provide a methodology to compare and contrast statistics under several and different realistic data conditions. The results provide probabilities of different outcomes and scenarios especially for small sample data and thus provide important ecological interpretation of the organism (i.e. insect) population dynamics.

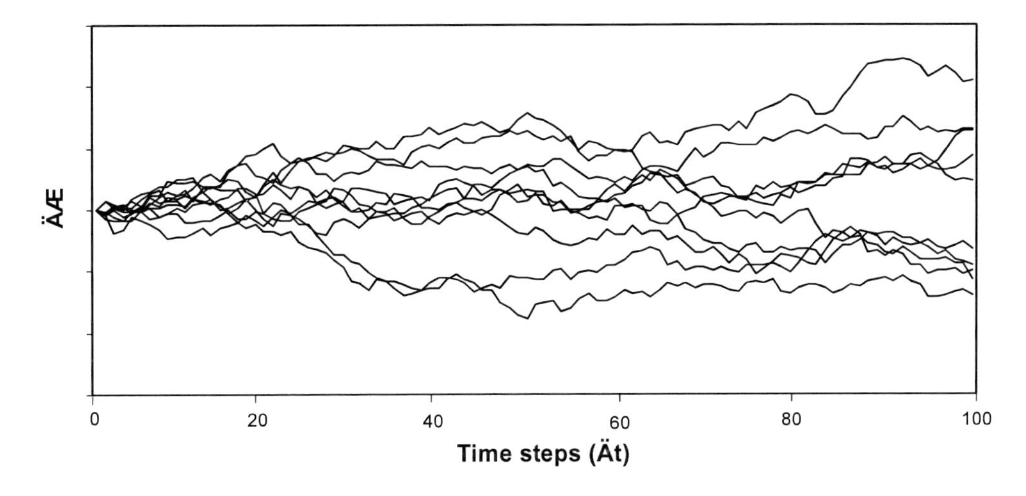

Figure 18. Brownian Motion Monte Carlo simulations (BMMC) of moth population dynamics. The values of ΔZ have a normal distribution with mean of $\mu\Delta t$ and variance of $\sigma^2\Delta t$. The model was generated for 10 hypothetical trials and 100 iteration time steps for a representative drift and variance rates.

5. Concluding Remarks

In the theoretic conceptual structures we used mathematical syntax, borrowed from probability theory, in order to gain new interpretations of certain biological phenomena during the life cycle of an insect when regarded as a semantic entity. The interest was to see whether the process of insect seasonality, which one might characterise as the probability of the appearance of certain developmental events, could be described by following the universal probability laws.

The Brownian motion is a continuous time stochastic process with continuous values and independent and stationary increments and is therefore obviously a Markov process. In this work some fundamental biological variables are modeled as Markov processes or Brownian motion movement. However, the range of application of Brownian motion goes far beyond this chapter and the fundamental study of microscopic particles in suspensions and related diffusion processes, including modelling of stock prices, thermal noise in electrical circuits, random perturbation and limiting behaviors in a variety of biological, physical, economic and management systems. It should also be noted that the Brownian motion is characterized by generality, since it is the asymptotic limit of many random variables and diffusion processes.

A practical utility in following the presented approaches, is to predict the future behavior of the system by calculating the cumulative density functions and further simulate, using Monte Carlo simulations, its trajectories under certain conditions. That is to predict the time evolution of certain developmental events under different boundary conditions. However, a key part of solving such kind of problems is to find the state description of the system that has the Markov properties. Thus, an accurate system definition is a prerequisite in modeling such kind of biological data, which may be for instance Markovian in respect to one state, but non Markovian to another. In addition, one major disadvantage of this kind of modeling procedures is oversimplification and subjective, to some degree, assumption of stationarity and time independence in order to fit the models. However, these trends, also evident in the given examples, can be handled to some degree. More sophisticated (from a mathematical point of view) Markov chain models are developed for processes in which deviations concerning stationarity and time independence are more or less present. They can also be used to test and handle the data for normality assumptions. Hence, stationarity deviations can be resolved to some extent, after the relative transformations and relative normality tests, while the Markov memory chain models have the potential to be applied where strong autocorrelations are present.

Under the above assumptions, the corresponding processes are modelled as a Markov chain, which for each initial state gives the probability that the defined system will undergo a transition following a certain attractor in time. In addition, under natural field conditions these environmental factors might be expected (as also indicated by the generated models) to influence the time of diapause induction and its termination. They may also influence the emergence of adults on the forthcoming season. It is also conceivable to indicate that the ability to utilize different natural variables by the species has evolved to such a degree that it predetermines its geographical distribution. Hence, the developmental physiology includes periodical fluctuations which control the times in which successive developmental events occur, in order to synchronize the emergence of populations with host specific growth season.

Diapause is the basic means by which insects and related arthropods cope with unfavourable environmental conditions in temperate zones, but also with other factors such as food-host quality [30] [79]. Thereby, diapause course is critical for ensuring species survival during different kinds of unfavourable conditions. For insects in the temperate zone, winter is the most unfavourable season. An early emergence from diapause might expose insects to harsh conditions, whereas a delayed emergence might result in the insects being out of synchrony with their hosts [78] [79]. Consequently, end timing of diapause is a very important adaptation on the ecology of insects, which assures successful resumption during the growth season.

Abiotic factors, such as chilling temperatures in field or laboratory and photoperiod, influence the potential outcome of the process significantly. Thus, cold storage either in the field or in a laboratory, acts as a driving force in regulating this phenomenon. In practical terms, that means that if temperatures are not favourable during autumn, the species will not enter diapause and the possibility of a fulfilment of one additional generation is an event with high possibility to occur. Whereas, if temperatures during winter are not suitable (effective chilling time intervals) for triggering the process of diapause termination, the phenology of the species, including synchronized spring emergence, will be strongly affected [59]. Hence, from an ecological standpoint, the time at which the motion will have reached its limit has consequences on the success of mating opportunities and host allocation. Furthermore, the adult forms of many insect species have been observed to emerge later than others. This pattern of adult emergence of a given species tends to be correlated with the diel photoperiod patterns and temperature fluctuations and also affects the number of generations during the growth season [32] [36] [38] [53] [69] [78].

These facts can act as driving forces in the insect's population dynamics affecting evolution and survivor. In addition, slight temperature alterations on a micro or macro environmental scale, in relation to regional photoperiodic lifecycles, also have a strong influence on the regional dispersion, geographical distributions and host allocation by a species. It is also worth noting that for many specialist (i.e. oligophagous) organisms, timing and synchrony are much more important when compared to general feeders. Thus, for successful mating and reproduction, an entire population of potential mates must emerge and be at similar developmental stages simultaneously. In the same context, the timing and development of resource-using life stages must coincide with the appearance of ephemeral resources for successful foraging in natural environments [21] [32] [33] [36] [53].

The behavioral and developmental rhythms of the insect's developmental events, as discussed in the previous sections, are governed strongly by external environmental factors. Moreover, the presented approaches in modeling some of the developmental events and seasonal dynamics of a moth, provide a relatively flexible and new method for handling discrete-valued biological data and modeling their time evolution. In addition, the method has the potential to describe the system's behavior under certain future conditions, in order to make some ecological interpretations in modified habitats. It also appears to be quite useful for detecting abrupt or steady changes in the structure and the time evolution of the process as a result of seasonal and long term variations of environmental factors such as temperature and photoperiod.

Hence, the functional unit of a population as a superorganism (i.e. system) and the underling stochastic processes used for its description have the potential to be formally described and therefore can also be predicted under certain boundary conditions. Under a holistic framework, this fact can add a high degree of determinisity and certainty in the system's description, which from a component point of view explains a mostly random behavior.

References

[1] Alt, W. (1980). Biased random walk models for chemotaxis and related diffusion approximations. *Journal of Mathematical Biology*, 9, 147-177

[2] Anderson, A. R. A., Chaplain, M. A. J., Newman, E. L., Steele, R. J. C. and Thompson, A. M. (2000). Mathematical modelling of tumour invasion and metastasis. *Journal of Theoretical Medicine*, 2, 129–154.

[3] Balzter, H. (2000). Markov chain models for vegetation dynamics. *Ecological Modelling*, 126, 139-154.

[4] Baum, L. and Petrie, T. (1966). Statistical inference for probabilistic functions of finite state Markov chains. *The Annals of Mathematical Statistics*, 37, 1554-1563.

[5] Baum, L. and Egon, J. (1967). An inequality with applications to statistical estimation for probabilistic functions of a Markov process and to a model for ecology. *Bulletin of the American Mathematical Society*, 73, 360-363.

[6] Beall, G. (1939). Methods for estimating the population of insects in a field. *Biometrica*, 30, 422-439.

[7] Beck, S. D. (1980) *Insect photoperiodism*, 2nd ed. Academic, New York.

[8] Benhamou, S. and Bovet, P. (1989). How animals use their environment: a new look at kinesis. *Animal Behaviour*, 38, 375–383.

[9] Benjamini, Y. (1988). Opening the box of a box plot. *The American Statistician*, 42, 257-262.

[10] Beran. J. (1994). *Statistics for long-memory processes*. Chapman and Hall, New York.

[11] Berg H. C. (1983). *Random walks in Biology*. Princeton University Press, N.J.

[12] Bins M. R., Nyrop J. P. and van der Werf. W. (2000). Sampling and monitoring in Crop Protection. In: *The theoretical basis for developing practical decision guides*. CABI Publishing, UK.

[13] Borodin, A. N. and Salimnen P. (2002). Handbook of Brownian motion: Facts and Formulae, 2nd ed., Basel: Birkhäuser Box, G. E. P., and Jenkins, G. (1976). *Time Series Analysis: Forecasting and Control*. Holden-Day.

[14] Braner, M. and Hairston Jr. N. G (1989). From cohort data to life table parameters via stochastic modelling. In: *Estimation and analysis of insect populations* (Lect. Notes Statist.), Berlin Heidelberg New York: Springer

[15] Brown, R. (1828). A brief account of microscopical observations made in the months of June, July and August, 1827, on the particles contained in the pollen of plants; and the general existence of active molecules in organic and inorganic bodies. *Philos. Mag.* 4, 161–173.

[16] Byers, J. A. (2001). Correlated random walk equations of animal dispersal resolved by simulation. *Ecology*, 82, 1680-1690.

[17] Chambers, Mrs. E. and Fowlkes, E. B. (1966). "*A Dictionary of Distributions: I.Comparisons With the Standard Normal.*" Paper presented at the Eastern Regional Meeting of the Institute of Mathematical Statistics, Upton, Long Island, New York.

[18] Chambers, J.; Cleveland, W.; Kleiner, B.; and Tukey, P. 1983. *Graphical Methods for data Analysis*. Belmont, CA: Wadsworth,

[19] Chen, E. H. (1971). The Power of the Shapiro-Wilk W Test for Normality in Samples from Contaminated Normal Distributions. *Journal of the American Statistical Association,* 66, 760–762.

[20] Chippendale, G. M. (1982). Insect diapause, the seasonal synchronization of life cycles, and management strategies. *Entomologia Experimentalis et Applicata*, 88, 1-7.

[21] Codling E. A., M. J. Plank and Benhamou S. (2008). Random walk models in Biology. *Journal of Royal Society of Interface*, 5, 813-834.

[22] Cox, D. R. and Miller H. D. (1965). *The Theory of Stochastic Processes*, Chapman and Hall, London.

[23] Crank, J. (1975). *The Mathematics of Diffusion*, 2nd ed. Oxford University Press, London, England.

[24] Cuthbert D. and, Fred W. S. (1971). *Fitting Equations to Data*. New York:John Wiley and Sons.

[25] Cleveland,W. (1985). The Elements of Graphing Data, Monterey, CA: Wadsworth Advanced Books and Software.

[26] Damos P. and Savopoulou-Soultani M. (2008). Temperature dependent bionomics and modeling of *Anarsia lineatella* (Lepidoptera: Gelechiidae) in the Laboratory. *Journal of Economic Entomology*, 101, 1557-1567.

[27] Damos P. and Savopoulou-Soultani M. (2010). Synchronized diapause termination of the peach twig borer *Anarsia lineatella* (Lepidoptera: Gelechiidae): Brownnian motion with drift? *Physiological Entomology*, 35, 64-75.

[28] Damos P. and Savopoulou-Soultani M. (2010). Development and validation of models in forecasting major lepidopterous peach pest complex for Integrated pest management programs. *Crop Protection*, 29, 1190-1199.

[29] Danks, H. V., (1987) Insect dormancy: An ecological perspective. *Biological Survey of Canada*, Ottawa, Canada.

[30] Dennis, B., Kemp, P. W. and Beckwith C. (1986). Stochastic Model of Insect: Estimation and Testing. *Environmental Entomology*, 15, 540-546.

[31] Denlinger, D. L., Giebultowiez, J. and Saunders D.S. (2001). *Insect Timing: Circadian Rhytmicity to Seasonality*. Elsevier, New York.

[32] Denlinger, D. L., (2002). Regulation of diapause. *Annual Review of Entomology*, 47, 93-122.

[33] Einstein, A. (1905). Eine neue Bestimmung der Moleküldimensionen, Inaugural-Dissertation, Universität Zürich, K.J.Wyss, Bern (1905); printed with some changes in Annalen der Physik (Leipzig), 19, (1906), 289-306.

[34] Einstein, A. (1905). Űber die von der molekularkinetischen Theorie der Wärme geforderte Bewegung von in ruhenden Flüssigkeiten suspendierten Teilchen. *Annals of Physics*, 17, 549–560.

[35] Ellers, J. and van Alpen, J. J. M.,(2002). A trade-off between diapause duration and fitness in female parasitoids. *Ecological Entomology*, 27, 279-284.

[36] Evans S. N. Sturmfels B. and Uhler. C. 2010. Commuting birth and death processes. *The Annals of Applied Probability*, 20, 238-266.
[37] Forkner. E. R., Marquis, R.J., Lill, J. T. and Le Corff, J. (2008) Timing is everything Phenological synchrony and population variability in leaf-chewing herbivores of Quercus. *Ecological Entomology*, 33, 276-285.
[38] Frigge, M., D. Hoaglin, and Iglewicz, B. (1989). Some implementations of the box plot. *The American Statistician*, 43, 50-54.
[39] Gihman, I. I., Skorohod, A. V.(1972). *Stochastic differential equations*. Berlin Heidelberg-New York: Springer.
[40] Grimmett, G. R. (1992). *Probability and random processes*, 2nd Edt., Clavedon Press, Oxford.
[41] Helland I. S. (1983). Diffusion models for the dispersal of insects near an attractive center. *Journal of Mathematical Biology*, 18, 103-122.
[42] Hill, N. A. and Häder, D. P. (1997). A biased random walk model for the trajectories of swimming micro-organisms. *Journal of Theoretical Biology*, 186, 503–526.
[43] Hod S. and Keshet. U. (2004). "Phase transition in random walks with long-range correlations", *Physical Reviews. E*, 70, p. 015104.
[44] Hodek, I. and Hodková M. (1988). Multiple role of temperature during insect diapause: a review. *Entomologia Experimentalis et Applicata*, 49, 53-165.
[45] Hulst, R. (1979). On the dynamics of vegetation: Markov chains as models of succession. *Vegetatio*, 40, 3-14.
[46] Horn, H. (1975). Markovian properties of forest succession. pages 196-211, In: *Ecology and Evolution of Communities*. Harvard University Press, Massachusetts, USA.
[47] Iosifescu M., Limnios N., and Oprisan G. (2010). *Introduction to Stochastic Models*, Wiley, London.
[48] Karatzas I. and Shreve S. E. (1991). *Brownian motion and stochastic calculus*. 2nd Ed. Springer-Verlag, N.Y.
[49] Kareiva, P. M., and Shigesada N. (1983). Analyzing insect movement as a correlated random walk. *Oecologia*, 56, 234- 238.
[50] Kenkel, N. (1993). Modelling Markovian dependence in populations of *Aralia nudicaulis*. *Ecology*, 74, 1700-1706.
[51] Korotkov, V., Logofet, D. and Loreau, M. (2001). Succession in mixed boreal forest of Russia: Markov models and non-Markov effects. *Ecological Modelling*, 142, 25-38.
[52] Leather, S. R., Walters K. F. A. and Ball, J. S. (1993). *The Ecology of Insects Overwintering*. Cambridge University Press, Cambridge.
[53] Legg, C. (1980). A Markovian approach to the study of heath vegetation dynamics. *Bulletin de Ecologie*, 11, 393-404.
[54] Logan, J. A. and Bentz B. J. (1999). Model analysis of mountain pine beetle seasonality. *Environmental* Entomology, 28, 924–934.
[55] Li, B.-L. (1995). Stability analysis of a non-homogenous Markovian landscape model. *Ecological Modelling*, 82, 247-256.
[56] Lippe, E., DeSmidt, J. and Glenn-Lewin, D. (1985). Markov models and succession: a test from a heathland in the Netherlands. *Journal of Ecology*, 73, 775-791.
[57] Logofet, D. and Lesnaya, E. (2000). The mathematics of Markov models: what Markov chains can really predict in forest successions. *Ecological Modelling*, 126, 258-298.

[58] Massaki, S., (2002) Ecophysiological consequences of variability in diapause intensity. *European Journal of Entomolgy*, 99, 143-154.

[59] Markov. A. A. (1907) "Extension of the limit theorems of probability theory to a sum of variables connected in a chain". Reprinted in Appendix B of: R. Howard. *Dynamic Probabilistic Systems, volume 1: Markov Chains*. John Wiley and Sons, (1971).

[60] McCulloch, C. E. and Cain M. L. (1989). Analyzing discrete movement data as a correlated random walk. *Ecology*, 70, 383-388.

[61] Melnyk S.S., Usatenko O.V., and Yampol'skii V.A. (2006). "Memory functions of the additive Markov chains: applications to complex dynamic systems", *Physica A*, 361, 405–415.

[62] Melnyk, S. S., Usatenko O. V., Yampol'skii V. A., Apostolov S. S. and Maiselis Z. A. (2006). Memory functions and correlations in additive binary Markov chains. *Journal of Physics A: Mathematical and General*, 39, 14289-14301.

[63] Milikowski, M. (1995). *Knowledge of numbers: a study of the Psychological Representation of the numbers 1-100*. Univesrity of Amsterdam, The Netherlands.

[64] Morgan, B. J. T. (2000). *Applied stochastic modeling*, Arnold, London.

[65] Narasimhan S.L., Nathan J.A., and Murthy K.P.N. (2005). Can coarse-graining introduce long-range correlations in a symbolic sequence?, *Europhysics Letters*, 69, 22-28.

[66] Okubo, A. (1980). Diffusion and ecological problems: mathematical models. *Biomathematics*, vol. 10, Berlin-Heidelberg-New York: Springer.

[67] Oster, G. and Takahashi Y. (1974). Models for age-specific interactions in a periodic environment. *Ecological Monographs*, 44, 483–501.

[68] Powell, J. A., Jenkins J., Logan J. A. and B. J. Bentz (2000). Seasonal temperature alone can synchronize life cycles. *Bulletin of Mathematical Biology*, 62, 977–998.

[69] Perrin J.P. (1965). *Discontinuous Structure of Matter*, In: Nobel Lectures 1922-1941, Elsevier, Amsterdam.

[70] Parzen, E. (1962). *Stochastic processes*. Holden-day, Inc. San Fransisco, Calif.

[71] Peterson S. C. and Noble P. B. (1972). A two-dimensional random-walk analysis of human granulocyte movement. *Biophysical Journal*, 12, 1048-1054.

[72] Ross M. S. (1996). *Stochastic Processes*. 2nd Ed. John Wiley ans Sons. Inc. N.Y.

[73] Ryan, T. A. and Joiner, B. L. (1973). Minitab: A Statistical Computing System for Students and Researchers. *The American Statistician*, 27, 222–225.

[74] Shapiro, S. S. and Wilk, M. B. (1965). An Analysis of Variance Test for Normality (Complete Samples). *Biometrika*, 52, 591–611.

[75] Skalski, G.T., Gilliam, J.F., (2000). Modeling diffusive spread in a heterogeneous population: a movement study with stream fish. *Ecology*, 81, 1685–1700.

[76] Skellam, J. G. (1951). Random dispersal in theoretical populations. *Biometrika*, 38, 196-218.

[77] Tauber, M. J., and Tauber C. A. (1976). Insect seasonality: diapause maintenance, termination, and post diapause development. *Annual Review of Entomology*, 21, 81-107.

[78] Tauber, M. J., Tauber C. A., and Masaki S. (1986). *Seasonal adaptations of insects*. Oxford University Press. New York

[79] Taylor, R. A. J. (1980). A family of regression equations describing the density distribution of dispersing organisms. *Nature*, 286, 53–55.

[80] Turchin, P. (1996). Fracatal analyses of animal movement: A critique. *Ecology*, 77, 2086-2090.

[81] Tufte, E. (1983). *The Visual Display of Quantitative Information*, Cheshire, CT: Graphics Press.

[82] Tucker, B. and Anand, M. (2003). The use of matrix models to detect natural and pollution-induced forest gradients. *Community Ecology*, 4, 89-110.

[83] Tucker B. C. and Anand M. (2004). The application of Markov Models in recovery and restoration. *International Journal of Ecology and Environmental Sciences*, 30, 131-140.

[84] Tukey, J.W. (1977). Exploratory Data Analysis, Reading, MA: Addison Wesley.

[85] Vogl, G. (2005). Diffusion and Brownian motion analogies in the migration of atoms, animals, men and ideas. In: "*Diffusion Funtamentals*", J. Kärger, F. Grinberg, P. Heitjans (Editors), Leipziger Universitätsverlag. International Conference "Diffusion Funtamentals I", Leipzig Sept. 2005.

[86] Waggoner, P. and Stephens, G. (1970). Transition probabilities for a forest. *Nature*, London 225, 1160-1161.

[87] Watkins, J. C. (1991). Diffusion models for Chemotaxis: A statistical analysis of non interactive unicellular movement. *Mathematical Biosciences*, 104, 271-303.

[88] Wilk, M. B. and Gnanadisikan, R. (1968). Probability Plotting Methods for the Analysis of Data. *Biometrika*, 55, 1–17.

In: Brownian Motion: Theory, Modelling and Applications ISBN: 978-1-61209-537-0
Editors: R.C. Earnshaw and E.M. Riley

Chapter 3

APPLICATIONS OF MULTIFRACTIONAL BROWNIAN MOTION TO INTERNET TRAFFIC

***Ming Li*[1,*] *and S. C. Lim*[2]**

[1]School of Information Science and Technology,
East China Normal University, Dong-Chuan Road, Shanghai, China
[2]Sutton Place, Singapore

ABSTRACT

Measuring fluctuations of the Internet traffic (traffic for short) is a challenge and complicated issue in computer communication networks. There are two types of fluctuations of traffic. One is local irregularity or local roughness and the other long-range dependence (LRD). The former is a local behavior of traffic at small time scales while the latter a global property of traffic at large time scales. In the world of computer science, people often use the terms of small time scaling phenomenon and the large time scaling one to represent the fluctuations of traffic locally and globally, respectively. We follow this preference in this chapter without confusions.

Tools to separately describe the scaling phenomena of traffic are desired in several aspects of computer communications, such as performance analysis of network systems. As for this tough issue, standard fractional Brownian motion (fBm) and or its increment process, fractional Gaussian noise (fGn), may be never enough even they are on an interval-by-interval basis, because two types of scaling phenomena may not be characterized by the Hurst parameter H singly. Though the multifractional fBm (mBm) that is indexed by the local Hölder function $H(t)$ may be used to capture the scaling phenomena of those two but $H(t)$ describes two types of scaling phenomena on a point-by-point basis only. In addition, $H(t)$ linear relates to the time-dependent fractal dimension $D(t) = 2 - H(t)$. To overcome the limitation of $H(t)$, we propose to describe the local irregularity of traffic, i.e., small time scaling phenomenon, by using $D(t)$ on a point-by-point basis and to measure the LRD of traffic by using $E[H(t)]$ on an interval-by-interval basis. Since $E[H(t)]$ is a constant within an observation interval while $D(t)$ is random in general, they are uncorrelated with each other. Thus, our proposed method can be used to separately characterize the small time scaling phenomenon and the large one

[*] E-mail address: ming_lihk@yahoo.com

of traffic in the global sense, providing a new outlook or tool to investigate the fluctuations of traffic.

1. Introduction

Denote by $x(t)$ the aggregated Internet traffic (traffic for short), implying the packet size of traffic at time t. Then, we have the first property (P1) of traffic expressed by

$$x(t) \geq 0, t \in (0, \infty). \tag{1.1}$$

The above holds because $x(t)$ is arrival traffic. The second property (P2) of traffic is expressed by

$$x_{\min} \leq x(t) \leq x_{\max}, \tag{1.2}$$

where $x_{\min}(t)$ and $x_{\max}(t)$ are constants restricted by the IEEE standard without technical reasons except the need to limit delays. For instance, the Ethernet protocol forces all packets of $x(t)$ to have $x_{\min}$ = 64 bytes and $x_{\max}$ = 1518 bytes without considering the Ethernet preamble and header (Stalling [1]).

The Poisson type models properly fit in with the traffic on old telephony networks, which are circuit-switched, see e.g., Gibson [2], Gall [3], Lin et al. [4], Manfield and Downs [5], Reiser [6], Cooper [7]. That type of models, however, fails to well characterize the traffic in the Internet, which is packet-switched. By using the techniques of packet switching, a message is marshaled into a set of packets (Coulouris et al. [8]). The size of each packet is random with the restriction expressed by (1.2). Each packet has the head message containing the source and destination addresses. Thus, different packets of a message may go through different routes from the source to the destination as routers examine the network resources and assign each packet a proper hop. A message can be an email file, a bulk of software package, audio or video. A packet usually goes along a route with light load. Consequently, traffic appears "burstiness" (see Tobagi et al. [9]) or intermittency and non-Poisson (Jain and Routhier [10]).

The following measure introduced by Cruz [11] characterizes the bound of the burstiness of traffic

$$0 \leq \lim_{t \to t_0} \int_{t_0}^{t} x(t)dt \leq \sigma, \tag{1.3}$$

see Raha et al. [12], Jiang and Liu [13], Jiang [14], Boudec and Patrick [15], Li and Zhao [16]. The integral expressed in (1.3) does not make sense if $\lim_{t \to t_0} \int_{t_0}^{t} x(t)dt \neq 0$ for the continuous $x(t)$ even in the field of Lebesgue's integrals, see Bartle and Sherbert [17], Trench [18]. However, it makes sense when it is considered in the domain of generalized functions (Kanwal [19], Gelfand and Vilenkin [20]). A simple way to explain (1.3) is

$$\lim_{t \to t_0} \int_{t_0}^{t} x(t)dt = \int_{t_0}^{t} \sigma_1 \delta(t - t_0)dt,$$

where $\sigma_1 \le \sigma$ and $\delta(t)$ is the Dirac-δ function. We take (1.3) as P3, i.e., the third property of traffic $x(t)$, representing the burstiness bound of $x(t)$, which is a local behavior of traffic.

Note that σ is t_0 dependent. Therefore, we may rewrite (1.3) by the following expression

$$0 \le \lim_{t \to t_0} \int_{t_0}^{t} x(t)dt \le \sigma(t_0). \tag{1.4}$$

The above exhibits that traffic has highly local irregularity or high burstiness as observed by Feldmann et al. [21], Willinger et al. [22], Jiang and Dovrolis [23], Papagiannaki et al. [24], Estan and Varghese [25], Paxson and Floyd [26]. Such a local irregularity considerably affects the polices or performances of telecommunication systems, such as queuing (see e.g., Nain [27], Draief and Mairesse [28]), end-to-end delay, see e.g., Nemeth et al. [29], Li and Zhao [30], Jiang and Yin [31], Wang et al. [32], Starobinski and Sidi [33], resource allocation (see e.g., Gravey et al. [34]), anomaly detection (Tian and Li [35]), admission control (Knightly and Shroff [36], Raha et al. [37], Jia et al. [38]), just naming a few.

Another measure below introduced by Cruz [11] describes the bound of the average rate of traffic

$$0 \le \lim_{t \to \infty} \frac{\int_{t_0}^{t} x(t)dt}{t - t_0} \le \rho = \text{constant}. \tag{1.5}$$

Note that the bound of the average rate expressed above describes a global property of traffic. It implies that the bound of the average rate of traffic is robust as ρ is a constant. This is agreement with the experimental observations stated by Feldmann et al. [21], Willinger et al. [22], Paxson and Floyd [26]. We regard (1.5) as the property 4 (P4) of traffic.

Taking into account P3 and P4 together, the accumulated traffic within $[t_0, t]$ is bounded by

$$\int_{t_0}^{t} x(u)du \le \sigma(t_0) + \rho(t - t_0). \tag{1.6}$$

Two properties, namely, P3 that is a local property and P4 which is a global one, play a role in teletraffic engineering. In terms of scaling, we say that the above implies that traffic has scaling phenomena in two folds. One is small time scaling and the other large one.

Denote the autocorrelation function (ACF) of traffic by

$$r_x(\tau) = \mathrm{E}[x(t)x(t + \tau)], \tag{1.7}$$

where E is the mean operator and τ the lag. Then, $r_x(\tau)$ for small lags, more precisely, for $\tau \to 0$, if $r_x(\tau)$ is sufficiently smooth on $(0, \infty)$, is given by

$$r_x(0) - r_x(\tau) \sim c|\tau|^{\alpha}, \tag{1.8}$$

where c is a constant and α the fractal index of $x(t)$. The fractal dimension of $x(t)$, denoted by D, is given by

$$D = 2 - \alpha/2, \tag{1.9}$$

see Adler [39], Hall and Roy [40], Chan et al. [41], Kent and Wood [42], Gneiting and Schlather [43], Lim and Li [44], Lim and Teo [45]. The parameter D is used to describe the local irregularity of traffic from a view of fractals. It is in terms of small time scaling of traffic, Li [46,47,48,49], Li and Lim [50,51,52].

Traffic $x(t)$ is said to be long-range dependence (LRD) if for $\beta \in (0, 1)$

$$r_x(\tau) \sim c|\tau|^{-\beta}, \text{ for } \tau \to \infty, \tag{1.10}$$

where β is the index of LRD. By using the Hurst parameter H, one has

$$H = 1 - \beta/2, \tag{1.11}$$

see Beran [53,54], Beran et al. [55], Mandelbrot [56,57], Willinger et al. [58]. The parameter H is utilized to characterize the global property, more precisely, LRD, of traffic from a view of fractals. Naturally, one needs two measures to describe two scaling phenomena as can be seen from our previous work [50-52]. However, conventionally fractional Brownian motion (fBm) as well as its increment process, i.e., fractional Gaussian noise (fGn), which are widely used in traffic engineering, is only indexed by H, see e.g., [1], [21-28], [53-55], [58-101], just citing a few. Therefore, there is a limitation of the standard fGn or fBm to characterize two scaling phenomena, namely, small time scaling and large one, which may be independent each other in general (see e.g., Li [46-49], Li and Lim [50-52], Li et al. [102,103]). Recall that the standard Brownian motion (Bm) (Hida [104]) can be taken as the root of stochastic processes in theory, see e.g., Press et al [105-138], simply mentioning a few. However, fGn as well as fBm is restricted for directly describing two independent scaling phenomena. Therefore, there is a research niche how to represent two independent scaling phenomena of traffic by using fBm, which differs from the generalized Cauchy (GC) process [47] as well as the generalized fGn explained by Li [139]. One thing contributed by Peltier and Levy-Vehel [140,141] is the generalization of the standard fBm by replacing the constant H with the Hölder function $H(t)$, which is also called the local Hurst function. The function $H(t)$ captures the multi-fractality of a sample path on a point-by-point basis. Hence, it is used for either the small scaling phenomena [140,141] or the large ones but both are on a point-by-point basis, see Ayache et al. [158], Lim and Teo [159]. The awkward thing to use $H(t)$ to describe two scaling phenomena is that $H(t)$ is linearly correlated with the fractal dimension $D(t)$, which equals to $2 - H(t)$ [140,141]. To overcome the difficulty to capture the large scaling phenomena of traffic in the global sense, we in this chapter introduce the measure expressed by E[$H(t)$]. In this way, we are able to use $D(t)$ to represent the small scaling phenomena of traffic on a point-by-point basis and E[$H(t)$] to characterize the large scaling phenomena of traffic in the global sense, respectively. The rest of the chapter is organized as follows. Preliminaries are briefed in Section 2. The local Hölder function $H(t)$ and its application to traffic modeling are explained in Section 3. Finally, discussions and conclusions are given in Section 4.

2. PRELIMINARIES

Traffic on old telephony networks is of the Poisson type. It has been successfully used in the design of infrastructure of old telephony networks for years (see e.g., Gibson [2]). It is such a success on old telephony networks that it has almost been taken as an axiom for modelling traffic in communication systems. Due to unsatisfactory performances of the Internet, such as traffic congestions, people began doubting about the Poisson model. To re-evaluate the Internet traffic models, people began measuring and analyzing the traffic at different sites in the Internet during different periods of times (see Paxson [142,143] and Traffic Archive at www.sigcomm.org/ITA/). Experimental processing real-traffic traces reveals that traffic has fractal properties. The early fractal model used for traffic modelling is fGn that was introduced in mathematics by Mandelbrot and van Ness [144]. In this section, we shall first brief the basics of conventional time series. Then, fBm and fGn are discussed.

2.1. Conventional Time Series

Let $\{x_l(t)\}$ (l = 1, 2, ...) be a 2-order stationary random process, where $x_l(t)$ is the lth sample function of the process. We use $x_l(t)$ to represent the process without confusion causing. Its mean in the wide sense can be expressed by

$$\mu_x^s(t) = \lim_{N\to\infty} \frac{1}{N} \sum_{l=1}^{N} x_l(t) = \text{const.} \tag{2.1}$$

Its autocorrelation function (ACF) can be written by

$$R_x^s(t, t+\tau) = \lim_{N\to\infty} \frac{1}{N} \sum_{l=1}^{N} x_l(t) x_l(t+\tau) = R_x^s(\tau). \tag{2.2}$$

In (2.1) and (2.2), the superscript s implies that the mean and the ACF are computed by using spatial average. The mean and the ACF of a process expressed by time average are written by

$$\mu_x^t(t) = \lim_{T\to\infty} \frac{1}{T} \int_0^T x_l(t) dt = \text{const,} \tag{2.3}$$

$$R_x^t(\tau) = \lim_{T\to\infty} \frac{1}{T} \int_0^T x_l(t) x_l(t+\tau) d\tau, \tag{2.4}$$

where the superscript t indicates that the mean and the ACF are computed by time average.

The process $x_l(t)$ is said to be ergodic if (2.5) and (2.6) hold,

$$\mu_x^s(t) = \mu_x^t(t) = \mu_x = \text{const}, \tag{2.5}$$

$$R_x^s(\tau) = R_x^t(\tau) = R(\tau). \tag{2.6}$$

Note that a real-traffic trace is a series of single history. In what follows, consequently, we just use $x(t)$ to represent a traffic process or a random function in general.

Denote by $p(x)$ the probability density function (PDF) of traffic $x(t)$. Then, the probability is given by

$$P(x_2) - P(x_1) = \mathrm{Prob}[x_1 < \xi < x_2] = \int_{x_1}^{x_2} p(\xi) d\xi. \tag{2.7}$$

The mean and the ACF of $x(t)$ based on PDF are written by (2.8) and (2.9), respectively,

$$\mu_x = \int_{-\infty}^{\infty} x p(x) dx, \tag{2.8}$$

$$R_x(\tau) = \int_{-\infty}^{\infty} x(t) x(t+\tau) p(x) dx. \tag{2.9}$$

Let V_x be the variance of x. Then,

$$V_x = \mathrm{E}[x(t) - \mu_x]^2 = \int_{-\infty}^{\infty} (x - \mu_x)^2 p(x) dx. \tag{2.10}$$

By conventional processes, we mean that $p(x)$ is light-tailed. More precisely, exponentially decayed, e.g., the Gaussian distribution given by

$$p(x) = \frac{1}{\sqrt{2\pi V_x}} e^{-\frac{(x-\mu_x)^2}{2V_x}}. \tag{2.11}$$

The Poisson distribution is a discrete probability distribution that expresses the probability of a number of events occurring in a fixed period of time if these events occur with a known average rate and independently of the time since the last event. In communication networks, one is interested in the work focused on certain random variables N

that count, among other things, a number of discrete occurrences (sometimes called "arrivals") that take place during a time-interval of given length. Denote the expected number of occurrences in this interval by a positive real number λ. Then, the probability that there are exactly n occurrences ($n = 0, 1, 2, \ldots$) is given by the Poisson distribution below

$$p(x;\lambda) = \frac{\lambda^k e^{-\lambda}}{n!}. \tag{2.12}$$

One thing worth noting is that either (2.11) or (2.12) decays fast. Therefore, according to (2.8) and (2.10), μ_x and V_x are convergent, which is actually a defaulted assumption in the traditional theory of communication networks. In addition, R_x in (2.9) is exponentially decayed. Therefore, the power spectrum density (PSD) below of a conventional random function exits in the domain of ordinary functions

$$S_x(\omega) = \int_{-\infty}^{\infty} R_x(\tau) e^{-j\omega\tau} d\tau. \tag{2.13}$$

In summary, a conventional random function has the following properties.

- $p(x)$ is light-tailed.
- μ_x and V_x exist.
- R_x decays exponentially.
- S_x exists in the domain of ordinary functions.

However, actual traffic challenges the above properties.

2.2. FGn and FBm for Traffic with LRD

Computer scientists claim that a traffic time series is heavy-tailed, see e.g., Abry et al. [75], Borgnat et al. [76], Willinger et al. [145], Li [48,146], Cappe et al. [147]. The tail of the PDF of traffic may be so heavy that its μ_x and V_x do not exist. Due to this meaning of the heavy tails, the ACF of traffic decays so slow in a hyperbolical manner such that R_x is non-integrable. Consequently, a random variable that represents a traffic time series can be no longer considered to be independent. Hence, LRD or long memory. On the other hand, the PSD of traffic with LRD has to be considered in the domain of generalized functions and it obeys a power law according to the theory of the Fourier transform, see Kanwal [19], Gelfand and Vilenkin [20]. Hence, $1/f$ noise.

- Let $B(t)$ be a random process. Then, $B(t_{n+1}) - B(t_n)$ ($n = 0, 1, 2, \ldots$) is its increment process. If $B(t)$ has the following properties, it is called Brownian motion (Bm) (Hida [104]).
- The increments $B(t + t_0) - B(t_0)$ are Gaussian.
- $\mathrm{E}[B(t + t_0) - B(t_0)] = 0$ and $\mathrm{Var}[B(t + t_0) - B(t_0)] = Vt$,

- where $V = \mathrm{E}\{[B(t+1) - B(t)]^2\} = \mathrm{E}\{[B(1) - B(0)]^2\} = \mathrm{E}\{[B(1)]^2\}$.
- In non-overlapping intervals $[t_1, t_2]$ and $[t_3, t_4]$, $B(t_4) - B(t_3)$ and $B(t_2) - B(t_1)$ are independent.
- $B(0) = 0$ and $B(t)$ is continuous at $t = 0$.

Let $B_H(t)$ be the fBm of the Weyl integral type with the Hurst parameter $H \in (0, 1)$. Let $\Gamma(\cdot)$ be the Gamma function. Then,

$$B_H(t) - B_H(0) = \frac{1}{\Gamma(H+1/2)}\left\{\int_{-\infty}^{0}[(t-u)^{H-0.5} - (-u)^{H-0.5}]dB(u) + \int_{0}^{t}(t-u)^{H-0.5}dB(u)\right\}. \quad (2.14)$$

The function $B_H(t)$ has the following properties.

- $B_H(0) = 0$.
- The increments $B_H(t+t_0) - B_H(t_0)$ are Gaussian.
- $\mathrm{Var}[B_H(t+t_0) - B_H(t_0)] = V_H t^{2H}$, where $V_H = \mathrm{E}\{[B_H(1)]^2\}$.
- $\mathrm{E}\{[B_H(t_2) - B_H(t_1)]^2\} = \mathrm{E}\{[B_H(t_2 - t_1) - B_H(0)]^2\} = \mathrm{E}\{[B_H(t_2 - t_1)]^2\} = V_H(t_2 - t_1)^{2H}$.
- $\mathrm{E}\{[B_H(t_2) - B_H(t_1)]^2\} = V_H(t_2)^{2H} + V_H(t_1)^{2H} - 2r[B_H(t_2), B_H(t_1)]$.

Thus, the ACF of $B_H(t)$, denoted by $r_{B_H,W}(t,s)$, is given by

$$r_{B_H,W}(t,s) = \frac{V_H}{(H+1/2)\Gamma(H+1/2)}\left[|t|^{2H} + |s|^{2H} - |t-s|^{2H}\right], \quad (2.15)$$

where

$$V_H = \mathrm{Var}[B_H(1)] = \Gamma(1-2H)\frac{\cos \pi H}{\pi H}. \quad (2.16)$$

Denote by $S_{B_H,W}(t,\omega)$ the PSD of $B_H(t)$. Then (Flandrin [148])

$$S_{B_H,W}(t,\omega) = \frac{1}{|\omega|^{2H+1}}(1 - 2^{1-2H}\cos 2\omega t). \quad (2.17)$$

From the above, we see that either the ACF or the PDF of $B_H(t)$ is time varying. Therefore, $B_H(t)$ is nonstationary.

Note that $B_H(t)$ is self-similar because it satisfies the definition of self-similarity. In fact,

$$B_H(at) \equiv a^H B_H(t), \quad a > 0, \quad (2.18)$$

where $\equiv$ denotes equality in the sense of probability distribution.

From (2.17), one sees that the PSD of $B_H(t)$ is divergent at $\omega = 0$, exhibiting a case of $1/f^{\alpha}$ noise, see Csabai [78] for the early work of $1/f$ noise in traffic theory. The relationship between the fractal dimension of fBm, denoted by D_{fBm}, and its Hurst parameter, denoted by H_{fBm}, is given by

$$D_{\text{fBm}} = 2 - H_{\text{fBm}}. \tag{2.19}$$

Note that the increment series, $B_H(t + s) - B_H(t)$, is fGn. Thus, one has

$$\begin{aligned}
&\mathrm{E}\{[B_H(t_4) - B_H(t_3)][B_H(t_2) - B_H(t_1)]\} = r\{[B_H(t_4) - B_H(t_3)], [B_H(t_2) - B_H(t_1)]\} \\
&= \mathrm{E}\{[B_H(t_4)B_H(t_2) - B_H(t_4)B_H(t_1) - B_H(t_3)B_H(t_2)] + B_H(t_3)B_H(t_1)\} \\
&= \mathrm{E}[B_H(t_4)B_H(t_2)] - \mathrm{E}[B_H(t_4)B_H(t_1)] - \mathrm{E}[B_H(t_3)B_H(t_2)] + \mathrm{E}[B_H(t_3)B_H(t_1)] \\
&= r[B_H(t_4), B_H(t_2)] - r[B_H(t_4), B_H(t_1)] - r[B_H(t_3), B_H(t_2)] + r[B_H(t_3), B_H(t_1)].
\end{aligned} \tag{2.20}$$

According to (2.14), therefore, one has

$$r[B_H(t_4), B_H(t_2)] = \frac{V_H}{2}\,[(t_4)^{2H} + (t_2)^{2H} - (t_4 - t_2)^{2H}], \tag{2.21}$$

$$r[B_H(t_4), B_H(t_1)] = \frac{V_H}{2}\,[(t_4)^{2H} + (t_1)^{2H} - (t_4 - t_1)^{2H}], \tag{2.22}$$

$$r[B_H(t_3), B_H(t_2)] = \frac{V_H}{2}\,[(t_3)^{2H} + (t_2)^{2H} - (t_3 - t_2)^{2H}], \tag{2.23}$$

$$r[B_H(t_3), B_H(t_1)] = \frac{V_H}{2}\,[(t_3)^{2H} + (t_1)^{2H} - (t_3 - t_1)^{2H}]. \tag{2.24}$$

Replacing the right hand of (2.20) by (2.21) ~ (2.24) yields

$$\begin{aligned}
&\mathrm{E}\{[B_H(t_4) - B_H(t_3)][B_H(t_2) - B_H(t_1)]\} = r\{[B_H(t_4) - B_H(t_3)], [B_H(t_2) - B_H(t_1)]\} \\
&= \frac{V_H}{2}\,[(t_4 - t_2)^{2H} + (t_3 - t_2)^{2H} - (t_4 - t_2)^{2H} - (t_3 - t_1)^{2H}].
\end{aligned} \tag{2.25}$$

In the discrete case, we let $t_1 = n$, $t_2 = n + 1$, $t_3 = n + k$, $t_4 = n + k + 1$. Then,

$$r\{[B_H(t_4) - B_H(t_3)], [B_H(t_2) - B_H(t_1)]\} = \frac{V_H}{2}\,[(k + 1)^{2H} - 2k^{2H} + (k - 1)^{2H}]. \tag{2.26}$$

Thus, the ACF of the discrete fGn (dfGn) is given by

$$r(k) = \frac{V_H}{2}\,[(k + 1)^{2H} - 2k^{2H} + (k - 1)^{2H}]. \tag{2.27}$$

Since the ACF is an even function, we have

$$r(k) = \frac{V_H}{2}\left[\left(|k|+1\right)^{2H} + \left||k|-1\right|^{2H} - 2|k|^{2H}\right], \tag{2.28}$$

where $k \in \mathbb{Z}$. Denote by $C_H(\tau;\ \varepsilon)$ the ACF of fGn in the continuous case. Then,

$$C_H(\tau;\ \varepsilon) = \frac{V_H \varepsilon^{2H-2}}{2}\left[\left(\frac{|\tau|}{\varepsilon}+1\right)^{2H} + \left|\frac{|\tau|}{\varepsilon}-1\right|^{2H} - 2\left|\frac{\tau}{\varepsilon}\right|^{2H}\right], \tag{2.29}$$

where $\varepsilon > 0$ is used by smoothing fBm so that the smoothed fBm is differentiable.

The PSD of dfGn was derived out quite early by Sinai [149]. It is given by

$$S_{\text{dfGn}}(\omega) = 2C_f(1-\cos\omega)\sum_{n=-\infty}^{\infty}\left|2\pi n + \omega\right|^{-2H-1}, \tag{2.30}$$

where $C_f = V_H(2\pi)^{-1}\sin(\pi H)\Gamma(2H+1)$ and $\omega \in [-\pi,\ \pi]$. The PSD of fGn is (see Li and Lim [150])

$$S_{\text{fGn}}(\omega) = V_H \sin(H\pi)\Gamma(2H+1)|\omega|^{1-2H}, \tag{2.31}$$

which exhibits that fGn belongs to the class of $1/f$ noises.

We say that $f(t)$ is asymptotically equivalent to $g(t)$ under the limit $x \to c$ if $f(t)$ and $g(t)$ are such that $\lim_{x\to c}\frac{f(t)}{g(t)} = 1$ (Murray [151]), i.e.,

$$f(t) \sim g(t)\ (t \to c) \text{ if } \lim_{x\to c}\frac{f(t)}{g(t)} = 1, \tag{2.32}$$

where c can be infinity. It has the property expressed by

$$f(t) \sim g(t) \sim h(t)\ (t \to c). \tag{2.33}$$

In this sense, $f(t)$ is called slowly varying function if $\lim_{u\to\infty}\frac{f(ut)}{f(u)} = 1$ for all t.

Let $x(t)$ and $r(\tau)$ be a random function and its ACF, respectively. Then, $x(t)$ is LRD if $r(\tau)$ is non-integrable while it is called short-range dependent (SRD) if $r(\tau)$ is integrable. Theoretically, any series whose ACF is non-integrable are LRD. In the field of telecommunications, however, the term of LRD traffic usually implies a hyperbolically decayed ACF given by

$$r(\tau) \sim c\tau^{2H-2} \; (\tau \to \infty),\, H \in (0.5, 1), \tag{2.34}$$

where $c > 0$ can be either a constant or a slowly varying function.

Note that $0.5[(\tau + 1)^{2H} - 2\tau^{2H} + (\tau - 1)^{2H}]$ can be approximated by $H(2H-1)(\tau)^{2H-2}$. In fact, that is the finite second-order difference of $0.5(\tau)^{2H}$ (Mandelbrot [56]). Approximating it with the second-order differential of $0.5(\tau)^{2H}$ yields

$$0.5[(\tau + 1)^{2H} - 2\tau^{2H} + (\tau - 1)^{2H}] \approx H(2H-1)(\tau)^{2H-2}. \tag{2.35}$$

From the above, one immediately sees that fGn contains three subclasses of time series. In the case of $H \in (0.5, 1)$, the ACF is non-summable and the corresponding series is of LRD. For $H \in (0, 0.5)$, the ACF is summable and fGn in this case is of SRD. FGn reduces to white noise when $H = 0.5$.

Among LRD processes, fGn has its advantage in traffic modeling. For example, it can be used to easily represent two types of traffic series, namely, self-similar process and processes with LRD. Note that LRD is a global property of traffic. However, in principle, self-similarity is a local property of traffic, which is measured by fractal dimension D.

Denote D_{fGn} and H_{fGn} the fractal dimension and the Hurst parameter of fGn, respectively. Then, according to the asymptotic expression (1.8), one has

$$r_{\mathrm{fGn}}(0) - r_{\mathrm{fGn}}(\tau) \sim c|\tau|^{2H_{\mathrm{fGn}}} \text{ for } |\tau| \to 0. \tag{2.36}$$

According to (1.8) and (1.9), therefore, one immediately gets

$$D_{\mathrm{fGn}} = 2 - H_{\mathrm{fGn}}. \tag{2.37}$$

Hence, for fGn type traffic, the local properties of traffic happen to be reflected in the global ones as noticed in mathematics by Mandelbrot [57, p. 27].

The above discussions exhibit that the standard fGn as well as fBm has its limitation in traffic modeling because it uses a single parameter H to characterize two different phenomena, that is, small time scaling and large one. The former is a local property and the latter global one.

3. Describing Fluctuations of Traffic Using H(t)

The above (2.37) implies that the local irregularity of a random function $X(t)$ is the same everywhere. This is too restricting to practical applications. As a matter of fact, if D of a traffic function $x(t)$ is a constant, σ of $x(t)$ in (1.4) is a constant too. This is a monofractal case, which is obviously in contradiction with real traffic as σ is time dependent, see (1.4).

One simple way to investigate the multi-fractality of traffic is to use the multifractional Brownian motion (mBm) that is the generalized fBm by replacing the constant H with a time-dependent function $H(t)$, where $t > 0$ and H: $[0, \infty] \to (a, b) \subset (0, 1)$, which is also called the local Hölder exponent, see Peltier and Levy-Vehel [141], Benassi et al. [152].

Following Peltier and Levy-Vehel [141], the mBm of a random function $X(t)$ is defined by

$$X(t) = \frac{1}{\Gamma(H(t)+1/2)} \int_{-\infty}^{0} \left[(t-s)^{H(t)-1/2} - (-s)^{H(t)-1/2} \right] dB(s) + \int_{0}^{t} (t-s)^{H(t)-1/2} dB(s), \quad (3.1)$$

where $B(t)$ is the standard Bm. The variance of $B_{H(t)}$ is given by

$$E\left[(X(t))^2 \right] = V_{H(t)} |t|^{2H(t)}, \quad (3.2)$$

where

$$V_{H(t)} = \frac{\Gamma(2-H(t))\cos(\pi H(t))}{\pi H(t)(2H(t)-1)}. \quad (3.3)$$

Since $V_{H(t)}$ is time-dependent, it will be desirable to normalize $B_{H(t)}$ such that $\mathrm{E}\left[(X(t))^2 \right] = |t|^{2H(t)}$ by replacing $X(t)$ with $X(t) / V_{H(t)}$.

For the subsequent discussion, $X(t)$ will be used to denote the normalized process. The explicit expression of the covariance of $X(t)$ can be calculated by

$$\mathrm{E}[X(t_1)X(t_2)] = N(H(t_1), H(t_2)) \left[|t_1|^{H(t_1)+H(t_2)} + |t_2|^{H(t_1)+H(t_2)} - |t_1 - t_2|^{H(t_1)+H(t_2)} \right], \quad (3.4)$$

where

$$N(H(t_1), H(t_2)) = \frac{\Gamma(2-H(t_1)-H(t_2))\cos\left(\pi \frac{H(t_1)+H(t_2)}{2} \right)}{\pi \left(\frac{H(t_1)+H(t_2)}{2} \right) (H(t_1)+H(t_2)-1)}. \quad (3.5)$$

With the assumption that $H(t)$ is β-Hölder function such that

$$0 < \inf(H(t)) \leq \sup(H(t)) < \min(1, \beta), \quad (3.6)$$

one may approximate $H(t + \lambda u) \approx H(t)$ as $\lambda \to 0$. Therefore, the local covariance function of the normalized mBm has the following limiting form for $\tau \to 0$

$$\mathrm{E}[X(t+\tau)X(t)] \sim \frac{1}{2} \left(|t+\tau|^{2H(t)} + |t|^{2H(t)} - |\tau|^{2H(t)} \right). \quad (3.7)$$

The variance of the increment process for $\tau \to 0$ becomes

$$E\left\{\left[X(t+\tau)-X(t)\right]^2\right\} \sim |\tau|^{2H(t)}, \tag{3.8}$$

which implies that the increment processes of mBm is locally stationary. It follows that the local Hausdorff dimension of the graphs of mBm is given by

$$\dim\{X(t),\ t\in[a,\ b]\} = 2-\min\{H(t),\ t\in[a,\ b]\} \tag{3.9}$$

for each interval $[a, b] \subset R^+$.

Due to the fact that the Hurst index H is time-dependent, mBm fails to satisfy the global self-similarity property and the increment process of mBm does not satisfy the stationary property. Instead, the standard mBm now satisfies the local self-similarity. Recall that fBm $B_H(t)$ is a self-similar Gaussian process with $B_H(at)$ and $a^H B_H(t)$ having identical finite-dimensional distributions for all $a > 0$. For a locally self-similar process, therefore, one may hope that the following expression can provide a description for the local self-similarity of $X(t)$:

$$X(at) \cong a^{H(t)} X(t),\ \forall a > 0, \tag{3.10}$$

where $\cong$ stands for equality in distribution. However, this definition of locally self-similar property would lead to a situation where the law of $X(s)$ depends on $H(t)$ when s is far away from $t: X(s) \cong (s/t)^{H(t)} X(t)$. A more satisfactory way of characterizing this property is the locally asymptotical self-similarity introduced by Benassi, et al. [152]. A process $X(t)$ indexed by the Hölder exponent $H(t) \in C^\beta$ such that $H(t)$: $[0, \infty] \to (a, b) \subset (0, 1)$ for $t \in R$ and $\beta >$ sup($H(t)$) is said to be locally asymptotically self-similar (lass) at point t_0 if

$$\lim_{\lambda \to 0_+} \left(\frac{X(t_0+\lambda u) - X(t_0)}{\lambda^{H(t_0)}} \right)_{u\in R} \cong \left(B_{H(t_0)}(u) \right)_{u\in R}, \tag{3.11}$$

where the equality in law is up to a multiplicative deterministic function of time and $B_{H(t_0)}$ is the fBm indexed by $H(t_0)$. It can be shown that mBm satisfies such a locally self-similar property, see Peltier and Levy-Vehel [141].

Based on the local growth of the increment process, one may write a sequence

$$S_k(j) = \frac{m}{N-1} \sum_{j=0}^{j+k} \left| X(i+1) - X(i) \right|, 1 < k < N, \tag{3.12}$$

where m is the largest integer not exceeding N/k. The local Hölder function $H(t)$ at point

$$t = j/(N-1) \tag{3.13}$$

is then given by (see Peltier and Levy-Vehel [141], Muniandy and Lim [153,154,155], Li et al. [156])

$$H(t) = -\frac{\log(\sqrt{\pi/2}S_k(j))}{\log(N-1)}. \quad (3.14)$$

The local box or Hausdorff dimension denoted by $D(t)$ is equal to

$$D(t) = 2 - H(t) = 2 + \frac{\log(\sqrt{\pi/2}S_k(j))}{\log(N-1)}. \quad (3.15)$$

The function $D(t)$ in (3.15) characterizes the local irregularity of traffic on a point-by-point basis.

Note that $H(t)$ may be used to describe the LRD of traffic on a point-by-point basis, see Ayache et al. [158]. From a view of applications, it is desired to represent the LRD, which is a global property of traffic at large time scales, on an interval-by-interval basis. As a matter fact, from a practical view of the Internet traffic, one is interested in the LRD measure, say H, to investigate how traffic at time t is persistently correlated with that at τ apart from t. Thus, the LRD at time t on a point-by-point basis, i.e., $H(t)$, may be difficult to be used in practice. In addition, since the local irregularity of traffic is independent of its LRD, see Li and Zhao [16], Li and Lim [50], while $H(t)$ linearly correlates to $D(t)$ (see (3.15)), $H(t)$ may be unsatisfactory to characterize the LRD property of traffic. Therefore, we propose the following expression to describe the LRD of traffic

$$H_{\mathrm{m}} = \mathrm{E}[H(t)], \quad (3.16)$$

where the subscript implies the mean.

Note 3.1. $\mathrm{E}[H(t)]$ should be understood on an interval-by-interval basis. □

Note 3.2. $\mathrm{E}[H(t)]$ is uncorrelated with $D(t)$. Denote by corr as a correlation operator. Then, considering that H_{m} is a constant, we have

$$\mathrm{corr}\{\mathrm{E}[H(t)], D(t)\} = 0. \quad (3.17)$$

□

Note 3.3. According to (3.15), we have

$$|\mathrm{corr}\{H(t), D(t)\}| = 1. \quad (3.18)$$

Eq. (3.18) exhibits that $H(t)$ is completely correlated with $D(t)$. □

We show two demonstrations of real-traffic traces named DEC-PKT-1.TCP and DEC-PKT-2.TCP that were recorded at Digital Equipment Corporation (DEC) in March 1995. Figure 1. plots its first 1025 data of traffic DEC-PKT-1.TCP, which is denoted by $x(i)$ to imply the size of the ith packet (i = 0, 1, …). Figure 2. shows its $D(i)$ of the first 8193 data points. The value of $H_{\mathrm{m}} = \mathrm{E}[H(i)]$ for DEC-PKT-1.TCP equals to 0.756 in the range of i = 0, …, 8192. Figures 3. and 4 are plots for DEC-PKT-2.TCP, where H_{m} = 0.754.

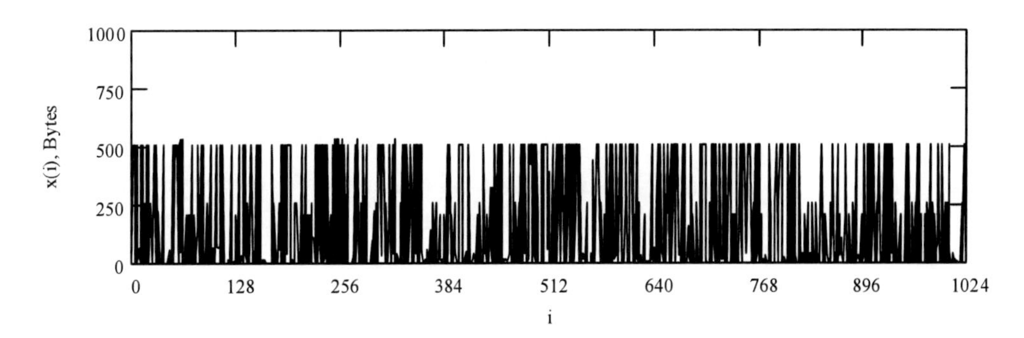

Figure 1. Traffic trace DEC-PKT-1.TCP in packet size.

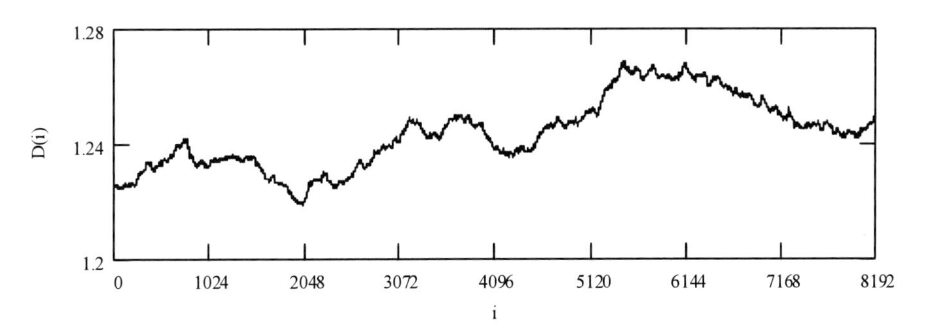

Figure 2. $D(i)$ of DEC-PKT-1.TCP for $i = 0, \ldots, 8192$. $H_m = 0.756$.

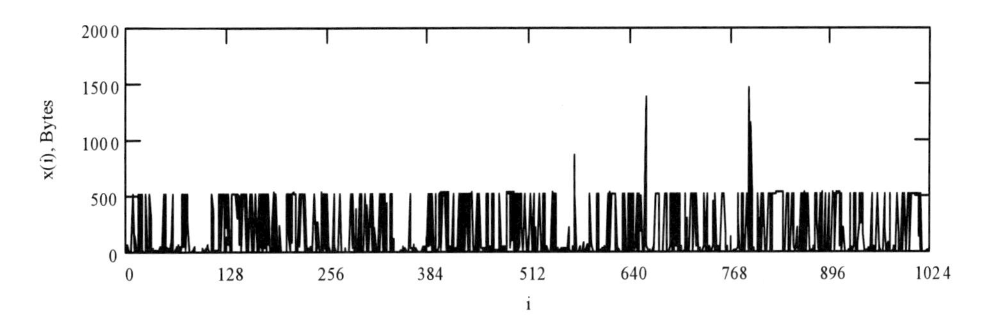

Figure 3. Traffic trace DEC-PKT-2.TCP in packet size.

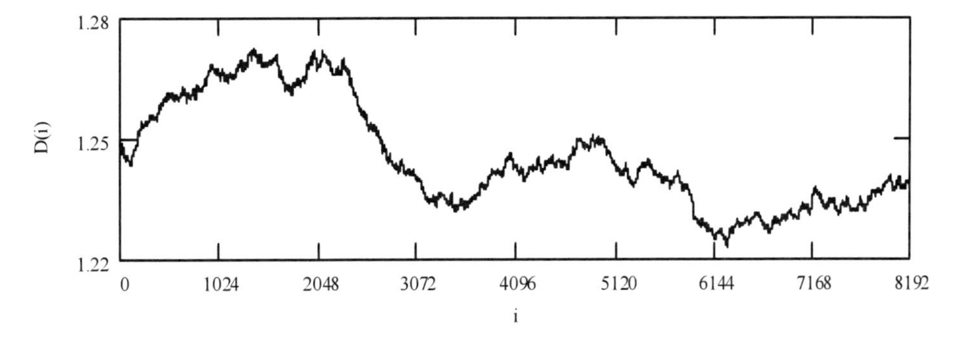

Figure 4. $D(i)$ of DEC-PKT-2.TCP for $i = 0, \ldots, 8192$. $H_m = 0.754$.

CONCLUSION

Scaling plays a role in describing the dynamics of traffic. For instance, traffic is stationary at small time scales but nonstationary at large time scales, see Li et. al [157]. The standard fBm or fGn is not enough as the Hurst parameter may not singly capture the two scaling phenomena that are in general independent of each other. The GC process has two parameters to separately measure the fractal dimension and the Hurst parameter but it describes the local irregularity on an interval-by-interval basis, see Li and Lim [50]. The mBm can be used to characterize the local irregularity and the LRD on a point-by-point basis [141,158,159] but it fails to separately describe the LRD and the local irregularity of traffic because $D(t)$ of mBm linearly relates to its $H(t)$. In this chapter, we have explained that $D(t)$ of mBm is independent of $E[H(t)]$. Thus, we suggest to use $D(t)$ to describe the local irregularity of traffic on a point-by-point basis for the small scaling phenomenon and propose to use $E[H(t)]$, instead of $H(t)$, to represent the LRD of traffic for the large scaling phenomenon on an interval-by-interval basis, providing a promising way to study the scaling phenomena of traffic.

ACKNOWLEDGMENTS

This work was supported in part by the National Natural Science Foundation of China under the project grant numbers 60573125, 60873264, 61070214, and the 973 plan under the project number 2011CB302800/2011CB302802.

REFERENCES

[1] Stalling, W. *High-Speed networks: TCP/IP and ATM design principles*; Prentice Hall, 1998.

[2] Gibson, J. D. editor-in-chief, *The Communications Handbook*; IEEE Press, 1997.

[3] Gall, F. Le. One Moment Model for Telephone Traffic. *Applied Mathematical Modelling*, (1982) 6(6): 415-423.

[4] Lin, P.; Leon, B.; Stewart, C. Analysis of Circuit-Switched Networks Employing Originating-Office Control with Spill-Forward. *IEEE Trans. Communications* (1978) 26(6): 754-766.

[5] Manfield, D.; Downs, T. On the One-Moment Analysis of Telephone Traffic Networks. *IEEE Trans. Communications* (1978) 27(8): 1169-1174.

[6] Reiser, M. Performance Evaluation of Data Communication Systems. *Proc. IEEE*, (1982) 70(2): 171-196.

[7] Cooper, R. B. *Introduction to Queueing Tehory*; Elsevier, 1981.

[8] Coulouris, G.; Dollimore, J.; Kindberg, T. *Distributed Systems: Concepts and Design*; 3rd Ed., Addison-Wesley, 2001.

[9] Tobagi, F. A.; Gerla, M.; Peebles, R. W.; Manning, E. G. Modeling and Measurement Techniques in Packet Communication Networks. *Proc. IEEE* (1978) 66(11): 1423-1447.

[10] Jain, R.; Routhier, S. Packet Trains-Measurements and a New Model for Computer Network Traffic. *IEEE Journal of Selected Areas in Communications* (1986) 4(6): 986-995.

[11] Cruz, R. L. A Calculus for Network Delay, Part I: Network Elements in Isolation. Part II: Network Analysis. *IEEE Trans. Information Theory* (1991) 37(1): 114-141.

[12] Raha, A.; Kamat, S.; Jia, X.; and Zhao, W. Using Traffic Regulation to Meet End-To-End Deadlines in ATM Networks. *IEEE Trans. Computers* (1999) 48(9): 917-935.

[13] Jiang, Y.-M. and Liu, Y. *Stochastic Network Calculus*; Springer, 2008.

[14] Jiang, Y.-M. Per-Domain Packet Scale Rate Guarantee for Expedited Forwarding. *IEEE/ACM Trans. Networking* (2006) 14(3): 630-643.

[15] Boudec, J.-Yves Le and Patrick, T. Network Calculus, *A Theory of Deterministic Queuing Systems for the Inter*net; Springer, 2001.

[16] Li, M.; Zhao, W. Representation of a Stochastic Traffic Bound. *IEEE Trans. Parallel and Distributed Systems* (2010) 21(9): 1368-1372.

[17] Bartle, R. G.; Sherbert, D. R. *Introduction to Real Analysis*; 3rd Ed., John Wiley and Sons, 2000.

[18] Trench, W. F. *Introduction to Real Analysis*; Pearson Education, 2003.

[19] Kanwal, R. P. *Generalized Functions: Theory and Applications*; 3rd Ed., Birkhauser, 2004.

[20] Gelfand, I. M.; Vilenkin, K. *Generalized Functions*; Vol. 1, Academic Press, New York, 1964.

[21] Feldmann, A.; Gilbert, A. C.; Willinger, W.; Kurtz, T. G. *The Changing Nature of Network Traffic: Scaling Phenomena. ACM SIGCOMM Comput. Commun. Rev.* (1998) 28: 5-29.

[22] Willinger, W.; Taqqu, M. S.; Sherman, R.; Wilson, D. V. Self-Similarity through High-Variability: Statistical Analysis of Ethernet LAN Traffic at the Source Level. *IEEE/ACM Trans. Netw.* (1997) 5(1): 71-86.

[23] Jiang, H.; Dovrolis, C. Why is the Internet Traffic Bursty in Short Time Scales? *ACM SIGMETRICS Performance Evaluation Review* (2005) 33(1): 241-252.

[24] Papagiannaki, K.; Cruz, R.; Diot, C. *Network Performance Monitoring at Small Time Scales*. IMC 2003, 2003, Miami, Florida, USA.

[25] Estan, C.; Varghese, G. New Directions in Traffic Measurement and Accounting: Focusing on the Elephants, Ignoring the Mice. *ACM Transactions on Computer Systems* (2003) 21(3): 270-313.

[26] Paxson, V.; Floyd, S. Wide Area Traffic: the Failure of Poison Modeling. *IEEE/ACM Trans. Networking* (1995) 3(3): 226-244.

[27] Nain, P. Impact of Bursty Traffic on Queues. *Stat. Infer. Stoch. Proc.* (2002) 5(3): 307-320.

[28] Draief, M.; Mairesse, J. Services within a Busy Period of an M/M/1 Queue and Dyck Paths. *Queueing Syst*. (2005) 49(1): 73-84.

[29] Nemeth, F.; Barta, P.; Szabo, R.; Biro, J. Network Internal Traffic Characterization and End-To-End Delay Bound Calculus for Generalized Processor Sharing Scheduling Discipline. *Performance Evaluation* (2005) 48(6): 910-940.

[30] Li, C.-Z.; Zhao, W. Stochastic Performance Analysis of Non-Feedforward Networks. *Telecommunication Systems* (2010) 43(3-4): 237-252.

[31] Jiang, Y.; Yin, Q.; Liu, Y.; Jiang, S. Fundamental Calculus on Generalized Stochastically Bounded Bursty Traffic for Communication Networks. *Computer Networks* (2009) 53(12): 2011-2019.

[32] Wang, S.; Xuan, D.; Bettati, R.; Zhao, W. Toward Statistical QoS Guarantees in a Differentiated Services Network. *Telecommunication Systems* (2010) 43(3-4): 253-263.

[33] Starobinski, D.; Sidi, M. Stochastically Bounded Burstiness for Communication Networks. *IEEE Trans. Information Theory* (2000) 46(1): 206-212.

[34] Gravey, A.; Boyer, J.; Sevilla, K.; Mignault, J. Resource Allocation for Worst Case Traffic in ATM Networks. *Performance Evaluation* (1997) 30(1-2): 19-43.

[35] Tian, K.; Li, M. A Reliable Anomaly Detector against Low-Rate DDOS Attack. *International Journal of Electronics and Computers* (2009) 1(1): 1-6.

[36] Knightly, E.; Shroff, N. Admission Control for Statistical QoS: Theory and Practice. *IEEE Network* (1999) 13(2): 20-29.

[37] Raha, A.; Kamat, S.; Zhao, W.; Jia, W. Admission Control for Hard Real-Time Connections in ATM LANs. IEE Proceedings - Communications, (2001) 148(4): 1-12.

[38] Jia, X.; Zhao, w.; Li, J. An Integrated Routing and Admission Control Mechanism for Real-Time Multicast Connections in ATM Networks. *IEEE Trans. Communications* (2001) 49(9): 1515-1519.

[39] Adler, A. J.; *The Geometry of Random Fields*; Wiley, New York, 1981.

[40] Hall, P.; Roy, R. On the Relationship between Fractal Dimension and Fractal Index for Stationary Stochastic Processes. *Ann. Appl. Probab.* (1994) 4(1): 241-253.

[41] Chan, G.; Hall, P.; Poskitt, D. S. Periodogram-Based Estimators of Fractal Properties. *Ann. Stat.* (1995) 23: 1684-1711.

[42] Kent, J. T.; Wood, A. T. Estimating the Fractal Dimension of a Locally Self-Similar Gaussian Process by Using Increments. *J. Roy Stat. Soc. B.* (1997) 59(3): 679-699.

[43] Gneiting, T.; Schlather, M. Stochastic Models that Separate Fractal Dimension and The Hurst Effect. *SIAM Review* (2004) 46(2): 269-282.

[44] Lim, S. C.; Li, M. Generalized Cauchy Process and Its Application to Relaxation Phenomena. *J. Phys. A: Math. Gen.* (2006) 39(12): 2935-2951.

[45] Lim, S. C.; Teo, L. P. Gaussian Fields and Gaussian Sheets with Generalized Cauchy Covariance Structure. *Stochastic Processes and Their Applications* (2009) 119(4): 1325-1356.

[46] Li, M. Teletraffic Modeling relating to Generalized Cauchy Process: Empirical Study; VDM Verlag, Germany, Nov. 6, 2009.

[47] Li, M. Generation of Teletraffic of Generalized Cauchy Type. *Physica Scripta* (2010) 81(2): 025007 (10pp).

[48] Li, M. Self-Similarity and Long-Range Dependence in Teletraffic. Proc., the 9th WSEAS Int. Conf. on Multimedia Systems and Signal Processing, Hangzhou, China, May 2009, 19-24.

[49] Li, M. *Essay on Teletraffic Models* (I). ACACOS'10, Hangzhou, China, April 2010, 130-135.

[50] Li, M.; Lim, S. C. Modeling Network Traffic Using Generalized Cauchy Process. *Physica A* (2008) 387(11): 2584-2594.

[51] Li, M.; Lim, S. C. Power Spectrum of Generalized Cauchy Process. *Telecommunication Systems* (2010) 43(3-4): 219-222.

[52] Li, M.; Lim, S. C. Modeling Network Traffic Using Cauchy Correlation Model with Long-Range Dependence. *Modern Physics Letters* B (2005) 19(17): 829-840.
[53] Beran, J. *Statistics for Long-Memory Processes*; Chapman and Hall, 1994.
[54] Beran, J. Statistical Methods for Data with Long-Range Dependence. *Statistical Science* (1992) 7(4): 404-416.
[55] Beran, J.; Shernan, R.; Taqqu, M. S.; Willinger, W. Long-Range Dependence in Variable Bit-Rate Video Traffic. *IEEE Trans. Communications* (1995) 43(2-3-4): 1566-1579.
[56] Mandelbrot, B. B. *Gaussian Self-Affinity and Fractals*; Springer, 2001.
[57] Mandelbrot, B. B. *The Fractal Geometry of Nature*; W. H. Freeman, New York, 1982.
[58] Willinger, W.; Paxson, V.; Riedi, R. H.; Taqqu, M. S. *Long-Range Dependence and Data Network Traffic, in Long-range Dependence: Theory and Applications.* Doukhan, P.; Oppenheim, G.; Taqqu, M. S. (eds.): Birkhäuser, 2002, 625-715.
[59] Willinger, W.; Taqqu, M. S.; Sherman, R.; Wilson, D. V. Self-Similarity through High-Variability: Statistical Analysis of Ethernet LAN Traffic at the Source Level. *IEEE/ACM Trans Networking* (1997) 5(1): 71-86.
[60] Willinger, W.; Taqqu, M. S.; Leland, W. E.; Wilson, D. V. Self-Similarity in High-Speed Packet Traffic: Analysis and Modeling of Ethernet Traffic Measurements. *Statistical Science* (1995) 10(1): 67-85.
[61] Tsybakov, B.; Georganas, N. D. Self-Similar Processes in Communications Networks. *IEEE Trans. Information Theory* (1998) 44(5): 1713-1725.
[62] Tsybakov, B.; Georganas, N. D. On Self-Similar Traffic in ATM Queues: Definitions, Overflow Probability Bound, and Cell Delay Distribution. *IEEE/ACM Trans. Networking* (1997) 5(3): 397-409.
[63] Adas, A. Traffic Models in Broadband Networks. *IEEE Communication Magazine* (1997) 35(7): 82-89.
[64] Michiel, H.; Laevens, K. Teletraffic Engineering in a Broad-Band Era. Proc. IEEE (1997) 85(12): 2007-2033.
[65] Li, M.; Zhao, W.; Jia, W.; Long, D.; Chi, C.-H. Modeling Autocorrelation Functions of Self-Similar Teletraffic in Communication Networks Based on Optimal Approximation in Hilbert Space. *Applied Mathematical Modelling* (2003) 27(3): 155-168.
[66] Li, M. An Approach to Reliably Identifying Signs of DDOS Flood Attacks Based on LRD Traffic Pattern Recognition. *Computers and Security* (2004) 23(7): 549-558.
[67] Li, M. Change Trend of Averaged Hurst Parameter of Traffic under DDOS Flood Attacks. *Computers and Security* (2006) 25(3): 213-220.
[68] Erramilli, A.; Narayan, O.; Willinger, W. Experimental Queuing Analysis with Long-Range Dependent Packet Traffic. *IEEE/ACM Trans. Networking* (1996) 4(2): 209-223.
[69] Erramilli, A.; Roughan, M.; Veitch, D.; Willinger, W. Self-Similar Traffic and Network Dynamics. *Proc. IEEE* (2002) 90(5): 800-819.
[70] Rolls, D. A.; Michailidis, G.; Hernández-Campos, F. Queueing Analysis of Network Traffic: Methodology and Visualization Tools. Comput Netw (2005) 48(3): 447-473.
[71] Carpio, K. J. E. Long-Range Dependence of Stationary Processes in Single-Server Queues. *Queueing Syst.* (2007) 55(2):123–130.
[72] Abry, P.; Borgnat, P.; Ricciato, F.; Scherrer, A.; Veitch, D. Revisiting an Old Friend: on the Observability of the Relation between Long Range Dependence and Heavy Tail. *Telecommunication Systems* (2010) 43(3-4): 147-165.

[73] Abry, P.; Veitch, D. Wavelet Analysis of Long-Range Dependent Traffic. *IEEE Trans. Inform Theory* (1998) 44(1): 2-15.

[74] Abry, P.; Baraniuk, R.; Flandrin, P.; Riedi, R.; and Veitch, D. Multiscale Nature of Network Traffic. *IEEE Sig. Proc. Mag.* (2002) 19(3): 28-46.

[75] Abrey, P.; Veitch, D. Wavelet Analysis of Long-Range Dependent Traffic. *IEEE Trans Inform Theory* (1998) 44(1): 2-15.

[76] Borgnat, P.; Dewaele, G.; Fukuda, K.; Abry, P.; Cho, K. Seven Years and One Day: Sketching the Evolution of Internet Traffic. Proc. the 28th IEEE INFOCOM 2009, pp. 711-719, Rio de Janeiro (Brazil), May 2009.

[77] Loiseau, P.; Gonçalves, P.; Dewaele, G.; Borgnat, P.; Abry, P. ; Primet, P. V.-B. Investigating Self-Similarity and Heavy-Tailed Distributions on a Large-Scale Experimental Facility. *IEEE/ACM Trans. Netw.* (2010) 18(4): 1261-1274.

[78] Csabai, I. 1/f Noise in Computer Network Traffic. *J. Phys. A: Math. Gen.* (1994) 27(12): L417-L421.

[79] Karagiannis, T.; Molle, M.; Faloutsos, M. Long-Range Dependence: Ten Years of Internet Traffic Modeling. *IEEE Internet Computing* (2004) 8(5): 57-64.

[80] Gong, W.-B.; Liu, Y.; Misra, V.; Towsley, D. Self-Similarity and Long Range Dependence on the Internet: a Second Look at the Evidence, Origins and Implications. *Comput. Netw.* (2005) 48(3): 377-399.

[81] Lee, Ian W. C.; Fapojuwo, A. O. Stochastic Processes for Computer Network Traffic Modeling. *Computer Communications* (2005) 29(1): 1-23.

[82] He, G.; Hou, J. C. On Sampling Self-Similar Internet Traffic. *Comput. Netw.* (2006) 50(16): 2919-2936.

[83] Stoev, S.; Taqqu, M. S.; Park, C.; Marron, J. S. On the Wavelet Spectrum Diagnostic for Hurst Parameter Estimation in the Analysis of Internet Traffic. *Comput. Netw.* (2005) 48(3): 423-445.

[84] Paxson, V. Fast Approximate Synthesis of Fractional Gaussian Noise for Generating Self-Similar Network Traffic. *ACM SIGCOMM Comput. Commun. Rev.* (1997) 27(5): 5-18.

[85] Jeong, H.-D. J.; Lee, J.-S. R.; McNickle, D.; Pawlikowski, P. Distributed Steady-State Simulation of Telecommunication Networks with Self-Similar Teletraffic. *Simulation Modelling Practice and Theory* (2005) 13(3): 233–256.

[86] Ledesma, S.; Liu, D. Synthesis of Fractional Gaussian Noise Using Linear Approximation for Generating Self-Similar Network Traffic. *ACM SIGCOMM Comput Commun. Rev.* (2000) 30(2): 4-17.

[87] Garrett, M. W.; Willinger, W. Analysis, Modeling and Generation of Self-Similar VBR Traffic. Proc ACM SigComm'94, London, 1994, 269-280.

[88] Li, M.; Chi, C.-H. *A Correlation-Based Computational Method for Simulating Long-Range Dependent Data.* J Franklin I (2003) 340(6-7): 503-514.

[89] Chakraborty, D.; Ashir, A.; Keeni, T. S. G. M.; Roy, T. K.; Shiratori, N. Self-Similar and Fractal Nature of Internet Traffic. *Int. J. Network Mgmt. (*2004) 14(2): 119–129.

[90] Lee, Ian W. C.; Fapojuwo, A. O. Analysis and Modeling of a Campus Wireless Network TCP-IP Traffic. *Computer Networks* (2009) 53(15): 2674-2687.

[91] Pitts, J. M.; Schormans, J. A. Introduction to ATM Design and Performance: with Applications Analysis Software; 2nd Ed., John Wiley, 2000.

[92] McDysan, D. QoS and *Traffic Management in IP and ATM Networks*; McGraw-Hill, 2000.

[93] Cleveland, W. S.; Sun, D. X. Internet Traffic Data. *Journal of the American Statistical Association* (2000) 95(451): 979–985.

[94] Cho, J.; Lee, C.; Cho, S.; Song, J. H.; Lim, J.; Moon, J. A Statistical Model for Network Data Analysis: KDD CUP 99' Data Evaluation and Its Comparing with MIT Lincoln Laboratory Network Data. *Simulation Modelling Practice and Theory* (2010) 18(4): 431-435.

[95] Feng, H.; Shu, Y.; Yang, O.; *Nonlinear Analysis of Wireless LAN Traffic*. Nonlinear Analysis: Real World Applications (2009) 10(2): 1021-1028.

[96] Darbha, S.; Rajagopal, K. R.; Tyagi, V. A Review of Mathematical Models for the Flow of Traffic and Some Recent Results. Nonlinear Analysis: Theory, Methods and Applications (2008) 69(3): 650-670.

[97] Masugi, M.; Takuma, T. Multi-Fractal Analysis of IP-Network Traffic for Assessing Time Variations in Scaling Properties. *Physica D* (2007) 225(2): 119-126.

[98] Nicol, D. M.; Yan, G. Discrete Event Fluid Modeling of Background TCP Traffic. ACM Trans. *Modeling and Computer Simulation* (2004) 14(3): 211-250.

[99] Leland, W. E.; Taqqu, M. S.; Willinger, W.; Wilson, D. V. On the Self-Similar Nature of Ethernet Traffic (Extended Version). *IEEE/ACM Trans. Networking* (1994) 2(1): 1-15.

[100] Crovella, E.; Bestavros, A. Self-Similarity in World Wide Web Traffic: Evidence and Possible Causes. *IEEE/ACM Trans. Networking* (1997) 5(6): 835-846.

[101] Nogueira, A.; Salvador, P.; Valadas, R.; Pacheco, A. Modeling Network Traffic with Multifractal Behavior. *Telecommunication Systems* (2003) 24(2-4): 339-362.

[102] Li, M.; Lim, S. C.; Zhao, W. Long-Range Dependent Network Traffic: A View from Generalized Cauchy Process. in Progress in Applied Mathematical Modeling; ed., F. Yang, Nova Science Publishers, USA, 2008, 319-336.

[103] Li, M.; Zhao, W.; Chen, S-Y. FGN Based Telecommunication Traffic Models. WSEAS Trans. *Computers* (2010) 7(9): 706-715.

[104] Hida, T. *Brownian Motion*; Springer, 1980.

[105] Press, W. H.; Teukolsky, S. A.; Vetterling, W. T.; Flannery, B. P. Numerical Recipes in C: the Art of Scientific Computing; 2nd Ed., Cambridge University Press, 1992.

[106] Li, M. Fractal Time Series - a Tutorial Review. *Mathematical Problems in Engineering* (2010).

[107] Ortigueira, M. D.; Batista, A. G. On the Relation between the Fractional Brownian Motion and the Fractional Derivatives. *Physics Letters A* (2008) 372(7): 958-968.

[108] Li, J.; Lu, X.; Li, M.; Chen, S-Y. Data Simulation of Matérn Type. *WSEAS Trans. Computers*, (2010) 7(9): 696-705.

[109] Shinozuka, M.; Deodatis, G. Simulation of Multi-Dimensional Gaussian Stochastic Fields by Spectral Representation. *Applied Mechanics Review* (1996) 49(1): 29-53.

[110] Harris, C. M.; ed., *Shock and Vibration Handbook*; 4th ed., McGraw-Hill, 1995.

[111] Akimaru, H.; Kawashima, K. *Teletraffic: Theory and Applications*; Springer-Verlag, 1993.

[112] Stanislaw, R. M.: *Ocean Surface Waves: Their Physics and Prediction*; World Scientific, 1996.

[113] Nichols, J. M.; Olson, C. C.; Michalowicz, J. V.; Bucholtz, F. A Simple Algorithm for Generating Spectrally Colored, Non-Gaussian Signals. *Probabilistic Engineering Mechanics* (2010) 25(3): 315-322.

[114] Pomerantz, A. E.; Tilke, P.; Song, Y.-Q. Generating Heterogeneity Spectra from Spatially Resolved Measurements. *Mathematical Geosciences* (2009) 41(7): 721-735.

[115] Taufer, E.; Leonenko, N. Simulation of Lévy-Driven Ornstein–Uhlenbeck Processes with Given Marginal Distribution. *Computational Statistics and Data Analysis* (2009) 53(6): 2427-2437.

[116] Rodriguez, E.; Echeverria, J. C.; Alvarez-Ramirez, J. $1/f^{\alpha}$ Fractal Noise Generation from Grünwald–Letnikov Formula. Chaos, Solitons and Fractals (2009) 39(2): 882-888.

[117] Cohen, S.; Lacaux, C.; Ledoux, M. A General Framework for Simulation of Fractional Fields. *Stochastic Processes and Their Applications* (2008) 118(9): 1489-1517.

[118] Michel, R. Simulation of Certain Multivariate Generalized Pareto Distributions. *Extremes* (2007) 10(3): 83-107.

[119] Plaszczynski, S. Generating Long Streams of $1/f^{\alpha}$ Noise. *Fluctuation and Noise Letters* (2007) 7(1): R1-R13.

[120] Horn, G.; Kvalbein, A.; Blomskøld, J.; Nilsen, E. An Empirical Comparison of Generators for Self Similar Simulated Traffic. *Performance Evaluation* (2007) 64(2): 162-190.

[121] Mallor, F.; Mateo, P.; Moler, J. A. A Comparison between Several Adjustment Models to Simulated Teletraffic Data. *Journal of Statistical Planning and Inference* (2007) 137(12): 3939-3953.

[122] Chakraborty, D.; Roy, T. K. Generation and Prediction of Self-Similar Processes by Surrogates. Fractals (2006) 14(1): 17-26.

[123] Restrepo, A.; Bovik, A. C. On the Generation of Random Numbers from Heavy-Tailed Distributions. *Proc. IEEE* (1988) 76(7): 838-840.

[124] Corsini, G.; Saletti, R. A $1/f^{\gamma}$ Power Spectrum Noise Sequence Generator. *IEEE Trans. Instrumentation and Measurement* (1988) 37(4): 615-619.

[125] Li, M. A method for requiring block size for spectrum measurement of ocean surface waves. *IEEE Trans. Instrumentation and Measurement* (2006) 55(6): 2207-2215.

[126] Gneiting, T.; Sevcikova, H.; Percival, D. B.; Schlather, M.; Jiang, Y. Fast and Exact Simulation of Large Gaussian Lattice Systems in R2. *Journal of Computational and Graphical Statistics* (2006) 15(3): 483-501.

[127] Makse, H. A.; Havlin, S.; Schwartz, M.; Stanley, H. E. Method for Generating Long-Range Correlations for Large Systems. *Physical Review E* (1996) 53(5): 5445-5449.

[128] Stein, M. L. Fast and Exact Simulation of Fractional Brownian Surfaces. *Journal of Computational and Graphical Statistics* (2002) 11(3): 587-599.

[129] Stein, M. L. Local Stationarity and Simulation of Self-Affine Intrinsic Random Functions. *IEEE Trans. Information Theory* (2001) 47(4): 1385-1390.

[130] Chan, G.; Wood, A. T. A. Simulation of Stationary Gaussian Vector Fields. *Statistics and Computing* (1999) 9(4): 265-268.

[131] Hänggi, P.; Marchesoni, F.; Nori, F. Brownian Motors. *Annalen der Physik* (2005) 14(1-3): 51-70.

[132] Hänggi, P.; Ingold, G.-L. Fundamental Aspects of Quantum Brownian Motion. *Chaos* (2005) 15(2): 026105 (*12pp)*

[133] Hänggi, P.; Marchesoni, F. Introduction: 100 Years of Brownian Motion. *Chaos* (2005) 15(2): 026101.

[134] Frey, E.; Kroy, K. Brownian Motion: a Paradigm of Soft Matter and Biological Physics. *Annalen der Physik* (2005) 14(1-3): 20-50.

[135] Sokolov, I. M.; Klafter, J. From Diffusion to Anomalous Diffusion: A Century after Einstein's Brownian Motion. *Chaos*, (2005) 15(2): 026103.

[136] Ankerhold, J.; Grabert, H.; Pechukas, P. Quantum Brownian Motion with Large Friction. *Chaos* (2005) 15(2): 026106 (*10 pp*).

[137] Luczka, J. Non-Markovian Stochastic Processes. Chaos (2005) 15(2): 026017 (13pp).

[138] Mitra, S. K.; Kaiser, J. F. *Handbook for Digital Signal Processing*; John Wiley and Sons, 1993.

[139] Li, M. Modeling Autocorrelation Functions of Long-Range Dependent Teletraffic Series Based on Optimal Approximation in Hilbert Space-a Further Study. *Applied Mathematical Modelling* (2007) 31(3): 625-631.

[140] Peltier, R. F.; Levy-Vehel, J. *A New Method for Estimating the Parameter of Fractional Brownian Motion*; INRIA RR 2696, 1994.

[141] Peltier, R. F.; Levy-Vehel, *J. Multifractional Brownian Motion: Definition and Preliminaries Results*; INRIA RR 2645, 1995.

[142] Paxson, V. *Measurements and Analysis of End-to-End Internet Dynamics*; Ph.D. Dissertation, University of California, 1997.

[143] Paxson, V. Growth Trends in Wide-Area TCP Connections. *IEEE Network* (1994) 8(4): 8-17.

[144] Mandelbrot, B. B.; van Ness, J. W. Fractional Brownian Motions, Fractional Noises and Applications. *SIAM Rev* (1968) 10(4): 422-437.

[145] Willinger, W.; Paxson, V.; Taqqu, M. S. Self-similarity and Heavy Tails: Structural Modeling of Network Traffic, in A Practical Guide to Heavy Tails: Statistical Techniques and Applications; Adler, R., Feldman, R., and Taqqu, M.S., eds., Birkh*ä*user, 1998.

[146] Li, M. Fractional Gaussian Noise and Network Traffic Modeling. Proc., the 8th WSEAS Int. Conf. on Applied Computer and Applied Computational Science, Hangzhou, China, May 2009, 34-39.

[147] Cappe, O.; Moulines, E.; Pesquet, J.-C.; Petropulu, A; Yang, X. S. Long-Range Dependence and Heavy-Tail Modeling for Teletraffic Data. *IEEE Signal Processing Magazine* (2002) 19(3): 14-27.

[148] Flandrin, P. On the Spectrum of Fractional Brownian Motion. *IEEE Trans. Information Theory* (1989) 35(1): 197-199.

[149] Sinai, Y. G. Self-Similar Probability Distributions. *Theory of Probability and Its Applications* (1976) 21(1): 64-80.

[150] Li, M.; Lim, S. C. A Rigorous Derivation of Power Spectrum of fractional Gaussian Noise. *Fluctuation and Noise Letters* (2006) 6(4): C33-C36.

[151] Murray, J. D. *Asymptotic Analysis*; Springer, 1984.

[152] Benassi, A.; Jaffard, S.; Roux, D. Elliptic Gaaussian Random Processes. *Revista Mathematica Iberoamericana* (1997) 13(1): 19-90.

[153] Muniandy, S. V.; Lim, S. C. On Some Possible Generalizations of Fractional Brownian Motion. *Physics Letters A* (2000) 266(2-3): 140-145.

[154] Muniandy, S. V.; Lim, S. C.; Murugan, R. Inhomogeneous Scaling Behaviors in Malaysia Foreign Currency Exchange Rates. *Physica A* (2001) 301(1-4): 407-428.

[155] Muniandy, S. V.; Lim, S. C. Modelling of Locally Self-Similar Processes Using Multifractional Brownian Motion of Riemann-Liouville Type. *Phys. Rev. E* (2001) 63(4): 046104.

[156] Li, M.; Lim, S. C.; Zhao, W. Investigating Multi-Fractality of Network Traffic Using Local Hurst Function. Advanced Studies in Theoretical Physics (2008) 2(10): 479-490.

[157] Li, M.; Chen, W.-S.; Han, L. Correlation Matching Method of the Weak Stationarity Test of LRD Traffic. *Telecommunication Systems* (2010) 43(3-4): 181-195.

[158] Ayache, A.; Cohen, S.; Vehel, J. L. The Covariance Structure of Multifractional Brownian Motion, with Application to Long Range Dependence. ICASSP, 2000 IEEE International Conference on Acoustics, Speech, and Signal Processing, vol. 6, 2000, 3810-3813.

[159] Lim, S. C.; Teo, L. P. Weyl and Riemann–Liouville Multifractional Ornstein–Uhlenbeck Processes. *Journal of Physics A: Mathematical and Theoretical* (2007) 40(23): 6035-6060.

In: Brownian Motion: Theory, Modelling and Applications ISBN: 978-1-61209-537-0
Editors: R.C. Earnshaw and E.M. Riley

Chapter 4

THE MULTIDIMENSIONAL BROWNIAN MOTION: NUMERICAL VALUATION OF AMERICAN AND BERMUDAN OPTIONS

Luis M. Abadie [*]
Basque Centre for Climate Change (BC3), Alameda Urquijo,
Bilbao, Spain

Closed-form analytical solutions for the pricing of options are available only in a limited number of simple cases. Beyond them, in more complex settings, the value of options is typically driven by several sources of risk. Other methods need to be developed in order to solve these cases. This paper develops several methods for pricing American and Bermudan options. In these options the exercise time is not restricted to the maturity of the options: rather, they can be exercised at any time prior to maturity or at pre-specified times before the expiration date. A number of numerical examples are shown.

First I address the analytical solution for a perpetual American option with two sources of risk. Both sources determine the value of a project, which serves as the underlying asset of an American option.

Second, I assess American and Bermudan options by means of multidimensional binomial lattices. A contribution of this chapter to the relevant literature is that in order to avoid negative probabilities the lattices are deployed to represent the behaviour of the corresponding futures contracts (whose stochastic equation has no drift), and futures prices are then used to estimate the spot price in a risk-neutral world. In this part I draw the optimal exercise boundary, i.e. the boundary between the "invest" region and the "wait" region. I also show the solution to an optimal control problem.

Another section is devoted to the Monte Carlo method. I obtain the price of path-dependent and American-style options, in the latter case using Least Squares Monte Carlo method along with the optimal exercise boundary under multiple sources of risk.

[*] E-mail: lm.abadie@bc3research.org

Keywords: Geometric Brownian motion, Numerical methods, Perpetual options, American options, Bermudan options, Binomial method, Least-squares Monte Carlo.

1. INTRODUCTION

There are a great many options in the financial world and indeed in the real world (Real Options), and their structures are seldom simple enough for them to be valued using an analytical solution. They can very often be exercised throughout a continuous period (American options) or at several discrete times within a period (Bermudan options). Unless the actual situation is simplified it is frequent in practice for there to be a high number of risk factors that affect valuation and need to be taken into account. The existence of a large number of risk factors results in the problem known as the "curse of dimensionality": the increase in calculation requirements is more than proportional to the increase in the number of risk factors. What numerical method is selected can be a significant factor in determining whether the processing time needed to obtain a result is acceptable. A great many stochastic models have been developed to take into account the behaviour of the underlying assets on which a specific derivative asset (such as an option or future) is to be valued. Some of those models are better suited than others to specific applications. One of the most widely used in finance is Geometric Brownian Motion (GBM), on which this study focuses. [1] Most of this study is given over to the valuation of American- and Bermudan-style options on multiple assets which exhibit GBM-type stochastic behaviour. To that end, numerical methods are mainly used. The increase in the power of computers in recent years facilitates the use of such methods, which sometimes require long computer processing times. All the examples given are concerned with GBM, but the numerical methods used can in many cases be extended with some additional development for use with processes of other types, e.g. mean-reverting processes.

In many cases this use of numerical methods entails the design of programs[2] that need to be debugged and checked. One way of checking them is to see whether they give the same result obtained with analytical solutions when programs are used to solve a simple case. Another way is to check whether results published in papers and books in the relevant body of financial literature can be reproduced using the same data. A third way is to use two different numerical methods to solve the same problem and check whether the results are very similar and converge as the number of steps into which the time to maturity is divided increases. Spreadsheets can also be used for some checks and to solve certain very simple cases.

All this enables gradual progress to be made towards the development of more complex programs, e.g. moving up from programs that solve two-dimensional binomial lattices to programs that solve three-dimensional binomial lattices. Checks such as those indicated above are covered in several of the sections in this chapter for the programming designed by the author, and the examples used to illustrate developments are solved. The rest of the chapter is structured as follows: Section 2 presents a review of the relevant literature, Section 3 briefly analyses the characteristics of the GBM model, Section 4 studies perpetual options

[1]For commodities, mean-reverting models are frequently used.

[2]In some cases – basically simple cases of one-dimensional stochastic processes – spreadsheet designs may also be used.

on assets whose prices follow GBM processes: in this case they are possible analytical solutions that may serve to check the results obtained from numerical methods, e.g. when the maturity of, for instance, a finite lifetime option is increased. Three main numerical methods are used: binomial and trinomial lattices, Monte Carlo simulation and finite difference methods. Section 5 covers the development of binomial trees, beginning with an analysis of a one-dimensional tree then moving on to two- and finally three-dimensional trees. Special care is taken to construct these trees in such a way as to avoid the existence of negative probabilities insofar as possible. To that end, the stochastic dynamics of futures contracts are used. Section 6 tackles the use of Monte Carlo methods to value path-dependent and American-style options (in the latter case via the Least Squares Monte Carlo method). Section 7 concludes.

2. Review of the Relevant Literature

There is a body of literature that deals with numerical methods for valuing options. Relevant works include those of Hull [11], Wilmott [16], Clewlow and Strickland [7] and Brandimarte [6][3], all of which discuss the three most widely used numerical methods: binomial and trinomial lattices, Monte Carlo simulation and finite difference methods. The use of binomial lattices stems from the paper by Cox et al. [9]. Binomial lattices are also studied in the papers by Copeland and Antikarov [8] and Luenberger [12]. A practical application of two-dimensional lattices is developed in Abadie and Chamorro [3], where a choice is made between a flexible technology and an inflexible one and the optimal investment option exercise curve is calculated. The three-dimensional lattice for both GBM and mean-reverting processes, avoiding the possibility of negative probabilities, is developed in Abadie et al. [2]. The n-dimensional binomial method for multiple GBMs is developed in Boyle et al. [5]; according to these authors this method does not guarantee that all probabilities will be positive. The Least Squares Monte Carlo (LSM) method was first put forward in a paper by Longstaff and Schwartz [13]. A practical application of the valuation of an investment when there are numerous sources of risk is developed in Abadie and Chamorro [4]. Finite difference methods are studied in Tavella [14]. Trigeorgis [15] studies the application of valuation methods, mainly binomial lattices, for the valuation of real options.

Valuation methods for perpetual American options are developed in Dixit and Pindyck [10] for both one and two dimensions. A practical application can be found in Abadie and Chamorro [1].

3 The GBM Stochastic Model

The GBM model behaves in the real world as specified in formula (1):

$$dS_t = \alpha S_t dt + \sigma S_t dz_t, \tag{1}$$

[3]This last study includes programs written in Matlab.

where S_t is the price of asset S at time t, α is the drift in the real world, σ is the instantaneous volatility and dz_t stands for the increment to a standard Wiener process.

The risk-neutral version of this model is:[4]

$$dS_t = (\alpha - \lambda)S_t dt + \sigma S dz_t, \quad (2)$$

where λS_t is the market price of risk (MPR)[5], which, as can be observed, is modelled as proportional to S_t. In this case the value at time t of a futures contract with maturity at T is obtained based on the following equation:

$$F(t,T) = S_t e^{(\alpha-\lambda)(T-t)} \quad (3)$$

Equation (3) shows that an asset whose spot price follows a GBM should appear on the futures market with quotes that increase in absolute value by greater amounts as maturity times increase. This behaviour serves to identify underlying assets as candidates for modelling with a GBM. On the other hand, in the real world the time$-t$ expectation of the spot price at time T is given by:

$$E_t(S_T) = S_t e^{\alpha(T-t)}. \quad (4)$$

The risk premium thus looks like this:

$$RP(t,T) = F(t,T) - E_t(S_T) = S_t e^{\alpha(T-t)}[e^{-\lambda(T-t)} - 1]. \quad (5)$$

The next step is to obtain the differential equation followed by the price of futures contracts over time. This results in:

$$F_S \equiv \frac{\partial F}{\partial S} = e^{(\alpha-\lambda)(T-t)}; F_{SS} = 0; F_t \equiv \frac{\partial F}{\partial t} = -(\alpha - \lambda)S_t e^{(\alpha-\lambda)(T-t)}.$$

Note that what is under study here is the variation in a futures contract characterised by maturity T as time t increases, and therefore the time remaining until maturity $\tau = T - t$ decreases.

If the differential equation for the future is to be used to draw up valuations, then equation (2) must be employed; thus, applying Ito's lemma, the following results:

$$dF_t = \sigma S_t e^{(\alpha-\lambda)(T-t)} dz_t^* = \sigma F_t dz_t^*. \quad (6)$$

[4]This version enables valuations to be drawn up, given that the results obtained when deducing the market price of risk from the trend can be discounted from the risk-free interest rate r.

[5]An alternative specification is $\alpha - \lambda = r - \delta$, where r is the risk-free interest rate and δ is the yield of convenience. Wilmott [16] features several models of the market price of risk.

In this case the trend has disappeared, which may be useful for use in numerical methods such as binomial trees since the possibility of obtaining negative probabilities in one-dimension calculations disappears.[6] As obtained here, in the risk-neutral world there can be no drift in futures prices: otherwise positive results could be obtained without investing anything. Equation (6) is used in Section 5 below to draw up valuations with the binomial lattice method.

4. THE PERPETUAL OPTION

4.1 The One-Dimensional Perpetual Option

The price of a derivative $F(S)$ based on an asset S that follows a GBM process such as that shown in equation (2) must satisfy differential equation (7):[7]

$$dF = \left[\frac{\partial F}{\partial S}(\alpha - \lambda)S + \frac{\partial F}{\partial t} + \frac{1}{2}\frac{\partial^2 F}{\partial S^2}\sigma^2 S^2 \right] dt + \frac{\partial F}{\partial S}\sigma S dz, \tag{7}$$

plus the corresponding boundary conditions. When a perpetual option is involved it must satisfy $\frac{\partial F}{\partial t} = 0$. In this case the differential equation is reduced to a form that does not depend on t, which facilitates the task of seeking an analytical solution:

$$dF = \left[\frac{\partial F}{\partial S}(\alpha - \lambda)S + \frac{1}{2}\frac{\partial^2 F}{\partial S^2}\sigma^2 S^2 \right] dt + \frac{\partial F}{\partial S}\sigma S dz.$$

Building a risk-free portfolio[8] results in:

$$\frac{1}{2}\sigma^2 S^2 \frac{\partial^2 F}{\partial S^2} + (\alpha - \lambda)S\frac{\partial F}{\partial S} - rF = 0. \tag{8}$$

When a perpetual call option and a perpetual put option with an exercise price K are involved there are three boundary conditions. These conditions and the general solution are shown in Table 1. Given the GBM process, if $S_t = 0$ then $S_\tau = 0$ for any value of $\tau \geq t$. Therefore in this case the value is nil for a perpetual call and K for a perpetual put.

[6]The implications of this approach for the two- and three-dimensional cases are analysed later.
[7]This equation is obtained by applying Ito's lemma.
[8]See Dixit and Pindyck [10].

The value-matching condition reflects the fact that at the optimal exercise time for the option (when the value of the underlying asset is S^*) the immediate exercise value must be equal to the value of keeping the option alive. The smooth-pasting condition, which is obtained by deriving the value-matching condition, must also be met.

Table 1: Boundary conditions and general solution

	Perpetual Call	Perpetual Put
$F(\cdot)$	$F(0)=0$	$F(\infty)=0$
value-matching	$F(S^*)=S^*-K$	$F(S^*)=K-S^*$
smooth-pasting	$F^{'}(S^*)=1$	$F^{'}(S^*)=-1$
general solution	$A_1S^{\beta_1}$	$A_2S^{\beta_2}$

The general solution of equation (8) is the following:

$$F(S)=A_1S^{\beta_1}+A_2S^{\beta_2}, \tag{9}$$

where $\beta_1>0$ and $\beta_2<0$. In the case of a perpetual call, $A_2=0$ must hold for the condition $F(0)=0$ to be met. In the case of a perpetual put, $A_1=0$ must hold for $F(\infty)=0$ to be met.

If AS^{β} is substituted in equation (8) the quadratic form (10) is obtained:

$$\frac{1}{2}\sigma^2\beta(\beta-1)+(\alpha-\lambda)\beta-r=0. \tag{10}$$

It is then straightforward to obtain the values of β_1 and β_2, which correspond to the positive and negative roots of equation (10) respectively. Therefore, the values of β will depend on the volatility σ, on the risk-free interest rate r and on the drift $\alpha-\lambda$ in the risk-neutral world.

The values of A_1 and A_2 must be determined together with the optimal exercise price S^* (the so-called "trigger price"), using the value-matching and smooth-pasting conditions. The formulae which enable the values of A_1, A_2 and S^* to be calculated are shown in Table 2.

Table 2. Coefficients A_1, A_2, and trigger price S^*

	Perpetual Call	Perpetual Put
S^*	$S^* = \frac{\beta_1}{\beta_1 - 1} K$	$S^* = \frac{\beta_2}{\beta_2 - 1} K$
A	$A_1 = \frac{1}{\beta_1 (S^*)^{\beta_1 - 1}}$	$A_2 = -\frac{1}{\beta_2 (S^*)^{\beta_2 - 1}}$

Example 1. Perpetual call and put options

Consider the following cases of American perpetual options. The results obtained using the above formulae are shown in Table 3.

Table 3. Example of perpetual American options

	American Call	American Put
Drift (risk-neutral): $\alpha - \lambda$	0.03	0.03
Spot price at initial time: S_0	100	100
Exercise price: K	102	98
Volatility: σ (per annum).	0.01	0.1
Risk-free interest rate: r	0.04	0.04
β_1	1.3326	1.2749
β_2	-600.3326	6.2749
S^*	408.6796	84.5291
A_1	0.101571	0.0000
A_2	0.0000	1.66418×10^{13}
$F(S)$	46.9850	4.6921

4.2. The Two-Dimensional Perpetual Model

This model is described in Dixit and Pindyck [10] under price and cost uncertainty.[9] The value of the option of investing in a project depends on income S and costs K; both are assumed to follow GBM processes. The optimal decision will depend on the ratio $x \equiv \frac{S}{K}$: if the value of income and costs double, the decision will remain the same.

Assume the following stochastic processes in a risk-neutral world:

$$dS_t = (\alpha_S - \lambda_S) S_t dt + \sigma_S S_t dz_t^S = (r - \delta_S) S_t dt + \sigma_S S_t dz_t^S,$$

[9]The only difference is that in Dixit and Pindyck [10] the immediate exercise price is $S/\delta - K$, while here it is $S - K$.

$$dK_t = (\alpha_K - \lambda_K)K_t dt + \sigma_K K_t dz_t^K = (r - \delta_K)K_t dt + \sigma_K K_t dz_t^K,$$

with $E(dz_t^S dz_t^K) = \rho dt$.

The value of the perpetual option $F(S,K)$ must satisfy the differential equation

$$\frac{1}{2}\sigma_S^2 S^2 F_{SS} + \frac{1}{2}\sigma_K^2 K^2 F_{KK} + \rho SK\sigma_S\sigma_K F_{SK} + (r-\delta_S)SF_S + (r-\delta_K)KF_K - rF = 0. \quad (11)$$

It must also satisfy the following boundary conditions:

(i) value-matching: $F(S^*, K) = S^* - K$;

(ii) smooth-pasting: $F_S(S^*, K) = 1$ and $F_K(S^*, K) = -1$.

Making $F(S,K) = Kf(\frac{S}{K}) = Kf(x)$ the following partial derivatives are obtained:

$$F_S = f'(x); F_K = f(x) - xf'(x); F_{SS} = \frac{1}{K}f''(x); F_{KK} = \frac{x^2}{K}f''(x); F_{SK} = -\frac{x}{K}f''(x).$$

Substituting in differential equation (11) and dividing by K the following is obtained:

$$(\frac{1}{2}\sigma_S^2 + \frac{1}{2}\sigma_K^2 - \rho\sigma_S\sigma_K)x^2 f''(x) + (\delta_K - \delta_S)xf'(x) - \delta_K f(x) = 0 \quad (12)$$

If $f(x) = A_1(x)^{\beta}$, substituting in equation (12) gives:

$$(\frac{1}{2}\sigma_S^2 + \frac{1}{2}\sigma_K^2 - \rho\sigma_S\sigma_K)\beta(\beta-1) + (\delta_K - \delta_S)\beta - \delta_K = 0,$$

from where the value of β can be obtained.

For a perpetual call option the following is obtained:

$$p^* \equiv \frac{S^*}{K^*} = \frac{\beta_1}{\beta_1 - 1}.$$

As in the one-dimensional case, the following is calculated:

$$A_1 = \frac{1}{\beta_1 (p^*)^{\beta_1 - 1}}.$$

Example 2. two-dimensional perpetual call option

Next, the following case is solved:

Spot price at initial time: $S = 100$ and $K = 100$; $p = 1$.

Drift (risk-neutral world): $\alpha_S - \lambda_S = 0.03$ and $\alpha_K - \lambda_K = 0.02$.

Risk-free interest rate: $r = 0.06$.

Volatility: $\sigma_S = 0.2$ and $\sigma_K = 0.3$.

Correlation: $\rho_{S,K} = 0.50$.

From these data the following is deduced: $\delta_S = r - (\alpha_S + \lambda_S) = 0.03$, $\delta_K = r - (\alpha_K + \lambda_K) = 0.04$.

Consequently: $\beta_1 = 1.4843$, $p^* = 3.064978$, $A_1 = \dfrac{1}{\beta_1 (p^*)^{\beta_1 - 1}} = 0.39167648$.

The value of the perpetual call option is: $KA_1 (p)^{\beta_1} = 39.167648$.

5. The Binomial Lattice

One of the most widely used numerical methods for valuing options in finance is the binomial lattice method: binomial lattices are binomial trees in which the branches are recombinant. The method is based on the use of a (discrete) binomial model of the behaviour of an asset, in such a way that the model converges towards a normal distribution (continuous model) when the number of steps into which the time to maturity is divided increases. Changes in value between a node in the lattice and the two subsequent nodes linked to it take place in accordance with certain probabilities. A major drawback of this method is that some of those probabilities could be negative, which would of course be unacceptable. Such negative probabilities can be avoided in many cases by constructing the lattice for the trend of futures quotes in the risk-neutral world, since there is no drift in their stochastic equation. In line with this idea, a one-dimensional lattice is first solved below, and the method is then extended to three dimensions.

5.1. The One-Dimensional Lattice

Some developments concerned with the one-dimensional binomial lattice which are used later in the multidimensional case are presented below. The first task is to construct the lattice using the dynamic of futures. Then its implementation is analysed using the dynamics of (the log of) futures.

5.1.1. Solution Using Futures Prices

After a time Δt the initial spot price S_0 can take two possible values: $S_0 u$ if the price increases and $S_0 d$ if it falls. To ensure that the branches of the tree recombine in the subsequent steps, the value chosen here is $d = \frac{1}{u}$.[10]

In calculating the parameters to be used in the binomial tree compatibility with the mean and the variance of the GBM stochastic process in the risk-neutral world must be maintained, i.e.:

$$E_0^Q(S_{\Delta t}) = S_0 e^{(\alpha-\lambda)\Delta t}, \tag{13}$$

$$Var(S_{\Delta t}) = S_0^2 e^{2(\alpha-\lambda)\Delta t}(e^{\sigma^2 \Delta t} - 1). \tag{14}$$

Given that there is no drift ($\alpha - \lambda = 0$), for the trend in futures of a GBM in the risk-neutral world these equations are:

$$E_0^Q(F_{\Delta t}) = F_0, \tag{15}$$

$$Var(F_{\Delta t}) = F_0^2(e^{\sigma^2 \Delta t} - 1). \tag{16}$$

Thus, the parameters to be used in the one-dimensional binomial tree must comply with the following:

(i) for the mean:

$$F = pFu + (1-p)Fd; \tag{17}$$

(ii) for variance:[11]

$$F^2(e^{\sigma^2 \Delta t} - 1) = pF^2u^2 + (1-p)F^2d^2 - F^2. \tag{18}$$

From this the following can be deduced:

$$p = \frac{1-d}{u-d}, \tag{19}$$

$$e^{\sigma^2 \Delta t} = pu^2 + (1-p)d^2. \tag{20}$$

---

[10] The combination of branches enables the calculation process to be considerably simplified.

[11] Using $Var(F_{\Delta t}) = E_0^Q(F_{\Delta t}^2) - [E_0^Q(F_{\Delta t})]^2$.

Parameters p, u and d are independent of F. The probabilities are therefore constant throughout the tree.[12] Since $u > 1$ and $d < 1$, $p > 0$ is always satisfied. Similarly, $1 - p \equiv q = \frac{u-1}{u-d} > 0$. Therefore, with this type of implementation the risk-neutral probabilities will always be positive.

The exact solution for the value of u can be found by solving the system of equations formed by (19) and (20). This can be seen to be:

$$u = \frac{e^{\sigma^2 \Delta t} + 1 + \sqrt{(e^{\sigma^2 \Delta t} + 1)^2 - 4}}{2}, \tag{21}$$

a solution which is valid regardless of the size of Δt.

Frequently, however, the simplified solution (22), deduced from equation (21), is used for small values of Δt:

$$u \cong e^{\sigma \sqrt{\Delta t}}. \tag{22}$$

The trigger price corresponds to the value of the underlying asset S^*, which means that the immediate exercise value is equal to the continuation value. In the one-dimensional case the optimal exercise value is a point on a line that separates the “invest” region from the “wait” region.

Example 3. valuation of an American call option using futures

Consider a call option on a commodity with the following conditions:

Drift in the risk-neutral world: $\alpha - \lambda = 0.03$.

Spot price at the initial time: $S_0 = 100$.

Time to maturity: $T = 1$ year.

Size of the step selected for the binomial tree: $\Delta t = 0.25$ years.

Exercise price: $K = 102$.

Volatility: $\sigma = 0.01$ (1% per annum).

Risk-free interest rate: $r = 0.04$.

If this information is used to calculate the parameters for valuation with the trend in the spot price in the risk-neutral world, the following values are obtained:[13]

[12] The same does not hold when the idea is to use a binomial lattice in valuing an option on an asset that follows a mean-reverting process and the spot price dynamic is used.

[13] Observe that $\alpha - \lambda = r - \delta$, i.e. the drift minus the market price of risk equals the risk-free interest rate minus the yield of convenience. If $\delta = 0$ then $\alpha - \lambda = r$, which is what appears in many binomial trees described in financial literature. However the representation $\alpha - \lambda$ or its equivalent $r - \delta$ is more general, and is used very frequently for commodity markets.

$$u = e^{\sigma\sqrt{\Delta t}} = 1.0050\ ,d = 0.9950\ ,p = \frac{e^{(\alpha-\lambda)\Delta t} - d}{u - d} = 1.2516\ ,1 - p = -0.2516.$$

These parameter values are unacceptable for the construction of the binomial tree and the valuing of the American call option.

Using the approach developed for the futures trend, the following values are obtained:

$$u = e^{\sigma\sqrt{\Delta t}} = 1.00501252\ \ ,d = 0.9950\ ,p = \frac{1 - d}{u - d} = 0.4988\ ,1 - p = 0.5012.$$

These values are used in the calculation shown in Table 4.

Table 4. shows an initial value for the future $F(t,T)$ at time $t = 0$, when there is one year $(T - t = 1)$ remaining before maturity, of $F(t,T) = S_0 e^{(\alpha-\lambda)(T-t)} = 103.0455$. Then, with 0.75 remaining before maturity, the value may shift to $103.0455u = 103.5620$ with a probability of 0.4988 or drop to $103.0455d = 102.5315$ with a probability of 0.5012. The tree representing the dynamic of the futures over time in the risk-neutral world is thus gradually constructed.

Based on the trend of the future, the equivalent spot values are reconstructed using

$$S_t = F(t,T)e^{-(\alpha-\lambda)(T-t)}.$$

Table 4. Valuation of American call with futures dynamics

	Time to maturity $(T-t)$				
	1.00	0.75	0.50	0.25	0.00
Futures	103.0455	103.5620	104.0811	104.6028	105.1271
		102.5315	103.0455	103.5620	104.0811
			102.0201	102.5315	103.0455
				101.5113	102.0201
					101.0050
	Time to maturity $(T-t)$				
	1.00	0.75	0.50	0.25	0.00
Spot	100.0000	101.2578	102.5315	103.8212	105.1271
		100.2503	101.5113	102.7882	104.0811
			100.5013	101.7654	103.0455
				100.7528	102.0201
					101.0050
	Time to maturity $(T-t)$				
	1.00	0.75	0.50	0.25	0.00
Call	1.0648	1.52	2.04	2.58	3.13
		0.64	1.02	1.55	2.08
			0.26	0.53	1.05
				0.01	0.02
					0.00

For example, for the value obtained of $F(0.75,1) = 103.5620$ following two rises and one fall, when 0.25 years remain before maturity the figure is $S_{0.75} =$

$103.5620^{-0.03\times0.25} = 102.7882$. To value the American call option, at the final nodes $W = max(S_T - K, 0)$ is obtained, where $K = 102$ is the exercise price. The process then works back towards the beginning. At previous times the following is selected:

$$\max(S_t - K, e^{-r\Delta t}(pW^+ + (1-p)W^-),$$

i.e. the maximum between exercising the option at that node or continuing. W^+ and W^- are the values obtained for the call in the previous step in the cases of price increase and price fall, respectively. The solution is obtained recursively, working backwards. The value obtained for the initial time is $c = 1.0648$. If the exact value of u had been used, this whole process would have been carried out with $u = 1.00501255$ instead of the approximate value of $u = 1.00501252$. The approximation usually used is therefore quite acceptable in this case. For purposes of illustration, a binomial tree with four steps per year is shown. The result comes closer to the correct value when a larger number of steps is used. For instance, for 100 steps ($\Delta t = 1/100$) the value of the call is $c = 1.0831$; for $1{,}000$ steps it is $c = 1.0835$. See Table 5.[14] The convergence of the American call and the perpetual call is checked below when the exercise period for the option increases. Twelve steps per year are used in all cases, along with futures dynamics. See Table 6. As can be seen in Table 6, as the interval in which the American call can be exercised increases its value converges with the value of the perpetual option.

Table 5. Valuation of American call with futures dynamics

Steps (per year)	4	12	100	500	1,000
Value (c)	1.0648	1.0847	1.0831	1.0836	1.0835

Table 6: Option value and life of option

Years	Steps	Value (c)
1	12	1.0847
5	60	11.6124
10	120	22.1111
25	300	40.3564
50	600	46.9832
75	900	46.9850
100	1200	46.9850
∞	-	46.9850

The numerical method provides us with a value that converges towards a known analytical solution. The trigger price S^* must satisfy the following condition:

[14]For a large number of steps it is preferable to use a suitable program as an alternative to the spreadsheet. The spreadsheet could be used as a checking feature in the design of the program. The calculations performed in this chapter were programmed in Matlab.

$S^* - K = e^{-r\Delta t}(pW^+ + (1-p)W^-)$.

Using the four-step binomial lattice the value obtained is $S^* = 406.47$. Using the program with 1000 steps the result obtained is $S^* = 408.4$.

Example 4. Valuation of an American put option using futures

Now consider the case of an American put with the following data:
Drift in the risk-neutral world: $\alpha - \lambda = 0.03$.
Spot price at the initial time: $S_0 = 100$.
Time to maturity: $T = 1$ year.
Size of the step selected for the binomial tree: $\Delta t = 0.25$ years.
Exercise price: $K = 98$.
Volatility: $\sigma = 0.1$ (10% per annum).
Risk-free interest rate: $r = 0.04$.
The trees for the future and spot prices change because in this case the value of σ is different. The decisions at the nodes also change. At maturity:

$W = max(K - S_T, 0)$.
At earlier stages:
$\max(K - S_t, e^{-r\Delta t}(pW^+ + (1-p)W^-)$.
The results shown in Table 7 are obtained, in line with the number of steps per year used.

Table 7. Valuation of American put with futures dynamics

Steps	4	12	100	500	1,000
Value (p)	2.2029	2.1081	2.1103	2.1064	2.1053

The results in Table 8 show the convergence towards the correct value as maturity increases; we use a consistent number of 100 steps per year.

Table 8. Option value and life of option

Years	Steps	Value (p)
1	100	2.1103
5	500	3.7750
10	1,000	4.3019
25	2,500	4.6320
50	5,000	4.6834
75	7,500	4.6872
100	10,000	4.6876
∞	-	4.6921

As can be seen, the numerical solution converges with the analytical solution for the perpetual put option when the exercise period is extended. The more steps per year are used the more accurate the calculations obtained become.

The trigger price S^* must meet the following condition:

$$K - S^* = e^{-r\Delta t}(pW^+ + (1-p)W^-).$$

Using the four-step binomial lattice the value obtained is $S^* = 90.42$. Using the $1{,}000$ - step program it is $S^* = 89.0$.

Example 5. valuation of a Bermudan put option using futures

Now consider the case of a Bermudan put with the following data:

Drift in the risk-neutral world: $\alpha - \lambda = 0.03$.

Spot price at initial time: $S_0 = 100$.

Time to maturity: $T = 1$ year.

Size of the step selected for the binomial tree: $\Delta t = 0.25$ years.

Exercise price: $K = 98$.

Volatility: $\sigma = 0.1$ (10% per annum).

Risk-free interest rate: $r = 0.04$.

Assume that the put can only be exercised half-yearly, i.e. at $T = 1$, $T = 0.5$, and $T = 0.0$. The values obtained for the underlying asset in the binomial lattice are exactly the same as those in Example 4. It is the decisions for valuing the option that change. As in the previous case, since the option can be exercised and continuation is not possible at the last time the decision is:

$$W = max(K - S_T, 0).$$

At earlier times when the Bermudan option cannot be exercised ($T = 0.75$ and $T = 0.25$):

$$W = e^{-r\Delta t}(pW^+ + (1-p)W^-).$$

At earlier times when the Bermudan option can be exercised:

$$\max(K - S_t, e^{-r\Delta t}(pW^+ + (1-p)W^-). \qquad (23)$$

This whole calculation process for the four-step case is shown in Table 9.

In this case the result obtained is $p = 2.1833$. This is somewhat lower than the figure obtained for an American put, since now there is less optionality. If the number of steps is increased a more precise result is obtained. Table 10 shows that for $1{,}000$ steps the result is 1.9988.

Table 9. Valuation of Bermudan put with futures dynamics

	Time to maturity $(T-t)$				
	1.00	0.75	0.50	0.25	0.00
Futures	103.0455	108.3287	113.8828	119.7217	125.8600
		98.0199	103.0455	108.3287	113.8828
			93.2394	98.0199	103.0455
				88.6920	93.2394
					84.3665
	Time to maturity $(T-t)$				
	1.00	0.75	0.50	0.25	0.00
Spot	100.0000	105.9185	112.1873	118.8272	125.8600
		95.8390	101.5113	107.5193	113.8828
			91.8512	97.2875	103.0455
				88.0293	93.2394
					84.3665
	Time to maturity $(T-t)$				
	1.00	0.75	0.50	0.25	0.00
Put	2.1833	0.62	0.00	0.00	0.00
		3.71	1.23	0.00	0.00
			6.15	2.42	0.00
				9.22	4.76
					13.63

Table 10. Valuation of Bermudan put with futures dynamics

Steps	4	12	100	500	1,000
Value (p)	2.1833	2.0056	2.0038	1.9999	1.9988

The trigger price S^* must meet the following condition:

$$K - S^* = e^{-r\Delta t}(pW^+ + (1-p)W^-).$$

The difference with the American put is that here the values W^+ and W^- are obtained differently. With the four-step binomial lattice the value obtained is $S^* = 92.07$. With the $1{,}000$ -step program it is $S^* = 92.2$.

5.1.2. Solution Using Log Futures

The transformation $x = \ln F$ results in equation (24).

$$dx = -\frac{\sigma^2}{2}dt + \sigma dz_t. \tag{24}$$

In this case the equations which must be satisfied when building the binomial tree are:

$$E(\Delta x) = p\Delta x - (1-p)\Delta x = -\frac{\sigma^2}{2}\Delta t, \tag{25}$$

$$E(\Delta x^2) = p\Delta x^2 + (1-p)\Delta x^2 = \sigma^2\Delta t + \frac{\sigma^4}{4}(\Delta t)^2 \cong \sigma^2\Delta t. \tag{26}$$

In this way the tree is recombinant. From equation (26) it is straightforward to obtain $\Delta x = \sigma\sqrt{\Delta t}$. Therefore:

$$p = \frac{1}{2} - \frac{\sigma\sqrt{\Delta t}}{4},$$

$$1 - p = \frac{1}{2} + \frac{\sigma\sqrt{\Delta t}}{4}.$$

In this case the value of Δt chosen must be low enough to ensure that negative probabilities are not obtained.

Example 6. valuation of an American call option using log futures prices

For the same data as in Example 3 the following is found: $\Delta x = \sigma\sqrt{\Delta t} = 0.005$; $p = 0.4988$ and $1 - p = 0.5012$. As can be checked, the probabilities are unchanged, as are the values obtained for the binomial tree with the spot price (which are shown in Table 11 and are identical to those in Table 4). As is logical, the same results are obtained with both methods.

Table 11. Valuation of American call with log futures dynamics

	Time to maturity $(T-t)$				
	1.00	0.75	0.50	0.25	0.00
Log Futures	4.6352	4.6402	4.6452	4.6502	4.6552
		4.6302	4.6352	4.6402	4.6452
			4.6252	4.6302	4.6352
				4.6202	4.6252
					4.6152

Table 11. (Continued)

	Time to maturity $(T-t)$				
	1.00	0.75	0.50	0.25	0.00
Spot	100.0000	101.2578	102.5315	103.8212	105.1271
		100.2503	101.5113	102.7882	104.0811
			100.5013	101.7654	103.0455
				100.7528	102.0201
					101.0050
	Time to maturity $(T-t)$				
	1.00	0.75	0.50	0.25	0.00
Call	1.0648	1.52	2.04	2.58	3.13
		0.64	1.02	1.55	2.08
			0.26	0.53	1.05
				0.01	0.02
					0.00

5.2. The Two-Dimensional Lattice

Consider two correlated stochastic processes in which the risk-neutral version of the futures dynamics is:

$$dF_i = \sigma_{F_i} F_i dz_i, \tag{27}$$

where $i = 1,2$. Equation (27) is a GBM process with no drift.

First take natural logarithms of futures prices:

$$x_i \equiv \ln F_i.$$

Applying Ito's Lemma the following is obtained:

$$dx_i = -\frac{1}{2}\sigma_i^2 dt + \sigma_i dz_i. \tag{28}$$

Observe that regardless of the process followed by the spot price, if the future follows equation (27) it is possible, based on equation (28), to construct a two-dimensional binomial lattice in which the risk-neutral probabilities are constant throughout the tree, and negative probabilities can be avoided in many cases.

With two dimensions at each node of the lattice, it is possible to move to $2^2 = 4$ different states of nature. Thus there are four probabilities to be computed, in addition to two incremental values (Δx_1 and Δx_2). For this purpose there are six equations.

The first equation establishes that the probabilities must sum to one:

$$p_{uu} + p_{ud} + + p_{du} + p_{dd} = 1. \quad (29)$$

The next two set the conditions for consistency regarding the second moment:

$$E(\Delta x_i^2) = (p_{uu} + p_{ud} + p_{du} + p_{dd})\Delta x_i^2 =$$
$$= \sigma_i^2 \Delta t + \frac{\sigma^4}{4}(\Delta t)^2 \cong \sigma_i^2 \Delta t.$$

When the increments Δt in the lattice are small, the term $(\Delta t)^2 \cong 0$. These equations allow the increments to be computed directly:

$$\Delta x_i = \sigma_i \sqrt{\Delta t}.$$

The next equation requires the probabilities to be consistent with observed correlation:

$$E(\Delta x_1 \Delta x_2) = (p_{uu} - p_{ud} - p_{du} + p_{dd})\Delta x_1 \Delta x_2 =$$
$$= \rho_{1,2}\sigma_1\sigma_2 \Delta t + \frac{\sigma_1^2\sigma_2^2}{4}(\Delta t)^2 \cong \rho_{1,2}\sigma_1\sigma_2\Delta t.$$
c

Taking into account the values of Δx_1 and Δx_1 the following is obtained:

$$p_{uu} - p_{ud} - p_{du} + p_{dd} = \rho_{1,2}. \quad (30)$$

The last two equations establish the conditions for consistency with the first moment:

$$E(\Delta x_1) = (p_{uu} + p_{ud} - p_{du} - p_{dd})\Delta x_1 = -\frac{\sigma_1^2}{2}\Delta t,$$

$$E(\Delta x_2) = (p_{uu} - p_{ud} + p_{du} - p_{dd})\Delta x_2 = -\frac{\sigma_2^2}{2}\Delta t.$$

These equations must be presented in the following form:

$$p_{uu} + p_{ud} - p_{du} - p_{dd} = -\frac{\sigma_1}{2}\sqrt{\Delta t} = -\frac{\Delta x_1}{2}, \quad (31)$$

$$p_{uu} - p_{ud} + p_{du} - p_{dd} = -\frac{\sigma_2}{2}\sqrt{\Delta t} = -\frac{\Delta x_2}{2}. \quad (32)$$

The solutions to the four equations (29), (30), (31) and (32) can be obtained straightforwardly:

$$p_{uu} = \frac{1}{4}[1 + \rho_{1,2} + \sqrt{\Delta t}(-\frac{\sigma_1}{2} - \frac{\sigma_2}{2})],$$

$$p_{ud} = \frac{1}{4}[1 - \rho_{1,2} + \sqrt{\Delta t}(-\frac{\sigma_1}{2} + \frac{\sigma_2}{2})],$$

$$p_{ud} = \frac{1}{4}[1 - \rho_{1,2} + \sqrt{\Delta t}(+\frac{\sigma_1}{2} - \frac{\sigma_2}{2})],$$

$$p_{ud} = \frac{1}{4}[1 + \rho_{1,2} + \sqrt{\Delta t}(+\frac{\sigma_1}{2} + \frac{\sigma_2}{2})].$$

As can be observed, a small enough Δt suffices to keep the probabilities positive. Only in the extreme cases of $\rho_{1,2} = +1$ and $\rho_{1,2} = -1$ could a negative probability arise, and in those cases the problem can be solved using two one-dimensional binomial trees.

In this case there is an optimal exercise boundary with the combinations of (S_1^*, S_2^*) based on which it is optimal to exercise the option. For each value of S_1^* the value of S_2 is sought for which the immediate exercise value is equal to the continuation value. These values form a point (S_1^*, S_2^*) on the optimal exercise boundary which separates the "invest" and "wait" regions.

Example 7. valuation of an American spread call option

The same exercise is performed as in Clewlow and Strickland [7] for an American Spread Call Option[15] with the following data:

Spot price at the initial time: $S_{1,0} = 100$ and $S_{2,0} = 100$.

Time to maturity: $T = 1$ year.

Size of the step selected for the binomial tree: $\Delta t = \frac{1}{3}$ years.

Drift (risk-neutral world): $\alpha_1 - \lambda_1 = 0.03$ and $\alpha_2 - \lambda_2 = 0.02$

Exercise price: $K = 1$.

Volatility: $\sigma_1 = 0.2$ and $\sigma_2 = 0.3$.

Risk-free interest rate: $r = 0.06$.

Correlation: $\rho_{1,2} = 0.50$.

---

[15] Exercising will result in $(S_1 - S_2) - K$.

For this option the following value is calculated at the final nodes:

$$W = max(S_{1,T} - S_{2,T} - K, 0).$$

At times $t < T$ when the American spread call option can be exercised:

$$\max(S_{1,t} - S_{2,t} - K, e^{-r\Delta t}(p_{uu}W^{++} + p_{ud}W^{+-} + p_{du}W^{-+} + p_{dd}W^{--}). \quad (33)$$

Using the conventional construction of the tree based on spot dynamics, the program gives the same result as in Clewlow and Strickland [7] for three steps. The probabilities calculated for the three-step case are the following:

$$p_{uu} = 0.3702, p_{ud} = 0.1442, p_{du} = 0.1058, p_{dd} = 0.3798.$$

The results corresponding to the different numbers of steps are shown in Table 12.

By contrast, if the method based on futures dynamics proposed here is used, the probabilities calculated for the three-step case are:

$$p_{uu} = 0.3389, p_{ud} = 0.1322, p_{du} = 0.1178, p_{dd} = 0.4111.$$

Since the number of steps is changed, in this case the results shown in Table 13 are obtained.

As can be checked, the values obtained are similar. The advantage is that now the model can be used in the certainty that the probabilities will be positive (unless the correlation is exactly $+1$ or -1).

Table 12. Valuation of American spread call with spot dynamics

Steps	3	12	100	500	1,000
Value	10.0448	10.1239	10.1527	10.1523	10.1528

Table 13. Valuation of American spread call with futures dynamics

Steps	3	12	100	500	1,000
Value	10.0779	10.1091	10.1494	10.1526	10.1529

The optimal exercise boundary is made up of those values for which the following holds at the initial time:

$$S^*_{1,0} - S^*_{2,0} - K = e^{-r\Delta t}(p_{uu}W^{++} + p_{ud}W^{+-} + p_{du}W^{-+} + p_{dd}W^{--}). \quad (34)$$

For this it is sufficient to take values of $S_{1,0}^*$ and find the values of $S_{2,0}^*$ for which equation (34) is satisfied. For example, taking $S_{1,0}^* = 100$, the figure obtained for the three-step case is $S_{2,0}^* = 59.8$. Thus, $(100,59.8)$ is a point on the optimal exercise boundary (with three steps per year).

Example 8. American call with stochastic exercise price

Consider the same data as for the perpetual American call with the stochastic exercise price in Example 2. The valuation is carried out with a two-dimensional lattice, using 12 steps per year and futures dynamics.

At the last times:

$$W = max(S_T - K_T, 0).$$

At times $t < T$ when the American spread call option can be exercised:

$$\max(S_t - K_t, e^{-r\Delta t}(p_{uu}W^{++} + p_{ud}W^{+-} + p_{du}W^{-+} + p_{dd}W^{--}). \quad (35)$$

The results shown in Table 14 are obtained, in line with maturity. For the 12-step case the probabilities calculated are:

$$p_{uu} = 0.3570, p_{ud} = 0.1286, p_{du} = 0.1214, p_{dd} = 0.3930.$$

Table 14. Option value and Option life

Years	Steps	Value
1	12	10.6549
5	60	22.3156
10	120	28.8330
25	300	36.0462
50	600	38.5864
75	900	39.0371
100	1,200	39.1330
∞	-	39.1676

The optimal exercise boundary must satisfy the following:

$$S_0^* - K_0^* = e^{-r\Delta t}(p_{uu}W^{++} + p_{ud}W^{+-} + p_{du}W^{-+} + p_{dd}W^{--}). \quad (36)$$

For instance, if $S_0^* = 100$ the figure obtained for the three-step case is $K_0^* = 58.3$. $(100,58.3)$ is therefore a point on the optimal exercise boundary (with three steps).

Example 9. an optimal control problem

Consider the following problem (which will be solved here initially with a two-dimensional lattice). A firm can produce a monthly output q which, logically, is limited to $q_{\min} \leq q \leq q_{\max}$, where it is assumed that $q_{\min} = 0$ and $q_{\max} = 100$. Production entails costs that depend on the quantity q_t and the unit cost c_t of the input used, as per the following equation:

$$C(q_t, c_t) = a_0 + a_1 c_t q_t + a_2 (q_t)^2.$$

There are therefore fixed costs a_0 which are independent of the output level, and costs $a_1 c_t q_t$ which are dependent on the output level, where a_1 represents the units of input required to produce one unit of output. The cost function is completed with the addend $a_2 (q_t)^2$, which means that the unit cost of producing one unit of output increases as output increases. On the other hand there is also income given by $p_t q_t$, where p_t is the market price of the output. Output and input prices follow GBM processes of the following form:

$$dp_t = (\alpha_p - \lambda_p) p_t dt + \sigma_p p_t dz_t^p,$$

$$dc_t = (\alpha_c - \lambda_c) c_t dt + \sigma_c c_t dz_t^c,$$

with $E(dz_t^p dz_t^c) = \rho dt$.

At the beginning of each month the company considers its input and output prices at that time and carries out an optimisation process to determine how much it wishes to produce (the quantity that maximises its profit), subject to its maximum output limitation:

$$p_t - a_1 c_t - 2a_2 q_t = 0 \Rightarrow$$

$$\Rightarrow q_t = \frac{p_t - a_1 c_t}{2a_2} \text{ with } q_{\min} \leq q_t \leq q_{\max}.$$

The data for performing this valuation exercise are the following:

$p_0 = 100$; $c_0 = 30$; $a_0 = 500$; $a_1 = 2$; $a_2 = 0.25$. Thus, the quantity to be produced at the initial time is $q_0 = 80$. Therefore the result for the first month is 264,000. For stochastic processes the following values are used in the example: $\alpha_p - \lambda_p = \alpha_c - \lambda_c = 0.05$; $\sigma_p = 0.15$; $\sigma_c = 0.10$; $\rho = 0.65$; $r = 0.06$. These values assume that $\delta_p = \delta_c = 0.01$. It is also assumed that the useful lifetime of the company is 20 years.

The solution is reached as follows:

a) First the two-dimensional binomial tree is constructed.
b) At the final nodes the value is zero.
c) At the intermediate nodes (once the quantity to be produced, q_t, has been chosen) the value obtained is:

$$p_t q_t - a_0 - a_1 c_t q_t - a_2 (q_t)^2 + C,$$

where the continuation value C is:

$$C = e^{-r\Delta t}(p_{uu}W^{++} + p_{ud}W^{+-} + p_{du}W^{-+} + p_{dd}W^{--}), \tag{37}$$

which in this case means $\Delta t = \frac{1}{12}$; the number of steps is therefore 240.

The result is a current profit of $539{,}350$. Bear in mind that losses per month are limited to 500 if nothing is produced but there is no limit on profits, since the price p_t may rise without constraint and c_t may decrease. If $q_t = 100$ is always produced each month, rather than the firm choosing the optimal quantity, then the binomial tree gives a result of 450.090 . This figure is 16.5% less than the result with the optimal output level. With $q_t = 100$ in all months there is an analytical solution that can be used to check the working of the binomial lattice, which is described below.

If annual output is 1200 and $p_0 = 100$, the current cumulative income is:

$$1200 \times 100 \int_0^{20} e^{(\alpha_p - \lambda_p)t} e^{-rt} dt = 120{,}000 \int_0^{20} e^{-\delta_p t} dt = 2{,}175{,}231.$$

Likewise, the fixed costs are as follows:

$$500 \times 12 \int_0^{20} e^{-rt} dt = 69{,}881.$$

The variable costs in proportion to the cost of input are :

$$1200 \times 2 \times 30 \int_0^{20} e^{-\delta_c t} dt = 1{,}305{,}139.$$

The costs proportional to $(q_t)^2$ are:

$$0.25 \times 12 \times 100^2 \int_0^{20} e^{-rt} dt = 349{,}403.$$

The overall net result obtained analytically is $450{,}809$. The figure of $450{,}090$ obtained with the binomial tree is just 0.16% below the exact value. Bear in mind that in the binomial lattice one step per month is used: logically, the accuracy of the result will increase if the number of steps is increased, but so will the computer processing time required to obtain the solution.[16]

There are usually several possible ways of finding a solution. Even at the initial stage it may be advisable to check results using a different valuation method. This same problem is solved via a different procedure in the section on Monte Carlo methods below.

5.3. The Three-Dimensional Lattice

Consider three correlated stochastic processes in which the risk-neutral version of the futures dynamic is:

$$dF_i = \sigma_{F_i} F_i dz_i, \tag{38}$$

where the index $i = 1,2,3$. First take the natural logarithms of the future prices:

$$x_i \equiv \ln F_i.$$

Applying Ito's Lemma the following is obtained:

$$dx_i = -\frac{1}{2}\sigma_i^2 dt + \sigma_i dz_i. \tag{39}$$

Observe that whichever process the spot price follows, if the future follows equation (38) then equation (39) can be used as the basis for constructing a three-dimensional binomial tree in which the risk-neutral probabilities are constant throughout the tree and negative probabilities can be avoided. With three dimensions at each node of the lattice, it is possible to move to $2^3 = 8$ different states of nature. Thus there are eight probabilities to be computed, in addition to three incremental values (Δx_1; Δx_2; Δx_3). For this purpose there are ten equations. The first equation establishes that the probabilities must sum to one:

$$p_{uuu} + p_{uud} + p_{udu} + p_{udd} + p_{duu} + p_{dud} + p_{ddu} + p_{ddd} = 1.$$

The next three impose the conditions for consistency regarding the second moment:

$$E(\Delta x_i^2) = (p_{uuu} + p_{uud} + p_{udu} + p_{udd} + p_{duu} + p_{dud} + p_{ddu} + p_{ddd})\Delta x_i^2 =$$
$$= \sigma_i^2 \Delta t + \frac{\sigma^4}{4}(\Delta t)^2 \cong \sigma_i^2 \Delta t.$$

[16] If more than 12 steps per year are used the stochastic process is better represented but a decision is not possible at all nodes. This is similar to a two-dimensional Bermudan option.

When the increments Δt in the lattice are small, the term $(\Delta t)^2 \cong 0$. These equations allow the increments to be computed directly:

$$\Delta x_i = \sigma_i \sqrt{\Delta t}.$$

The next three equations require the probabilities to be consistent with observed correlations:

$$E(\Delta x_1 \Delta x_2) = (p_{uuu} + p_{uud} - p_{udu} - p_{udd} - p_{duu} - p_{dud} + p_{ddu} + p_{ddd})\Delta x_1 \Delta x_2 =$$
$$= \rho_{1,2}\sigma_1\sigma_2\Delta t + \frac{\sigma_1^2\sigma_2^2}{4}(\Delta t)^2 \cong \rho_{1,2}\sigma_1\sigma_2\Delta t,$$

$$E(\Delta x_1 \Delta x_3) = (p_{uuu} - p_{uud} + p_{udu} - p_{udd} - p_{duu} + p_{dud} - p_{ddu} + p_{ddd})\Delta x_1 \Delta x_3 =$$
$$= \rho_{1,3}\sigma_1\sigma_3\Delta t + \frac{\sigma_1^2\sigma_3^2}{4}(\Delta t)^2 \cong \rho_{1,3}\sigma_1\sigma_3\Delta t,$$

$$E(\Delta x_2 \Delta x_3) = (p_{uuu} - p_{uud} - p_{udu} + p_{udd} + p_{duu} - p_{dud} - p_{ddu} + p_{ddd})\Delta x_2 \Delta x_3 =$$
$$= \rho_{2,3}\sigma_2\sigma_3\Delta t + \frac{\sigma_2^2\sigma_3^2}{4}(\Delta t)^2 \cong \rho_{2,3}\sigma_2\sigma_3\Delta t.$$

Remembering that $(\Delta t)^2 \cong 0$ and bearing in mind the values for Δx_1, Δx_2, and Δx_3, the following is obtained:

$$p_{uuu} + p_{uud} - p_{udu} - p_{udd} - p_{duu} - p_{dud} + p_{ddu} + p_{ddd} = \rho_{1,2}$$

$$p_{uuu} - p_{uud} + p_{udu} - p_{udd} - p_{duu} + p_{dud} - p_{ddu} + p_{ddd} = \rho_{1,3}$$

$$p_{uuu} - p_{uud} - p_{udu} + p_{udd} + p_{duu} - p_{dud} - p_{ddu} + p_{ddd} = \rho_{2,3}$$

The last three equations establish the conditions for consistency with the first moment:

$$E(\Delta x_1) = (p_{uuu} + p_{uud} + p_{udu} + p_{udd} - p_{duu} - p_{dud} - p_{ddu} - p_{ddd})\Delta x_1 = -\frac{\sigma_1^2}{2}\Delta t,$$

$$E(\Delta x_2) = (p_{uuu} + p_{uud} - p_{udu} - p_{udd} + p_{duu} + p_{dud} - p_{ddu} - p_{ddd})\Delta x_2 = -\frac{\sigma_2^2}{2}\Delta t,$$

$$E(\Delta x_3) = (p_{uuu} - p_{uud} + p_{udu} - p_{udd} + p_{duu} - p_{dud} + p_{ddu} - p_{ddd})\Delta x_3 = -\frac{\sigma_2^2}{2}\Delta t.$$

From them the following is derived:

$$p_{uuu}+p_{uud}+p_{udu}+p_{udd}-p_{duu}-p_{dud}-p_{ddu}-p_{ddd}=-\frac{\sigma_1\sqrt{\Delta t}}{2},$$

$$p_{uuu}+p_{uud}-p_{udu}-p_{udd}+p_{duu}+p_{dud}-p_{ddu}-p_{ddd}=-\frac{\sigma_2\sqrt{\Delta t}}{2},$$

$$p_{uuu}-p_{uud}+p_{udu}-p_{udd}+p_{duu}-p_{dud}+p_{ddu}-p_{ddd}=-\frac{\sigma_3\sqrt{\Delta t}}{2}.$$

There are thus seven equations and eight unknowns. In principle, several solutions are possible. However, the method suggested by Boyle et al. [5] is adopted here. This gives the following probabilities, which satisfy the above equations:

$$p_{uuu}=\frac{1}{8}[1+\rho_{1,2}+\rho_{1,3}+\rho_{2,3}+\sqrt{\Delta t}(-\frac{\sigma_1}{2}-\frac{\sigma_2}{2}-\frac{\sigma_3}{2})],$$

$$p_{uud}=\frac{1}{8}[1+\rho_{1,2}-\rho_{1,3}-\rho_{2,3}+\sqrt{\Delta t}(-\frac{\sigma_1}{2}-\frac{\sigma_2}{2}+\frac{\sigma_3}{2})],$$

$$p_{udu}=\frac{1}{8}[1-\rho_{1,2}+\rho_{1,3}-\rho_{2,3}+\sqrt{\Delta t}(-\frac{\sigma_1}{2}+\frac{\sigma_2}{2}-\frac{\sigma_3}{2})],$$

$$p_{udd}=\frac{1}{8}[1-\rho_{1,2}-\rho_{1,3}+\rho_{2,3}+\sqrt{\Delta t}(-\frac{\sigma_1}{2}+\frac{\sigma_2}{2}+\frac{\sigma_3}{2})],$$

$$p_{duu}=\frac{1}{8}[1-\rho_{1,2}-\rho_{1,3}+\rho_{2,3}+\sqrt{\Delta t}(+\frac{\sigma_1}{2}-\frac{\sigma_2}{2}-\frac{\sigma_3}{2})],$$

$$p_{dud}=\frac{1}{8}[1-\rho_{1,2}+\rho_{1,3}-\rho_{2,3}+\sqrt{\Delta t}(+\frac{\sigma_1}{2}-\frac{\sigma_2}{2}+\frac{\sigma_3}{2})],$$

$$p_{ddu}=\frac{1}{8}[1+\rho_{1,2}-\rho_{1,3}-\rho_{2,3}+\sqrt{\Delta t}(+\frac{\sigma_1}{2}+\frac{\sigma_2}{2}-\frac{\sigma_3}{2})],$$

$$p_{ddd}=\frac{1}{8}[1+\rho_{1,2}+\rho_{1,3}+\rho_{2,3}+\sqrt{\Delta t}(+\frac{\sigma_1}{2}+\frac{\sigma_2}{2}+\frac{\sigma_3}{2})].$$

These probabilities have the same structure as those derived by Boyle et al. [5]. Also, when the correlation is ($\rho_{i,j}=0$ c) all eight probabilities are close to $\frac{1}{8}$ if a small size of Δt is chosen.

Example 10. three-dimensional European call option

Consider the problem solved in Boyle, Evnine and Gibbs [5] for a three-dimensional European call, valuing a call at the maximum for three assets with the following data:

Spot price at initial time: $S_i=100$.

Time to maturity: $T=1$ year.

Drift (risk-neutral world): $\alpha_i-\lambda_i=0.10$.

Exercise price: $K = 100$.

Volatility: $\sigma_i = 0.2$.

Risk-free interest rate: $r = 0.10$.

Correlation: $\rho_{i,j} = 0.50$.

The results obtained (in line with the number of steps) for a European call and a European put match those of Boyle, Evnine and Gibbs [5] for the 20, 40 and 80 steps shown in their paper. See Table 15. Table 15 also shows the results obtained here for American and Bermudan puts using the risk-neutral probability calculation method proposed by these authors, but with a different decision at the nodes. See Table 16: E denotes the value of exercising the option and C the value of continuing without exercising it.

Table 15. Three-dimensional option values with spot dynamics

Steps	Maximum European Call	Maximum European Put	Maximum American Put	Maximum Bermudan Put
20	22.2807	0.9192	2.2602	1.3520
40	22.4792	0.9254	2.2080	1.3287
80	22.5764	0.9290	2.1402	1.3092
160	22.6245	0.9308	2.1083	1.2966
320	22.6484	0.9318	2.0877	1.2907

Table 16. Decision rules at nodes

	European Call	European Put	American Put	Bermudan Put
Final nodes	Max(E_C ,0)	Max(E_P ,0)	Max(E_P ,0)	Max(E_P ,0)
Intermediate nodes	C	C	Max(E_P, C)	
With decision				Max(E_P, C)
With no decision				C

The value of exercising the call is given by:

$$Max(S_1, S_2, S_3) - K.$$

The value of exercising the put is:

$$K - Max(S_1, S_2, S_3).$$

and the continuation value is:

$$C = e^{-r\Delta t}(p_{uuu}W^{+++} + p_{uud}W^{++-} + p_{udu}W^{+-+} + \tag{40}$$

$$+ p_{udd}W^{+--} + p_{duu}W^{-++} + p_{dud}W^{-+-} + p_{ddu}W^{--+} + p_{ddd}W^{---}). \tag{41}$$

W^{+++} represents the value for the case when the price of all three assets increases in the following period; W^{++-} the value when the price of the first two assets increases but that of the third decreases in the following period, and so on. Bear in mind that, as usual, the tree is solved recursively starting from the last time and working back step by step to the initial time.

The probability values obtained for 20 and 40 steps are shown in Table 17.

Table 17. Risk-neutral probabilities

	Spot Dynamics		Futures Dynamics	
	20 steps	40 steps	20 steps	40 steps
p_{uuu}	0.3460	0.3362	0.3041	0.3066
p_{uud}	0.0737	0.0704	0.0597	0.0605
p_{udu}	0.0737	0.0704	0.0597	0.0605
p_{udd}	0.0513	0.0546	0.0653	0.0645
p_{duu}	0.0737	0.0704	0.0597	0.0605
p_{dud}	0.0513	0.0546	0.0653	0.0645
p_{ddu}	0.0513	0.0546	0.0653	0.0645
p_{ddd}	0.2790	0.2888	0.3209	0.3184

Using the risk-neutral probabilities for the futures dynamics and the corresponding conversions between future and spot prices the results shown in Table 18 are obtained.

Table 18. Three-dimensional option with futures dynamics

Steps	Maximum European Call	Maximum European Put	Maximum American Put	Maximum Bermudan Put
20	22.4670	0.9513	2.2425	1.3779
40	22.6059	0.9770	2.2031	1.3448
80	22.6306	0.9461	2.1257	1.3190
160	22.6492	0.9371	2.1003	1.2984
320	22.6640	0.9382	2.0851	1.2947

The optimal exercise boundary is formed by those points where the immediate exercise value at the initial time is equal to the continuation value. Thus, for the maximum American put, the following must hold:

$$Max(S_{1,0}^{*}, S_{2,0}^{*}, S_{3,0}^{*}) - K = e^{-r\Delta t}(p_{uuu}W^{+++} + p_{uud}W^{++-} + p_{udu}W^{+-+} +$$

$$+ p_{udd}W^{+--} + p_{duu}W^{-++} + p_{dud}W^{-+-} + p_{ddu}W^{--+} + p_{ddd}W^{---}). \quad (42);$$

Choosing two values, e.g. $S^*_{1,0} = 100$ and $S^*_{2,0} = 100$, the value of $S^*_{3,0}$ that satisfies equation (42) is found. A point $(S^*_{1,0}, S^*_{2,0}, S^*_{3,0})$ is obtained that belongs on the boundary between the "invest now" region and the "wait" region. In the three-dimensional case that boundary is a surface.

6. The Least-Squares Monte Carlo Method

6.1. Path-Dependent Options

Example 11: Path-dependent option

This section looks again at the same optimal control case solved by binomial trees in Example 9. There are two GBM-type correlated stochastic processes:

$$dp_t = (\alpha_p - \lambda_p) p_t dt + \sigma_p p_t dz_t^p,$$

$$dc_t = (\alpha_c - \lambda_c) c_t dt + \sigma_c c_t dz_t^c,$$

with: $E(dz_t^p dz_t^c) = \rho dt.$

In order to obtain a reasonably accurate valuation a sufficiently high number of paths must be simulated for each stochastic variable. Each path is generated step by step from the initial values. To obtain the value of a given step it is necessary at each time □ to have correlated random samples $u_{1,t}$ and $u_{2,t}$. These samples can be obtained from standard normal variates $e_{1,t}$ and $e_{2,t}$ as follows:

$$u_{1,t} = e_{1,t},$$

$$u_{2,t} = e_{1,t}\rho + e_{2,t}\sqrt{1-\rho^2}.$$

The useful lifetime of the plant is 20 years, so there are 240 steps for each path. If, for instance, $50{,}000$ paths are used for the input price and a further $50{,}000$ for the output price then 24 million independent random samples need to be generated, drawn from a normal standard distribution $N(0,1)$. The paths are generated with these values as follows:

$$p_{t+\Delta t} = p_t e^{(\alpha_p - \lambda_p - \frac{1}{2}\sigma_p^2)\Delta t + \sigma_p \sqrt{\Delta t} u_{1,t}},$$

$$c_{t+\Delta t} = c_t e^{(\alpha_c - \lambda_c - \frac{1}{2}\sigma_c^2)\Delta t + \sigma_c \sqrt{\Delta t} u_{2,t}}.$$

As with the binomial tree, the optimal value of q_t is calculated. This value must be between $q_{\min} = 0$ and $q_{\max} = 100$. With these values, the results for each month are calculated. Then they are discounted at the initial time at the risk-free interest rate. This gives a value V for each path. If n is the number of paths and $V(i)$ is the result obtained for a path i, the result of the valuation is:

$$\frac{1}{n}\sum_{i=1}^{i=n} V(i).$$

The results are shown in Table 19, in line with the number of paths used. The results obtained by this method are very close to the figure of $539{,}350$ obtained with the binomial lattice.

Table 19. Monte Carlo valuation in optimal control

Number of Paths	Value
25,000	545,190
50,000	542,540
75,000	544,370
100,000	542,690
125,000	539,930
150,000	540,310

6.2. American-Style Options

It was long believed that Monte Carlo methods were not well suited to valuing American-style options (they were seen as better suited to valuing path-dependent options).[17] By contrast, binomial and trinomial trees and finite difference methods were seen as most suitable for valuing American-style options. However, it is difficult to extend these methods when the number of risk factors is high, i.e. they suffer from the so-called "curse of dimensionality". Several papers, notably that of Longstaff and Schwartz [13], have now markedly changed that belief.

Example 12. The LSM method and the valuation of three-dimensional American options

Here valuation is to be run by this method using $N =$ 100,000 Monte Carlo simulations. The (investment) option has a finite life of one year, which is decomposed into 160 periods (i.e. $\Delta t = 1/160$). The rate of improvement of the quality of MC estimates (or

[17]See for instance Hull [11].

the rate of decrease of error) is known to be on the order of $1/\sqrt{N}$. In this respect, one can be relatively confident of the accuracy of the results.

To generate the paths correlated random samples are needed. These samples are generated as follows, from standard normal variates $e_{1,t}$, $e_{2,t}$ and $e_{3,t}$:

$$u_{1,t} = e_{1,t}, \tag{43}$$

$$u_{2,t} = e_{1,t}\rho_{12} + e_{2,t}\sqrt{1-\rho_{12}^2}, \tag{44}$$

$$u_{3,t} = e_{1,t}\rho_{13} + e_{2,t}\frac{\rho_{23} - \rho_{12}\rho_{13}}{\sqrt{1-\rho_{12}^2}} + e_{3,t}\sqrt{1-\rho_{13}^2 - \frac{(\rho_{23}-\rho_{12}\rho_{13})^2}{1-\rho_{12}^2}}. \tag{45}$$

The Monte Carlo simulations use:

$$S_{i,t+\Delta t} = S_{i,t} e^{(\alpha_i - \lambda_i - \frac{1}{2}\sigma_i^2)\Delta t + \sigma_i\sqrt{\Delta t}u_{i,t}},$$

where in the case considered here $i = 1,2,3$. The procedure can easily be generalised for a larger number of stochastic processes.

Given the values of V_t at any time and on each path, the Least Squares Monte Carlo (LSM) approach is used. At the last time T, the value of the three-dimensional American call option on the sum of the values of the three assets on each path is:

$$V_T = \max(S_{1,T} + S_{2,T} + S_{3,T} - K, 0). \tag{46}$$

For earlier times the method requires the computation of a series of parameters that allow a linear combination of basic functions to be constructed. This combination allows the continuation value to be estimated at each step. The specification adopted consists of a second-order expected continuation value function with ten regressors (since there are three sources of risk), namely:

$$E_t^Q[e^{-r\Delta t}V_{t+1}(S_{1,t+1}, S_{2,t+1}, S_{3,t+1})] \approx a_1 + a_2 S_{1,t} + a_3 S_{1,t}^2 +$$
$$+ a_4 S_{2,t} + a_5 S_{2,t}^2 + a_6 S_{3,t} + a_7 S_{3,t}^2 + a_8 S_{1,t} S_{2,t} + a_9 S_{1,t} S_{3,t} + a_{10} S_{2,t} S_{3,t}. \tag{47}$$

The value of the ten coefficients can be obtained at any time, considering the paths that are in-the-money and by applying ordinary least squares. The optimal exercise boundary is given by a surface formed with those values of $S_{1,0}$, $S_{2,0}$ and $S_{3,0}$ for which the present value of exercising the option at that time i equals the continuation value. Given the continuation value, the decisions at intermediate times on a path are very similar to those for binomial lattices. That is, the continuation value is compared with the immediate exercise value and the greater of the two is chosen.

Consider the following put option on the sum of the values of the three assets S_i, with $i = 1,2,3$, which follow a GBM stochastic process with the following parameters:

Spit price at initial time: $S_i = 100$.

Time to maturity: $T = 1$ year.

Drift (risk-neutral world): $\alpha_i - \lambda_i = 0.08$.

Exercise price: $K = 300$.

Volatility: $\sigma_i = 0.3$.

Risk-free interest rate: $r = 0.10$.

Correlation: $\rho_{i,j} = 0.50$.

A valuation is run with 160 steps and $100{,}000$ paths.

Once the paths are generated, the mean value at time $T = 1$ should be close to $E_0(S_{i,T}) = S_{i,0}e^{(\alpha_i - \lambda_i)T} = 108.3287$. The actual mean values obtained with $100{,}000$ paths for each asset at time T were 108.2921, 108.3893 and 108.3807 for $S_{1,T}$, $S_{2,T}$ and $S_{3,T}$ respectively.

A similar check was run on the correlations between the logarithms of the values of the assets at the last time. The figures obtained were 0.5006, 0.4970 and 0.4990 respectively.

An option price of 20.4122 was also obtained by LSM. The consistency of this value was checked by means of a three-dimensional binomial tree, using futures dynamics. The results obtained via the tree, in line with the number of steps used, are shown in Table 20.

Table 20. checking LSM value

Steps	Lattice Value
80	20.5818
160	20.5061
320	20.5083

The values of the parameters of the continuation function at the initial time are shown in Table 21.

At the optimal exercise boundary, the immediate exercise value equals the continuation value. Thus, when:

$$\max\left(S^*_{1,0} + S^*_{2,0} + S^*_{3,0} - K, 0\right) = a_1 + a_2 S^*_{1,0} + a_3 (S^*_{1,0})^2 + a_4 S^*_{2,0} + a_5 (S^*_{2,0})^2 + a_6 S^*_{3,0} + a_7 (S^*_{3,0})^2 + a_8 S^*_{1,0} S^*_{2,0} + a_9 S^*_{1,0} S^*_{3,0} + a_{10} S^*_{2,0} S^*_{3,0}. \quad (48)$$

Table 21. coefficients on the continuation function

a_1	510.8459
a_2	-12.2465
a_3	0.0228
a_4	2.2821
a_5	-0.0098
a_6	1.2859
a_7	-0.0159
a_8	0.0268
a_9	0.0478
a_{10}	-0.0340

Note that if values are given e.g. to $S^*_{1,0}$ to seek a point on the optimal exercise boundary that satisfies equation (48), different values are also obtained for the coefficients of the continuation function.

CONCLUSION

Numerical methods are essential in valuing derivative assets and managing the corresponding risks. The practical possibility of using such methods to solve complex problems has increased substantially in recent years thanks to the increase in calculating power of computers and to improvements in the products available for carrying out the relevant programming.

However, it is still essential to select the most suitable method for each specific problem to be solved. In particular the most suitable stochastic method and the degree of dimensionality (depending on the relevant risk factors) must be selected, the parameters of the model must be estimated or chosen and the most suitable numerical estimation method for the process must be selected. When calculations are being carried out not for an academic paper but for a program which is to be used frequently at a company where there is a need to obtain the correct result, other significant factors must also be taken into account, such as design, processing time and optimisation of resources on the computer that is to run the program.

Reality reveals that the sources of risk in an activity, a business or an asset portfolio are seldom one-dimensional. There is usually a broad range of risks, and all the most significant of them at least should be taken into account in any valuation.

This paper seeks to show how American and Bermudan type multi-dimensional options can be valued by numerical methods when there are several processes that show GBM type behaviour. It also shows how to run certain checks to ensure that the IT programs developed to solve problems of this type are behaving properly. Much of the development work in the case of binomial lattices is based on futures dynamics, in an effort to prevent the appearance of negative probabilities.

The procedures developed can be extended, with certain important changes, to processes of other types, such as mean-reverting processes.

REFERENCES

[1] Abadie,L.M.,and Chamorro,J.M. (2010). Toward sustainability through investments in energy efficiency. In W.H. Lee, V.G. Cho (Eds.), *Handbook of Sustainable Energy*, pp 735-774. NY, US: Nova Science Publishers Inc.

[2] Abadie, L.M., Chamorro J.M.,and González-Eguino, M. (2010). Optimal Investment in Energy Efficiency under Uncertainty. *BC3 Working Paper Series 2009-11.*

[3] Abadie, L.M.,and Chamorro, J.M. (2008). Valuing flexibility: The case of an integrated gasification combined cycle power plant. *Energy Economics 30,* pp 1850-1881.

[4] Abadie, L.M.,and Chamorro J.M. (2009) Monte Carlo Valuation of natural gas investments. *Review of Financial Economics 18,* pp 10-22.

[5] Boyle, P.P., Evnine, J.,and Gibbs S. (1989). Numerical Evaluation of Multivariate Contingent Claims. *The Review of Financial Studies 2(2),* pp. 241-250.

[6] Brandimarte, P. (2006). *Numerical Methods in Finance and Economics* Second Edition. Hoboken, New Jersey: John Wiley and Sons.

[7] Clewlow, L.,and Strickland, C. (1998). *Implementing Derivatives Models.* Sussex, UK: John Wiley and Sons.

[8] Copeland, T., and Antikarov, V. (2003). *Real Options: A practitioners guide.* US: Thomson.

[9] Cox, J., Ross, S., and Rubinstein, M. (1979). Option Pricing: A Simplified Approach. *Journal of Financial Economics 7*, pp. 229-264.

[10] Dixit, A. K., and Pindyck, R.S. (1994). *Investment under uncertainty.* Princeton University Press.

[11] Hull, J.C. (2003). *Options, Futures and Other Derivatives*, Fifth Edition. Delhi, India: Pearson Education, Inc.

[12] Luenberger, D.G. (1998). *Investment Sciehnce.* New York, US: Oxford University Press.

[13] Longstaff, F.A., and Schwartz, E.S. (2001). Valuing American Options by Simulation: A Simple Least Squares Approach. *Review of Financial Studies 14(1)*, pp 113-147.

[14] Tavella, D. A. (2002). *Quantitative Methods in Derivatives Pricing.* New York, US: John Wiley and Sons.

[15] Trigeorgis, L. (1996). *Real Options Managerial Flexibility and Strategy in Resource Allocations*. Cambridge, Massachusetts: The MIT Press.

[16] Wilmott, P. (2006). *Paul Wilmott on Quantitative Finance*. West Sussex,UK: John Wiley and Sons.

Reviewed by José M. Chamorro (Dpt. Financial Economics II, University of the Basque Country, Spain)

In: Brownian Motion: Theory, Modelling and Applications ISBN: 978-1-61209-537-0
Editors: R.C. Earnshaw and E.M. Riley

Chapter 5

BROWNIAN MOTION-BASED MODEL FOR ENHANCED THERMAL CONDUCTIVITY OF NANOFLUIDS

S. M. Sohel Murshed*
Department of Mechanical, Materials and Aerospace Engineering
University of Central Florida, Orlando, Florida, U.S.A.

ABSTRACT

Nanofluids are a new class of heat transfer fluids which are engineered by dispersing nanometer-sized solid particles in conventional fluids. This is a rapidly emerging interdisciplinary field where nanoscience, nanotechnology, and thermal engineering meet. Since the novel concept of *nanofluids* was coined in 1995, this research topic has attracted tremendous interest from researchers worldwide due to their exciting thermal properties and potential applications in numerous important fields. Although research works have shown that nanofluids exhibit significantly higher thermal conductivity compared to their base fluids, the underlying mechanisms for the enhancement are still debated and not thoroughly understood. Despite considerable theoretical efforts devoted to the development of model for the prediction of the effective thermal conductivity of nanofluids, there has been little agreement among different studies and no widely accepted model is also available due to inconclusive heat transfer mechanisms of nanofluids. Nevertheless, fundamental understanding of the underlying mechanisms and development of a unanimous theoretical model are crucial for exploiting potential benefits and applications of nanofluids. In this chapter, a new and improved Brownian motion-based model is introduced for the prediction of the enhanced thermal conductivity of nanofluids. In addition to the Brownian motion of nanoparticles, this model also takes into account several other important factors such as particle size and interfacial nanolayer that contribute to the enhancement of the effective thermal conductivity of nanofluids. The conventional kinetic theory-based Brownian motion term has been renovated using effective diffusion coefficient concept. The present model shows reasonably good agreement with the experimental results of various aqueous nanofluids and gives better predictions compared to classical and other recently developed models. Besides providing a brief review on theoretical studies and various heat transfer mechanisms of

* E-mail: drsmurshed@knights.ucf.edu

nanofluids, details of the present model development and its validation with the experimental results are also discussed in this chapter.

1. INTRODUCTION

1.1. Background of Nanofluids

With an ever-increasing thermal load due to smaller features of microelectronic devices and more power output, cooling for maintaining desirable performance and durability of such devices is one of the most important technical issues in many high-tech industries. The conventional method to increase the cooling rate is to use extended heat transfer surfaces. However, this approach requires an undesirable increase in the size of the thermal management systems. In addition, the inherently poor thermophysical properties of traditional heat transfer fluids such as water, ethylene glycol (EG) or engine oil (EO) greatly limit the cooling performance. Thus, these conventional cooling techniques are not suitable to meet the demand of these high-tech industries. There is always a need to develop advanced cooling techniques and innovative heat transfer fluids with improved heat transfer performance. It is also well known that at room temperature, metals possess at least an order-of- magnitude higher thermal conductivity than fluids. For example, the thermal conductivity of silver (429 W/m·K) at room temperature is around 700 times greater than that of water (0.607 W/m·K) and about 3000 times greater than that of engine oil (0.145 W/m·K). Therefore, the thermal conductivities of fluids that contain suspended metallic or nonmetallic (oxide) particles are expected to be significantly higher than those of conventional heat transfer fluids.

As thermophysical properties, particularly thermal conductivity of a fluid play a vital role in the development of energy-efficient heat transfer equipment, numerous theoretical and experimental studies on increasing thermal conductivity of liquids by suspending small particles have been conducted since the treatise by Maxwell more than a century ago [1]. However, all such studies on thermal conductivity of suspensions have been confined to millimeter- or micrometer-sized particles. The major problems of such suspensions are the rapid sedimentation of particles, clogging of flow channel, and increased pressure drop for pumping the fluid. In contrast, due to smaller size and larger interface area nanoparticles remain in suspension and result in reduced erosion and clogging. Nanoparticles are also suitable for use in microsystems because they are orders of magnitude smaller than the microsystems.

Over the last several decades, scientists and engineers have attempted to develop fluids, which offer better cooling or heating performance. However, it is only in 1995 that Steve Choi [2] at Argonne National Laboratory of USA coined the novel concept of "*nanofluid*" to meet the cooling challenges facing many high-tech industries. It should also be noted that nanoparticles-suspensions (i.e. nanofluids) were used in a boiling heat transfer study as early as 1984 by Yang and Maa [3] and another experimental study on thermal conductivity and viscosity of several types of nanoparticles-suspensions was performed by Masuda and co-workers [4] in 1993. Apart from the works of these two groups [3, 4], Arnold Grimm [5], a German researcher also won a German patent on the enhanced thermal conductivity of nano- and micro-sized particles suspensions in 1993. Aluminum particles of 80 nm to 1μm were suspended into a fluid and about 100% increase in the thermal conductivity of the fluid for loadings of 0.5 to 10 volume % was reported in his patent. As a new and innovative class of heat transfer fluids, nanofluids represent a rapidly emerging field where nanoscale science and thermal engineering meet. Nanofluids are engineered by dispersing nanometer-sized solid particles, rods or tubes in conventional heat transfer fluids such as water, ethylene glycol, or engine oil. From previous investigations, nanofluids were found to have significantly higher thermal conductivity than their base fluids [6-15] and they can offer numerous potential benefits and applications in many important fields such as microelectronics, microfluidics, transportation, manufacturing, HVAC (heating, ventilating and air-conditioning), and biomedical and thus have attracted great interest from the research community [15-17]. Although significant progress has been made, variability and controversies in the reported data and heat transport mechanisms still exist with nanofluids [15, 18, 19].

1.2. Synthesis and Potential Benefits of Nanofluids

Synthesis and preparation of sample nanofluids are the key steps for enhanced heat transfer performance of nanofluids. Techniques for good dispersion and production of uniform-sized nanoparticles in liquids or directly producing stable nanofluids are crucial. There are mainly two techniques for synthesizing nanofluids: the two-step process and the direct evaporation technique or single-step process [15, 16]. In the two-step process, dry nanoparticles are first produced by an inert gas condensation method and they are then dispersed into a fluid. On the other hand, the direct evaporation technique (i.e. single-step process) synthesizes nanoparticles and disperses them into a fluid in a single step. At present, most researchers used the two-step process to prepare nanofluids by dispersing commercial or self-produced nanoparticles in base liquids.

The optimization of thermal properties of nanofluids requires stable nanofluids, which can be ensured by proper synthesis and dispersion procedures. The impact of nanofluid technology is expected to be great considering that heat transfer performance of heat exchangers or cooling devices is vital in numerous industries. When the nanoparticles are properly dispersed, nanofluids can offer numerous benefits besides the anomalously high effective thermal conductivity.

Some of these benefits are improved heat transfer and stability, microchannel cooling without clogging, miniaturized systems, and reduction in pumping power. The better stability of nanofluids will prevent rapid settling and reduce clogging in the walls of heat transfer devices. The high thermal conductivity of nanofluids translates into higher energy efficiency, better performance, and lower operating costs. They can reduce energy consumption for pumping heat transfer fluids. Miniaturized systems require smaller inventories of fluids where nanofluids can be used. Thermal systems can be smaller and lighter with nanofluids.

In vehicles, smaller components result in better gasoline mileage, fuel savings, lower emissions, and a cleaner environment. With these highly desired thermal properties and potential benefits, nanofluids are thought to have a wide range of applications including transportation sector (because of the higher thermal conductivity nanofluids would allow for smaller, lighter engines, pumps, radiators, and other components) and micro-electromechanical systems (MEMS). Details of nanofluids potential benefits and applications can be found elsewhere [15].

2. Enhanced Thermal Conductivity of Nanofluids

2.1. Mechanisms

The classical models (e.g., Maxwell model) which were developed from the effective medium theory for the prediction of the effective thermal conductivity of composites have been verified by experimental data for mixtures with low concentrations of milli- or micro-meter sized particles. Experiments have shown that nanofluids exhibit anomalously high thermal conductivity which cannot be predicted accurately by the classical models. In early studies, Wang et al. [9] and Keblinski et al. [20] proposed several mechanisms to explain the observed anomalously high thermal conductivity of nanofluids and these mechanisms are not considered by classical models. Wang et al. [9] suggested that the microscopic motion of nanoparticles, surface properties, and the structural effects might be the reasons for the enhanced thermal conductivity of nanofluids. In nanofluids, the microscopic motion of the nanoparticles due to van der Waals force, stochastic force (causing Brownian motion) and electrostatic force can be significant. The surface properties and structural effects were not confirmed as potential mechanisms in their study. Keblinski et al. [20] later elucidated four possible mechanisms for the anomalous increase in thermal conductivity of nanofluids which are i) Brownian motion of the nanoparticles, ii) liquid layering at the liquid/particle interface, iii) nature of the heat transport in the nanoparticles, and iv) effect of nanoparticle clustering. These mechanisms are briefly discussed in the following paragraphs.

Due to the Brownian motion, nanoparticles move randomly through the liquid and may collide, thereby enabling direct solid-solid transport of heat from one to another, which can increase the effective thermal conductivity. By a simple analysis, Keblinski et al. [20] showed that the thermal diffusion of nanoparticles is much faster than Brownian diffusion and thus the contribution of Brownian motion to energy transport in nanofluids is not significant. Nevertheless, it is plausible that the Brownian motion can contribute in the thermal conductivity of nanofluids with small concentration and small-sized nanoparticles. Keblinski et al. [20] explained that when the size of the nanoparticles in a nanofluid becomes smaller than the phonon mean-free path, phonons no longer diffuse across the nanoparticle but move ballistically without any scattering. However, it is difficult to envision how ballistic phonon transport could be more effective than a very-fast diffusion phonon transport, particularly to the extent of explaining anomalously high thermal conductivity of nanofluids. No further work or analysis has been reported on the ballistic heat transport nature of nanoparticles. Instead, the continuum approach was adopted in most of the reported works [21-27].

Liquid layering around the nanoparticle i.e. nanolayer was considered as another mechanism responsible for the enhanced thermal conductivity of nanofluids. The basic idea is that liquid molecules can form a layer around the solid particles and thereby enhance the local ordering of the atomic structure at the interface region. Hence, the atomic structure of such liquid layer is significantly more ordered than that of the bulk liquid. Given that solids, which have much ordered atomic structure, exhibit much higher thermal conductivity than the liquids, the liquid layer at the interface would reasonably have a higher thermal conductivity than the bulk liquid. Thus, the nanolayer is considered as an important factor enhancing the thermal conductivity of nanofluids.

In analyzing effect of clustering, the effective volume of a cluster is considered much larger than the volume of the particles due to the lower packing fraction (ratio of the volume of the solid particles in the cluster to the total volume of the cluster) of the cluster. Since heat can be transferred rapidly within such clusters, the volume fraction of the highly conductive phase (cluster) is larger than the volume of solid, thus increasing its thermal conductivity [20, 28]. However, in general, clustering may exert a negative effect on heat transfer enhancement, particularly at a low volume fraction by settling small particles out of the liquid and creating a large region of "particle free" liquid with a high thermal resistance [20]. Besides these aforementioned mechanisms, the effects of nanoparticle surface chemistry and particles interaction could be significant in enhancing the thermal conductivity of nanofluids.

2.2. Existing Models for Nanofluids

In order to predict the effective thermal conductivity of solid particle suspensions (k_{eff}), several models have been developed since the treatise by Maxwell [1]. As mentioned previously, these classical models such as those attributed to Maxwell [1] and Hamilton-Crosser [29] were developed for predicting the effective thermal conductivity of a continuum medium with well-dispersed solid particles. The Maxwell model [1] predicts the effective electrical or thermal conductivity of liquid-solid suspensions for very low (dilute) concentration of spherical particles. Hamilton and Crosser [29] modified Maxwell's model for both spherical and non-spherical particles by applying a shape factor. For spherical particles, the Hamilton and Crosser (HC) model reduces to the Maxwell model.

Research showed that these classical models are unable to predict the anomalously high thermal conductivity of nanofluids. This is because these models do not include the nanoscale effects of particle size such as the interfacial layer at the particle/liquid interface, and motion of particles, which are considered as important factors for the enhancement of the thermal conductivity of nanofluids. Recently, many theoretical studies have been carried out to predict the anomalously increased thermal conductivity of nanofluids. Several models have been proposed by considering various mechanisms. A detailed review of numerous models proposed for the prediction of effective thermal conductivity of nanofluids is provided in recent review articles [15, 30]. However, some commonly used classical models and recently developed models used for the prediction of the enhanced thermal conductivity of nanofluids are given in Table 1. All the reported models can be categorized into two general groups, which are static and dynamic models. The static-based models assume stationary nanoparticles in the base fluid in which the thermal transport properties are predicted by effective medium approximations-based models such as Maxwell model and Hamilton-Crosser model. Most of the researchers [14, 21-28, 31, 32] in this group employed the concept of interfacial layer at liquid/particle interface to develop models and to explain the anomalous enhancement of the thermal conductivity of nanofluids. On the other hand, the dynamic models are based on the premise that nanoparticles experience random movement in the fluid. This motion is believed to be responsible for transporting energy directly through collision between nanoparticles or indirectly through microconvection of liquid that enhances the transport of thermal energy. Researchers in the dynamic model group emphasize the contribution of dynamic part related to particle motion in their model development [33-37]. Although Wang et al. [9] and Keblinski et al. [20] claimed that the contribution of Brownian motion to energy transport in nanofluids is not significant, other researchers [33-35] held contrary views. Among very few theoretical efforts on combined static and dynamic effects of nanoparticles, Prasher et al. [36] considered dynamic contribution of nanoparticles in their thermal conductivity model for nanofluids. However, their model contains three unknown parameters which use as empirical or fitting parameters. In addition, dynamic contribution was coupled with Maxwell's model without taking into account the effect of interfacial layer in their model. Besides particle volume fraction (ϕ_p), particles size (r_p), temperature, particles dispersions and particle movement should be taken into account in developing model for the effective thermal conductivity of nanofluids. Since nanoparticles in base fluids can easily experience Brownian force, it is plausible that the observed thermal conductivity of nanofluids is from the combined static (thermal properties and interfacial layer) and dynamic (e.g. Brownian motion) contributions of dispersed nanoparticles.

Table 1. Summary of models for the effective thermal conductivity of nanofluids

Researchers	Models/ Expressions and remarks
Maxwell [1]	$k_{eff}/k_f = \dfrac{k_p + 2k_f + 2\phi_p(k_p - k_f)}{k_p + 2k_f - \phi_p(k_p - k_f)}$ It depends on the thermal conductivities of particle (k_p) and base fluids (k_f) and volume fraction of solid. Valid for spherical particles only.
Hamilton and Crosser [29]	$k_{eff}/k_f = \left[\dfrac{k_p + (n-1)k_f - (n-1)\phi_p(k_f - k_p)}{k_p + (n-1)k_f + \phi_p(k_f - k_p)}\right]$ where shape factor $n = 3/\psi$ where ψ is the particle sphericity. Valid for both the spherical and cylindrical particles
Yu and Choi [22, 23]	1) $k_{eff}/k_f = \dfrac{k_{cp} + 2k_f + 2\phi_p(k_{cp} - k_f)(1+\beta)^3}{k_{cp} + 2k_f - \phi_p(k_{cp} - k_f)(1+\beta)^3}$ It is renovated Maxwell model where k_{cp} is the thermal conductivity of complex nanoparticles, $\beta = h/r_p$ and h is layer thickness. 2) $k_{eff}/k_f = 1 + \dfrac{n\phi_e A}{1 - \phi_e A}$ It is renovated HC model where $A = \dfrac{1}{3}\sum_{j=a,b,c} \dfrac{k_{pj} - k_f}{k_{pj} + (n-1)k_f}$, $n = 3\psi^{-\alpha}$, ϕ_e is the equivalent volume fraction of complex ellipsoids and α is an empirical parameter.
Kumar et al. [35]	$k_{eff}/k_f = 1 + c\dfrac{K_B T}{2\pi\eta r_p^3}\dfrac{\phi_p r_f}{k_f(1-\phi_p)}$ It is developed from the combination of kinetic theory and Fourier's law where η is the viscosity, c is a constant, r_f is the radius of liquid particle, T is the temperature and K_B is the Boltzmann's constant.

Researchers	Models/ Expressions and remarks
Jang and Choi [37]	$k_{eff}/k_f = 1+(k_p/k_f-1)\phi_p + 3c\phi_p(d_f/d_p)\mathrm{Re}_p^2\,\mathrm{Pr}$ It is the combination of convection and conduction heat transport where c is a constant and d_f is the diameter of base fluid molecule.
Prasher et al. [36]	$k_{eff}/k_f = (1+A\phi_p\,\mathrm{Re}^m\,\mathrm{Pr}^{0.333})\dfrac{(1+2\alpha)+2\phi_p(1-\alpha)}{(1+2\alpha)-\phi_p(1-\alpha)}$ where $\alpha_f = 2R_b k_f/d_p$. This model included interfacial resistance with empirical or fitting parameters A, m, R_b.
Xie et al. [24]	$k_{eff}/k_f = 1+3\Theta\phi_p + \dfrac{2\Theta^2\phi_p^2}{1-\Theta\phi_p}$ It is a static model and considered the presence of a nanolayer.
Leong et al. [26]	$k_{eff} = \dfrac{(k_p-k_{lr})\phi_p k_{lr}[2\gamma_1^3-\gamma^3+1]+(k_p+2k_{lr})\gamma_1^3[\phi_p\gamma^3(k_{lr}-k_f)+k_f]}{\gamma_1^3(k_p+2k_{lr})-(k_p-k_{lr})\phi_p[\gamma_1^3+\gamma^3-1]}$ It is a three-phase (particle, interfacial layer, and base fluid) static model and valid for spherical particle. Here $\gamma = 1+h/r_p$, $\gamma_1 = 1+h/2r_p$ and k_{lr} is the thermal conductivity of nanolayer.
Murshed et al. [14]	$k_{eff} = \dfrac{(k_p-k_{lr})\phi_p k_{lr}[\gamma_1^2-\gamma^2+1]+(k_p+k_{lr})\gamma_1^2[\phi_p\gamma^2(k_{lr}-k_f)+k_f]}{\gamma_1^2(k_p+k_{lr})-(k_p-k_{lr})\phi_p[\gamma_1^2+\gamma^2-1]}$ It is a three-phase static model and valid for cylindrically particle.

3. Modeling of Effective Thermal Conductivity of Nanofluids

3.1. Static-Based Modeling

As mentioned previously, interfacial nanolayer is considered as a key mechanism responsible for enhanced thermal conductivity of nanofluids. In the interfacial region between a nanoparticle and the base fluid, liquid molecules are bonded or otherwise oriented resulting in unique thermophysical properties that are different from those of the particles and the base fluid. Hence, it is important to consider the interfacial layer as a separate component in the particle-fluid mixture (nanofluids) in order to include its effects on the mixture. Here, pure nanofluids are therefore considered as a mixture of three components (regions)– the nanoparticle of radius r_p, the interfacial layer between particle/fluid of thickness h, and the fluid medium and a static thermal conductivity model for the nanofluids (non-interacting nanoparticles) was previously developed by the author [26]. In developing the model, following two-dimensional (r and θ), steady-state heat conduction equation was taken as base equation

$$\frac{\partial}{\partial r}\left(r^2\frac{\partial T}{\partial r}\right) + \frac{1}{\sin\theta}\frac{\partial}{\partial\theta}\left(\sin\theta\frac{\partial T}{\partial\theta}\right) = 0 \qquad (1)$$

With the proper assumptions and imposing right boundary conditions for an uniform external temperature gradient (e.g. E_∞) parallel to the z-axis, Equation (1) was solved for the spherical coordinate systems and the temperature gradients in each component were then determined. By making use of these temperature gradients and applying spatial averages of the heat fluxes in all components, a static model for the effective thermal conductivity of nanofluids with spherical nanoparticles was obtained as follows [26]:

$$k_{st} = \frac{(k_p - k_{lr})\phi_p k_{lr}[2\gamma_1^3 - \gamma^3 + 1] + (k_p + 2k_{lr})\gamma_1^3[\phi_p\gamma^3(k_{lr} - k_f) + k_f]}{\gamma_1^3(k_p + 2k_{lr}) - (k_p - k_{lr})\phi_p[\gamma_1^3 + \gamma^3 - 1]} \tag{2}$$

where, ϕ_p is the particle volumetric fraction, $\gamma = r_{cp} / r_p = 1 + h / r_p$, $\gamma_1 = 1 + h/2r_p$, r_{cp} is the radius of complex nanoparticle (nanoparticle with interfacial layer), and k_p, k_{lr} and k_f are the thermal conductivities of particle, interfacial layer and base fluid, respectively. The detailed discussion and mathematical derivations for this model containing spherical nanoparticles with interfacial nanolayer can be found elsewhere [26].

The orderness and orientation of fluid molecules absorbed on the surface of a nanoparticle (similar to surface adsorption) result in an intermediate value of thermal conductivity of nanolayer i.e., $k_f < k_{lr} < k_p$. Hence, the thermal conductivity of an interfacial layer can be written as $k_{lr} = \omega k_f$, where $\omega > 1$ is an empirical parameter which depends on the orderness of fluid molecules in the interface, nature and surface chemistry of nanoparticles.

Substituting $k_{lr} = \omega k_f$ into Equation (2), the static part of the effective thermal conductivity of nanofluids can be expressed as

$$k_{st} = k_f \frac{\phi_p\omega(k_p - \omega k_f)[2\gamma_1^3 - \gamma^3 + 1] + (k_p + 2\omega k_f)\gamma_1^3[\phi_p\gamma^3(\omega - 1) + 1]}{\gamma_1^3(k_p + 2\omega k_f) - (k_p - \omega k_f)\phi_p[\gamma_1^3 + \gamma^3 - 1]} \tag{3}$$

If there is no interfacial layer at the particle/liquid interface (i.e., $k_{lr} = k_f$ and $\gamma_1 = \gamma = 1$) Equation (3) reduces to the Maxwell model [1].

3.2. Brownian Motion-Based Modeling

3.2.1. Modifying the Brownian Motion

In order to determine the Brownian motion of nanoparticles in suspensions containing a specific volumetric loading of nanoparticles, it is more justifiable to employ the Brownian motion term which is also a function of the particle volume fraction than the conventional kinetic theory-based formulation for Brownian motion [35, 36]. Thus, a modified Brownian motion term (U_{MBM}) is deduced by incorporating the nanoparticle volume fraction.

The mean square displacement of the particle in a suspension along any given direction (e.g., x-direction), ξ^2 can be written as [38-39]

$$\xi^2 = 2D_{eff}t_c \tag{4}$$

where D_{eff} is the effective diffusion coefficient and t_c is the characteristic time.

For a suspension having finite concentration of particles, it is important to use the effective diffusion coefficient, which also depends on particle volume fraction. Based on suspension concentration at equilibrium, i.e. for the effective partition coefficient, $K_{eff} \equiv 1$, the effective diffusion coefficient of a suspension of spheres can be express as [40]

$$D_{eff} = D(1 - 1.5\phi_p) \tag{5}$$

where according to Stokes-Einstein formula [41], $D = \dfrac{K_B T}{6\pi\eta r_p}$ is the coefficient of diffusion of the particle (radius, r_p) suspended in an infinite liquid medium of viscosity (η), and K_B is the Boltzmann's constant. Thus, applying Equation (5) into Equation (4) gives

$$\xi = \sqrt{2D(1-1.5\phi_p)t_c} \tag{6}$$

where based on the Lengevin's equation, $t_c = \dfrac{m}{6\pi\eta r_p}$ and m is the mass of the particle.

With the effective diffusion coefficient, the modified Brownian motion (U_{MBM})) can be formulated from the mean displacement of the particle (ξ) and the characteristic time as

$$U_{MBM} = \frac{\xi}{t_c} = \frac{\sqrt{2D(1-1.5\phi_p)}}{\sqrt{t_c}} \tag{7}$$

Simplifying Equation (7) with the terms for D and t_c the final form of the modified Brownian motion becomes

$$U_{MBM} = \sqrt{\frac{2K_B T(1-1.5\phi_p)}{m}} \tag{8}$$

Equation (8) shows that the larger the particle volume fraction, the smaller the diffusion coefficient and thus, the weaker the Brownian motion.

3.2.2. Modeling the Dynamic Part of Thermal Conductivity

For uniform complex nanoparticles (cp) in a base fluid as shown in Figure 1, the net axial heat flux due to the movement of nanoparticles (U_{MBM}) resulting from the Brownian force (F_B) impacting on them can be written as [42-43]

$$q''_{net-dy} \approx -\frac{1}{2} n m_{cp} c_{p-cp} d_s U_{MBM} \nabla T \tag{9}$$

where n is the number density of particles, m_{cp}, c_{p-cp} and d_s are the mass, specific heat, and average separation distance of complex nanoparticles, respectively.

For $nm_{cp} = \rho_{cp}$, Equation (9) gives the effective thermal conductivity of suspensions due to Brownian motion of nanoparticles (k_{dy}) as

$$k_{dy} = \frac{1}{2}\rho_{cp}c_{p-cp}d_s U_{MBM} \tag{10}$$

From Equation (8), U_{MBM} for complex nanoparticles is expressed as

$$U_{MBM} = \sqrt{\frac{3K_B T(1-1.5\phi_{cp})}{2\pi\rho_{cp}r_{cp}^3}} \tag{11}$$

Substituting Equation (11) into Equation (10) and making use of $\phi_{cp} = \phi_p \gamma^3$, the following final form of Brownian motion-contributed thermal conductivity is obtained

$$k_{dy} = \frac{1}{2}\rho_{cp}c_{p-cp}d_s[\sqrt{\frac{3K_B T(1-1.5\gamma^3\phi_p)}{2\pi\rho_{cp}\gamma^3 r_p^3}} \tag{12}$$

where $d_s = 0.893 r_{cp}\phi_{cp}^{-1/3} = 0.893 r_p \phi_p^{-1/3}$ [39] and density (ρ_{cp}) and specific heat (c_{p-cp}) of complex nanoparticles can be determined from the formulations presented in the subsequent sections.

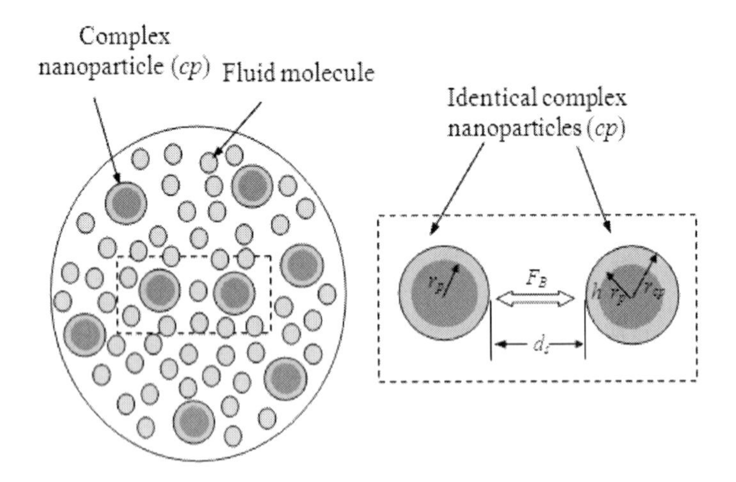

Figure 1. Schematic of dynamic mechanisms of nanoparticles in a base fluid.

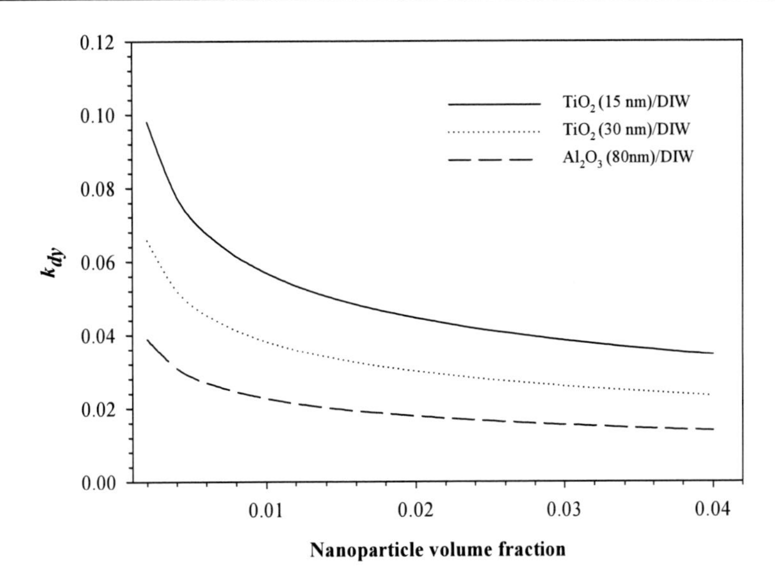

Figure 2. Dynamic thermal conductivity of nanofluids with nanoparticle volume fraction.

Figure 2. depicts the effect of particles size and volume fraction on the dynamic thermal conductivity component obtained using Equation (12) for three types of aqueous nanofluids. Since the specific heat and density of TiO_2 are very close to those of Al_2O_3, the differences between lines of k_{dy} for these nanofluids (Figure 2) are mainly due to the difference in particle size. As expected, it shows that the smaller the particle the larger the dynamic (Brownian motion-based) thermal conductivity (k_{dy}). It can also be seen that mainly for smaller size and low volume fraction of nanoparticles, the dynamic part of thermal conductivity is significant. The reason is that the smaller the particle size the greater the movement of particles in the fluid. It is also noted that this k_{dy} increases nonlinearly with decreasing particle volume fraction from 0.005 due to large interparticle separation distance (d_s) at such small particle volume fraction. For $\phi_p < 0.005$, the interparticle separation distance (d_s) is too large to cause any interaction through Brownian force of particles. Thus, the dynamic contribution of the thermal conductivity (k_{dy}) is not applicable for $\phi_p < 0.005$. However, such condition is not applicable to static contribution of thermal conductivity (k_{st}), which does not depend on Brownian motion or interparticle separation distance.

3.2.3. Determination of Interfacial Layer Thickness

As each nanoparticle is considered to have a nanolayer at the nanoparticle/fluid interface and its thickness, h remains in the model, the value of a nanolayer thickness is required to calculate the thermal conductivity of nanofluids. Based on the electron density profile at the interface, Hashimoto et al. [44] established a new definition of the interfacial layer thickness at the surface of spherical micro-domains, which is given as

$$h = \sqrt{2\pi}\sigma \tag{13}$$

where σ is a parameter characterizing the diffuseness of the interfacial boundary and its typical value falls in the range of 0.3 to 0.6 nm. By extension of the Debye equation, Li et al.

[45] introduced the same model for determining the interfacial layer thickness of a pseudo two-phase solid-liquid system. For $\sigma = 0.4$ nm, Equation (13) yields an interfacial layer thickness of 1 nm. In fact, the experimental results of Yu et al. [46] and the molecular dynamics simulations performed by Xue et al. [47] showed that the typical interfacial layer thickness between the solid and liquid phases is of the order of a few atomic distances namely, ≈ 1 nm. Hence, an interfacial layer thickness h of 1 nm can reasonably be used to predict the thermal conductivity of nanofluids by the present model.

3.2.4. Determination of Density and Specific Heat of Complex Nanoparticles

In order to determine the density and specific heat of complex nanoparticles, the average density and specific heat profiles for interfacial layer need to be known. Then, using these profiles of the interfacial layer in the volume fraction mixture rule, the average density and specific heat of complex nanoparticle can be obtained. Therefore, the profiles for average density and specific heat of the interfacial layer is determined first from an exponential distribution of concentration in radial direction in the interfacial layer [48]. Geometry of interfacial layer of a spherical nanoparticle in a base fluid is shown in Figure 3.

Considering an exponential distribution of concentration (radial direction) in the interfacial layer of particle-fluid interface (Figure 3), the density profile can be formulated as follows:

$$\rho_{lr}(r) = a\exp[-b(\frac{r-r_p}{h})] \quad r_p \le r \le r_p + h \tag{14}$$

where ρ_{lr} is density of interfacial layer and r_p is the radius of particle. The constants a and b are to be determined by using the boundary conditions which are

$$\rho_{lr}\big|_{r=r_p} = \rho_p \text{ and } \rho_{lr}\big|_{r=r_p+h} = \rho_f \tag{15}$$

where ρ_p and ρ_f are the particle and fluid densities, respectively.

By using Equation (15), the constants a and b in the Equation (14) are solved as follows

$$a = \rho_p \text{ and } b = \ln(\rho_p / \rho_f) \tag{16}$$

Thus, the density profile in the interfacial layer can be obtained from Equation (14) as

$$\rho_{lr}(r) = \rho_p \exp[-\ln(\rho_p / \rho_f)(\frac{r-r_p}{h})] \tag{17}$$

The average density of the interfacial layer has the form

$$\rho_{lr-avg} = \frac{1}{V_{lr}} \int \rho_{lr}(r) dV \tag{18}$$

where $V_{lr} = \frac{4}{3}\pi[(r_p + h)^3 - r_p^3] = \frac{4}{3}\pi h[3r_p^2 + 3r_p h + h^2] = \frac{4}{3}\pi ph$, $p = (3r_p^2 + 3r_p h + h^2)$, and $dV = 4\pi r^2 dr$.

Hence, Equation (18) becomes

$$\rho_{lr-avg} = \frac{3}{4\pi ph}\int_{r_p}^{r_p+h} 4\pi r^2[a\exp[-b(\frac{r-r_p}{h})]]dr \tag{19}$$

By solving Equation (19), the expression for the average density of the interfacial layer on a spherical particle becomes

$$\rho_{lr-avg} = \frac{3\rho_p}{pb^3}(2h^2 + 2br_p h + b^2 r_p^2) - \frac{3\rho_f}{pb^3}[(h^2(2 + 2b + b^2) + br_p(br_p + 2bh + 2h)] \tag{20}$$

With the average density of interfacial layer, the density of complex nanoparticles can be obtained from the volume fraction mixture rule as

$$\rho_{cp} = \frac{1}{V_{cp}}(V_p \rho_p + V_{lr}\rho_{lr-avg}) = [\frac{1}{\gamma^3}\rho_p + \{1 - \frac{1}{\gamma^3}\}\rho_{lr-avg}] \tag{21}$$

Substituting Equation (20) into Equation (21) and simplifying, the final expression of density of complex nanoparticles becomes

$$\rho_{cp} = \frac{1}{\gamma^3}\rho_p + (1 - \frac{1}{\gamma^3})\{\frac{3\rho_p}{pb^3}(2h^2 + 2br_p h + b^2 r_p^2) - \frac{3\rho_f}{pb^3}[h^2(2 + 2b + b^2) + br_p(br_p + 2bh + 2h)]\} \tag{22}$$

Similar to the above density expression, the average specific heat of complex nanoparticle can be derived as

$$c_{p-cp} = \frac{1}{\gamma^3}c_{p-p} + (1 - \frac{1}{\gamma^3})\{\frac{3c_{p-p}}{pb'^3}(2h^2 + 2b' r_p h + b'^2 r_p^2) - \frac{3c_{p-f}}{pb'^3}[h^2(2 + 2b' + b'^2) + b' r_p(b' r_p + 2b' h + 2h)]\} \tag{23}$$

where $b' = \ln(c_{p-p} / c_{p-f})$.

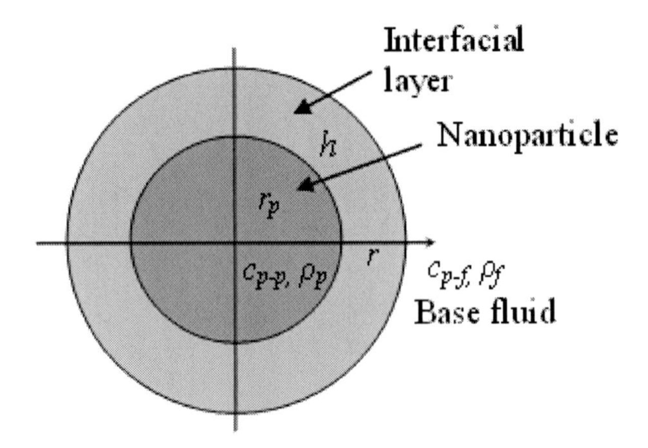

Figure 3. Interfacial layer geometry of a spherical nanoparticle in a base fluid.

3.3. Combined Model for the Effective Thermal Conductivity

The effective thermal conductivity of nanofluids is considered to be from both the static and dynamic mechanisms. The effects of these mechanisms are treated to be additive and by combining the static and dynamic mechanisms the effective thermal conductivity of nanofluids ($k_{eff\text{-}nf}$) has the form [43]

$$k_{eff-nf} = k_{st} + k_{dy} \tag{24}$$

Substituting k_{st} from Equation (3) and k_{dy} from Equation (12) into Equation (24) yields the final model for the effective thermal conductivity of nanofluids as

$$k_{eff-nf} = \left\{ k_f \frac{\phi_p \omega (k_p - \omega k_f)[2\gamma_1^3 - \gamma^3 + 1] + (k_p + 2\omega k_f)\gamma_1^3[\phi_p \gamma^3(\omega - 1) + 1]}{\gamma_1^3(k_p + 2\omega k_f) - (k_p - \omega k_f)\phi_p[\gamma_1^3 + \gamma^3 - 1]} \right\} + \left\{ \frac{1}{2}\rho_{cp} c_{p-cp} d_s \left[\sqrt{\frac{3K_B T(1 - 1.5\gamma^3\phi_p)}{2\pi\rho_{cp}\gamma^3 r_p^3}} \right\} \tag{25}$$

The significant features of this model are summarized as follows:

i. The model is developed by considering nanoparticles with a nano-interfacial layer together with their static and dynamic mechanisms in the base fluid. The particle size effect is also included in the model.
ii. The second term on the right hand side of Equation (25) is the dynamic contribution of thermal conductivity (k_{dy}), which takes into account the effect of particle Brownian motion as well as temperature. As mentioned previously, this term (k_{dy}) is applicable for $\phi_p > 0.005$.

iii. The conventional kinetic theory-based Brownian motion term has been renovated using effective diffusion coefficient concept to incorporate the effect of volume fraction on Brownian motion of nanoparticles.

iv. The model can also predict the temperature-dependent thermal conductivity of nanofluids.

v. If there is no interfacial layer, the static part of the model reduces to the Maxwell model [1] and when $\phi_p = 0$ (i.e. $\rho_{cp} = (3\phi_p m_{cp})/(4\pi r_p^3) = 0$) the entire model reduces to k_f.

Table 2. demonstrates the contributions of static and dynamic mechanisms to the effective thermal conductivity of TiO_2 (15 nm)/deionized water (DIW)-based nanofluids predicted by the present model given by Equation (25). As can be seen from Table 2, while the major contributions arise from static mechanisms such as volume fraction, particle size and interfacial nanolayer, Brownian motion-based dynamic mechanism can also play a significant role in enhancing the thermal conductivity of nanofluids particularly at low volume fraction of nanoparticles.

4. Results and Discussion

For spherical particles, the Hamilton-Crosser model [29] is the same as the Maxwell model [1]. Therefore, the Maxwell model is used as the representative of classical models. The present model is validated by comparing its results with own experimental results and data from the literature. The thermal conductivity of nanofluids was measured by using transient hot-wire technique which has been described elsewhere [12]. The predictions by the present model are also compared with the results from Maxwell model [1] as well as results from recent models specifically developed for nanofluids by Kumar et al. [35] and Prasher et al. [36].

Figures 4 to 8 demonstrate that the present model shows fairly good agreement with the experimental results and gives far better predictions compared to Maxwell's model as well as models developed for nanofluids by Kumar et al. [35] and Prasher et al. [36]. As shown in Figure 4, the predictions by the present model for TiO_2 (15 nm)/DIW-based nanofluids are in good agreement with the experimental results which are completely under-predicted by Maxwell's [1] and Prasher et al.'s [36] models.

Table 2. Contributions of static and dynamic mechanisms to the effective thermal conductivity of nanofluids predicted by Equation (25)

TiO_2 (15 nm)/DIW-based nanofluids			
Nanoparticle volume %	$k_{eff\text{-}nf}$(W/m·K)	k_{st} (%)	k_{dy} (%)
0.6	0.698	90.3	9.7
1	0.704	91.9	8.1
2	0.733	93.9	6.1
3	0.768	95	5

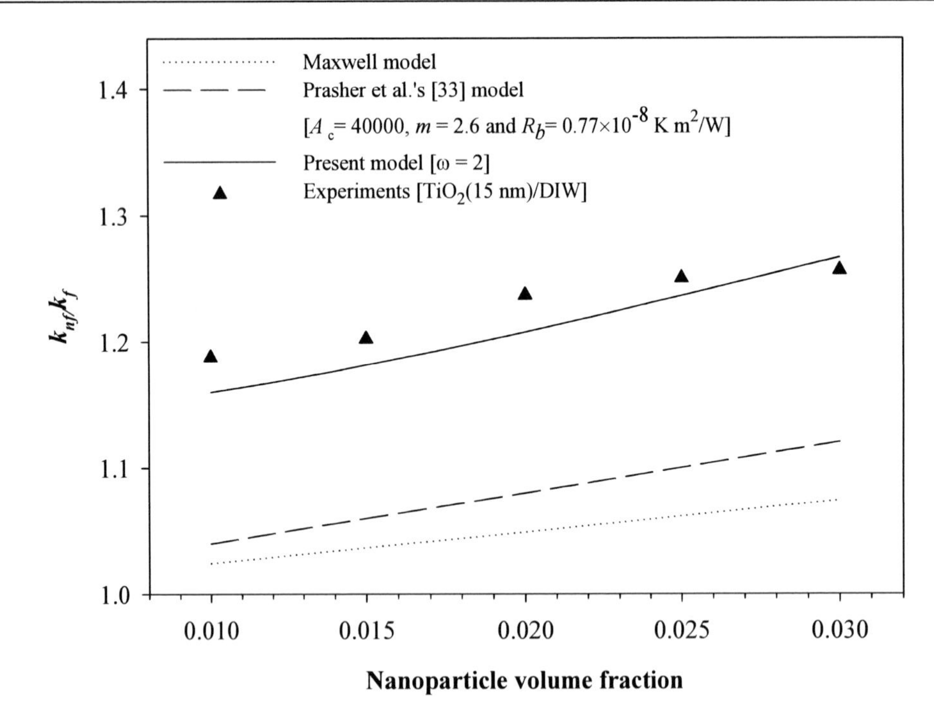

Figure 4. Comparison of present model's predictions with experimental results as well as predictions of other models for TiO_2/DIW-based nanofluids.

Like previous studies [8-9, 11-12, 14], Figures 5 to 6 demonstrate that for Al_2O_3/water-based nanofluids, the present model shows better agreement with the present experimental results and results from the literature [6]. It can also be seen that the thermal conductivities predicted by Prasher et al.'s model [36] are much lower than the experimental data. Although their model considers the effect of particle size and micro-convection, it could not predict the anomalously high thermal conductivity of nanofluids. This may be because their model includes effect of interfacial resistance and does not consider the effect of interfacial layer, particle movement and surface chemistry, which play significant roles in suspensions containing nanoparticles. It is also noted that suggested values of three unknown parameters i.e. R_b, A_c, and m were needed in the calculation of thermal conductivity when using Prasher et al.'s model [36]. Currently, these empirical parameters cannot be obtained by experimental or theoretical mean. Figures 4 to 8 clearly demonstrate that the classical Maxwell model is completely unable to predict the effective thermal conductivity of nanofluids. Figure 7. shows that for CuO/water-based nanofluids, the present model fits very well with the results obtained by Eastman et al. [6] and gives better predictions compared to other models. Like other types of nanofluids, Prasher et al.'s model [36] also under-predicts the thermal conductivity of this nanofluid. The predictions by the present model and other models are also compared with the experimental data for Fe_3O_4 (10 nm)/water-based nanofluids obtained from Zhu et al. [49] (Figure 8). When compared with experimental results, the present model gives better predictions of the effective thermal conductivity of these nanofluids compared to other models. Since the values of the empirical parameters in Prasher et al.'s [36] model are not given in their paper for this nanofluid, Kumar et al.'s model [35] is used to compare with the predictions by the present model. It can be seen that Kumar et al.'s model also severely under-predicts the results for the effective thermal conductivity of this nanofluids.

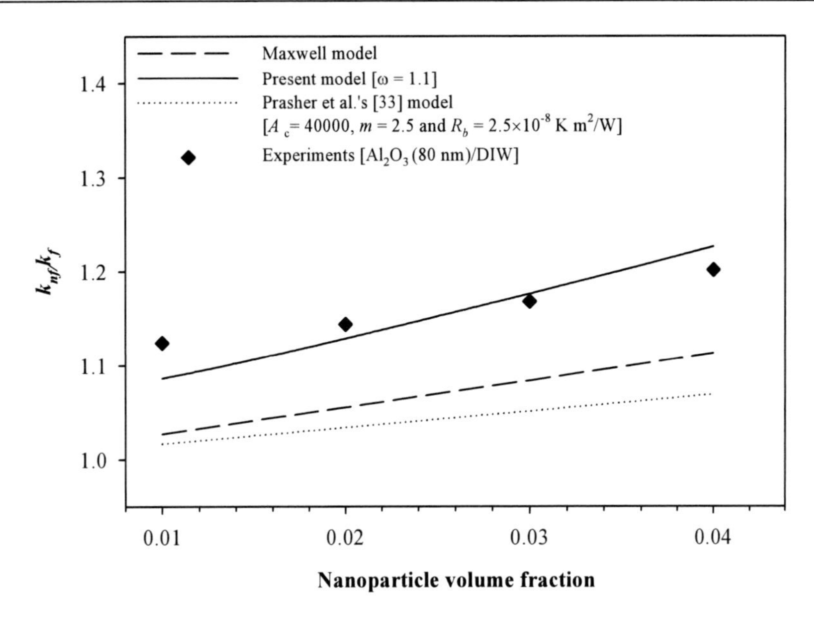

Figure 5. Comparison of present model's predictions with experimental results and predictions of other models for Al_2O_3 /DIW-based nanofluids.

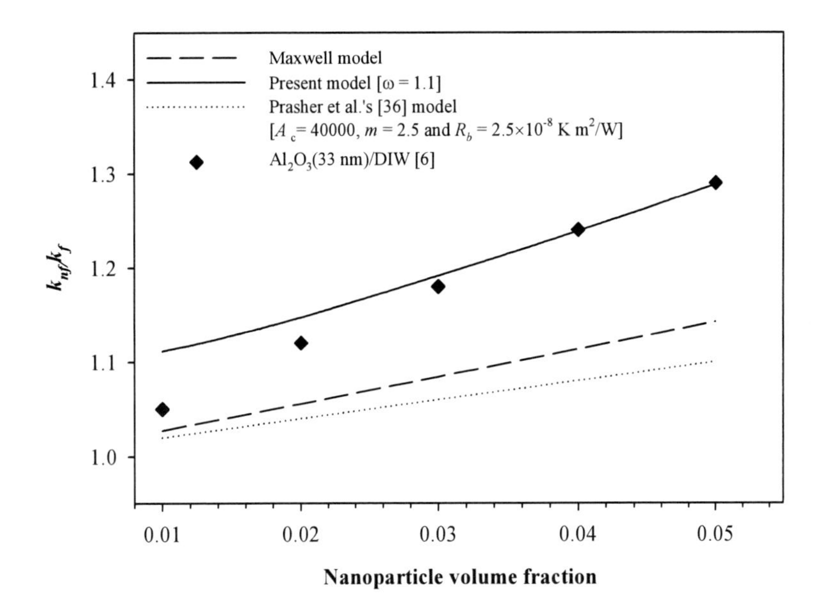

Figure 6. Comparison of present model's predictions with results from literature [6] and predictions by other models for Al_2O_3 /DIW-based nanofluids.

In this study, the value of the nanolayer thermal conductivity factor (ω) was chosen to be between 1.1 and 2.5. This is because the exact thermal conductivity of the interfacial layer is not known and cannot be obtained either experimentally or theoretically. However, it is presumed that the orderness and orientation of fluid molecules absorbed on the surface of particle (like surface adsorption) will result in a value of the nanolayer thermal conductivity which is intermediate between the thermal conductivities of nanoparticles and its base fluid.

The anomalous thermal conductivity of liquid film between mica plates was experimentally verified by Metsik [50]. A significant increase in the thermal conductivity of liquid film was found for water and ethyl alcohol when the thickness of liquid film reduces to 50 nm. Several researchers [22, 24, 51-52] also used a thermal conductivity value of the nanolayer which is between 2 to 4 times of thermal conductivity of base fluids.

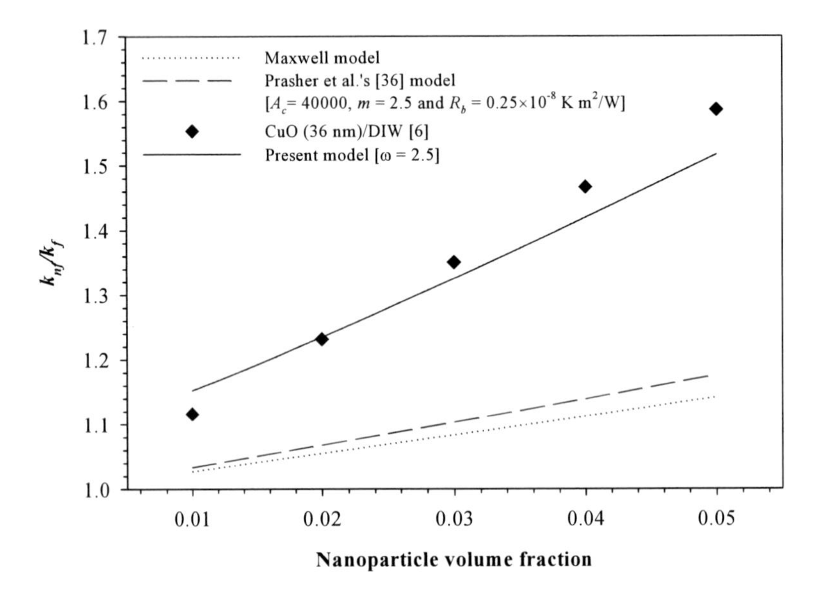

Figure 7. Comparison of present model's predictions with experimental data from literature [6] and other models' predictions for CuO/DIW-based nanofluids.

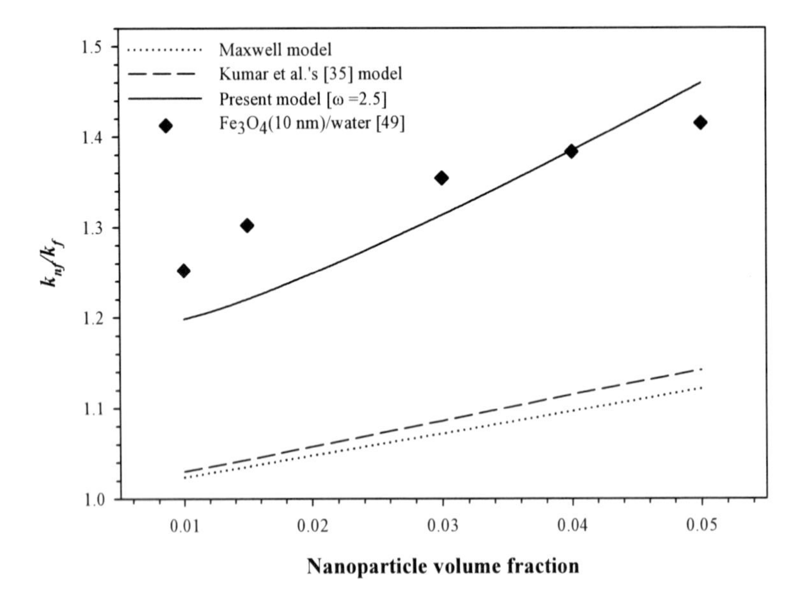

Figure 8. Comparison of present model's predictions with experimental results from literature [49] and other models' predictions for Fe_3O_4/water-based nanofluids.

CONCLUSION

The reported classical models have been found to be unable to predict the anomalously high thermal conductivity of nanofluids. This is not very surprising as these models do not include the nanoscale effects of dispersed nanoparticles such as particle size, the interfacial layer at the particle/liquid interface, and other dynamic micro-mechanisms (e.g., Brownian motion of nanoparticles), which are considered as important factors for the enhancement of the thermal conductivity of nanofluids.

Despite numerous theoretical studies devoted to model development for the prediction of the effective thermal conductivity of nanofluids, there is no widely accepted model available due to inconclusive heat transfer mechanisms of nanofluids. Most of the cases, these recent models also fail to explain and predict the observed thermal conductivity of nanofluids. In addition, these models cannot be validated with a wide range of experimental data.

An improved model for the prediction of the effective thermal conductivity of nanofluids is introduced in this chapter. The model is developed by incorporating both static and dynamic mechanisms that are behind the anomalous thermal conductivity of nanofluids. In addition to the effect of Brownian motion of nanoparticles, the model also takes into account several other important factors such as particle size and interfacial nanolayer. Furthermore, temperature-dependent thermal conductivity of nanofluids can be determined using this model. Compared to existing classical and other recently developed models, present model shows better agreement with experimental results. This is attributed to the incorporation of those key static and dynamic mechanisms in the present model. The static part of the complete model is directly developed from the steady-state heat conduction equation by considering nanofluids as a mixture of three components (i.e. nanoparticle, nanolayer, and base fluid). In the dynamic part, the conventional kinetic theory-based Brownian motion term has been renovated using the effective diffusion coefficient concept in order to formulate the Brownian motion of a suspension containing a certain volumetric loading of nanoparticles. It is found that the major contribution to the enhanced thermal conductivity of nanofluids originated from the static mechanisms. However, the Brownian motion-based dynamic mechanism is also significant for nanofluids with smaller-size and low concentration of nanoparticles. It can reasonably be concluded that the effective thermal conductivity of nanofluids is due to both static and dynamic mechanisms of nanoparticles.

Nevertheless, some challenges still remain as the thickness and thermal properties of an interfacial layer between a nanoparticle and base fluid cannot be determined at this time and more research efforts are to be made to deal with these issues together with any other potential mechanisms in future model development.

REFERENCES

[1] Maxwell, J. C. *A Treatise on Electricity and Magnetism*, Clarendon Press: Oxford, U.K., 1891.

[2] Choi, S. U. S. In *Developments and applications of non-Newtonian flows*; Siginer, D. A.; Wang, H. P.; Eds.; ASME Publishing: New York,USA,1995; FED-Vol. 231/MD-Vol. 66, pp 99-105.

[3] Yang, Y. M.; Maa, J. R. *Int. J. Heat Mass Transfer* 1984, 27, 145–147.
[4] Masuda, H; Ebata, A.; Teramae, K.; Hishinuma, N. *Netsu Bussei* 1993, 4, 227-233.
[5] Grimm, A. *Powdered Aluminum-containing Heat Transfer Fluids*, German patent DE 4131516 A, 1993.
[6] Eastman, J. A.; Choi, S. U. S.; Li, S.; Thompson, L. J. In *Proceedings of the Symposium on Nanophase and Nanocomposite Materials II*; Komarneni, S.; Parker, J. C.; Wollenberger, H.; Eds.; Materials Research Society: Boston, USA, 1997; Vol. 457, pp 3-11.
[7] Eastman, J. A.; Choi, S. U. S.; Li, S.; Yu, W.; Thompson, L. J. *Appl. Phys. Lett.* 2001, 78, 718-720.
[8] Lee, S.; Choi, S. U. S.; Li, S.; Eastman, J. A. *J. Heat Transfer* 1999, 121, 280-289.
[9] Wang, X.; Xu, X.; Choi, S. U. S. *J. Thermophys. Heat Transfer* 1999, 13, 474-480.
[10] Xuan, Y.; Li, Q. *Int. J. Heat Fluid Flow* 2000, 21, 58-64.
[11] Das, S. K.; Putra, N.; Thiesen, P.; Roetzel, W. *J. Heat Transfer* 2003, 125, 567-574.
[12] Murshed, S. M. S.; Leong, K. C.; Yang, C. *Int. J. Therm. Sci.* 2005, 44, 367-373.
[13] Hong, T.; Yang, H.; Choi, C. J. *J. Appl. Phys.* 2005, 97, 064311-1-064311-4.
[14] Murshed, S. M. S.; Leong, K. C.; Yang, C. *Int. J. Therm. Sci.* 2008, 47, 560-568.
[15] Murshed, S. M. S.; Leong, K. C.; Yang, C. *Appl. Therm. Eng.* 2008, 28, 2109-2125.
[16] Choi, S. U. S.; Zhang, Z. G.; Keblinski, P. In *Encyclopedia of Nanoscience and Nanotechnology;* Nalwa, H. S.; Ed.; American Scientific Publishers: Los Angeles, USA, 2004; Vol. 6, pp 757-773.
[17] Das, S. K. *Heat Transfer Eng.* 2006, 27, 1-2.
[18] Keblinski, P.; Prasher, R.; Eapen, J. *J. Nanopart Res.* 2008, 10, 1089-1097.
[19] Murshed, S. M. S. *J. Nanopart Res.* 2009, 11, 511-512.
[20] Keblinski, P.; Phillpot, S.; Choi, S. U. S.; Eastman, J. A. *Int. J. Heat Mass Transfer* 2002, 45, 855-863.
[21] Wang, B-X.; Zhou, L-P.; Peng, X-F. *Int. J. Heat Mass Transfer* 2003, 46, 2665-2672.
[22] Yu, W.; Choi, S. U. S., *J. Nanoparticle Res.* 2003, 5, 167-171.
[23] Yu, W.; Choi, S. U. S., *J. Nanoparticle Res.* 2004, 6, 355-361.
[24] Xie, H.; Fujii, M.; Zhang, X. *Int. J. Heat Mass Transfer* 2005, 48, 2926-2932.
[25] Ren, Y.; Xie, H.; Cai, A., *J. Phys. D: Appl. Phys.* 2005, 38, 3958-3861.
[26] Leong, K. C.; Yang, C.; Murshed, S. M. S. *J. Nanoparticle Res.* 2006, 8, 245-254.
[27] Murshed, S. M. S.; Leong, K. C.; Yang, C. *Int. J. Nanosci.* 2006, 5, 23-33.
[28] Hong, T; Yang, H.; Choi, C. J. *J. Appl. Phys.* 2005, 97, 064311-1–064311-4.
[29] Hamilton, R. L.; Crosser, O. K. *Ind. Eng. Chem. Fund.* 1962, 1, 187-191.
[30] Wang, X. Q.; Mujumdar, A. S., *Int. J. Therm. Sci.* 2007, 46, 1-19.
[31] Kang, H. U.; Kim, S. H.; Oh, J. M. *Exp. Heat Transfer* 2006, 19, 181-191.
[32] Sabbaghzadeh, J.; Ebrahimi, S. *Int. J. Nanosci.* 2007, 6, 45-49.
[33] Xuan, Y.; Li, Q.; Hu, W. *AIChE J.* 2003, 49, 1038-1043.
[34] Koo, J.; Kleinstreuer, C. *J. Nanoparticle Res.* 2004, 6, 577-588.
[35] Kumar, D. H.; Patel, H. E.; Kumar, V. R. R.; Sundararajan, T.; Pradeep, T.; Das, S. K. *Phys. Rev. Lett.* 2004, 93, 4301-4304.
[36] Prasher, R.; Bhattacharya, P.; Phelan, P. E. *Phys. Rev. Lett.* 2005, 94, 025901-1–025901-4.
[37] Jang, S. P.; Choi, S. U. S. *Appl. Phys. Lett.* 2004, 84, 4316-4318.

[38] Perrin, M. J. *Brownian Movement and Molecular Reality*, Taylor and Francis, London, 1910.

[39] Chandrasekhar, S. *Rev. Mod. Phys.* 1943, 15, 1-89.

[40] Deen, W. M. *Analysis of Transport Phenomena*, Oxford University Press, New York, 1998.

[41] Einstein, A. *Investigations on the Theory of the Brownian Movement*, Dover Publications, Inc.: New York, 1956.

[42] Bird, R. B.; Stewart, W. E.; Lightfoot, E. N. *Transport Phenomena*, John Wiley and Sons, Inc., New York, 2002.

[43] Murshed, S. M. S.; Leong, K. C.; Yang, C. *Appl. Therm. Eng.* 2009, 29, 2477-2483.

[44] Hashimoto, T.; Fujimura, M.; Kawai, H. *Macromolecules* 1980, 13, 1660-1669.

[45] Li, Z. H.; Gong, Y. J.; Pu, M.; Wu, D.; Sun, Y. H.; Wang, J.; Liu, Y.; Dong, B. Z. *J. Phys. D: Appl. Phys.* 2001, 34, 2085-2088.

[46] Yu, C.-J.; Richter, A. G.; Datta, A.; Durbin, M. K.; Dutta, P. *Physica B* 2000, 283, 27-31.

[47] Xue, L.; Keblinski, P; Phillpot, S. R.; Choi, S. U. S.; Eastman, J. A. *Int. J. Heat Mass Transfer* 2004, 47, 4277-4284.

[48] Liu, X.-Y. *Surf. Sci.* 1993, 290, 403-412.

[49] Zhu, H.; Zhang, C.; Liu, S.; Tang, Y.; Yin, Y. *Appl. Phys. Lett.* 2006, 89, 023123-1–023123-3.

[50] Metsik, M. S. In *Research in Surface Forces,* Deryagin, B. V.; Ed.; Vol. 4, Consultants Bureau, New York, 1975.

[51] Xue, Q-Z. *Phys. Lett. A* 2003, 307, 313-317.

[52] Xue, Q-Z.; Xu, W.-M. *Mater. Chem. Phys.* 2005, 90, 298-301.

In: Brownian Motion: Theory, Modelling and Applications ISBN: 978-1-61209-537-0
Editors: R.C. Earnshaw and E.M. Riley

Chapter 6

BROWNIAN MOTION OF CARBON NANOTUBE IN A MOLTEN POLYMER AND ITS APPLICATION

Masayuki Yamaguchi** and Howon Yoon
School of Materials Science, Japan Advanced Institute of Science and Technology
Asahidai, Nomi, Ishikawa, Japan

ABSTRACT

As increasing the demand for electronic application of polymeric materials, a polymer with good electrical conductivity has been desired and investigated intensively these days. Carbon nanotubes (CNTs) are widely used as one of the high-performance modifiers because of the unique properties such as high stability, high electric and thermal conductivity, and large aspect ratio. Therefore, the properties of polymer composites containing CNTs have been studied intensively over the last decade for applications in electrostatic dissipation, electromagnetic interference shielding, and radio frequency interference shielding. In this chapter, the effect of Brownian motion of CNTs in a polymer matrix on the distribution in a molten state and the electrical conductivity in a solid state is explained in detail.

It is found that CNT-filled polymer composites without post-processing annealing procedure could be a thermodynamically non-equilibrium system even for the sample prepared by a compression-molding, in which conductive network formation depends on the processing temperature and residence time. Further, post-processing annealing can change the distribution state of CNTs and thus the electrical and mechanical properties. The redistribution process of CNTs at the post-processing annealing is known as dynamic percolation, which was firstly reported by the research group of Sumita. When a composite is heated at a temperature above the melting point of a polymer matrix, it is expected that CNTs move to a stable condition by means of Brownian motion, and finally form a continuous network if CNT content exceeds a critical value.

Furthermore, the thermally activated transfer process of CNTs between surfaces of immiscible polymer pairs is investigated. Compression-molded sheets of polypropylene (PP) containing 20 wt% of CNTs are laid on another polymer sheet that is immiscible with PP. It is found that CNTs immigrate from PP to the surface of the immiscible

* Phone +81-761-51-1621, Facsimile +81-761-51-1625, E-mail m_yama@jaist.ac.jp

polymer such as polycarbonate (PC) during heat treatment by diffusion. The piled sheets are easily separated each other after cooling owing to the immiscible nature. The formation of a thin CNT-rich layer on the surface of the separated PC sheet produces electrical conductivity. Since CNT transfer is attributed to Brownian motion, heating conditions such as temperature and duration time are responsible for the amount of transferred CNTs. This transfer is influenced by the dispersion state of CNTs in the composite, the size of CNTs and their compatibility, and the chemical structure of the second polymer. Moreover, this method can be used to improve surface properties while minimizing CNT content in the bulk of the polymer composite and could be a feasible process to integrate CNTs into various devices.

INTRODUCTION

Blending technique of fillers into polymeric materials has been employed for a long time since the commercial application of polymers, because it enhances various properties of final products, such as rigidity, dimensional stability, and heat resistance. Furthermore, higher loading of fillers often leads to good cost-performance, because most of conventional fillers are inexpensive as compared with polymers. Therefore, it cannot be ignored especially for plastics, rubber, and paint industries. Typical fillers are listed in Table 1, in which various fillers are classified by materials and the shape. Fillers with low aspect ratio are good for large volume-filling, because their excluded volume is not large as compared with fibers. On the contrary, the maximum content of fibrous fillers is considerably lower than that of spherical ones because of large extruded volume of fibrous fillers. The fibrous fillers, however, can modify various properties such as thermal and electrical conductivity by the addition of a small amount if they form network structure in a continuous polymer matrix. Some functional fillers including fibrous ones, which are summarized in detail in the book by Xanthos (2010), are listed in Table 2.

Table 1. List of typical fillers

	Organic	Inorganic				
		Oxides	Hydroxides	Salts	Silicates	Others
Low Aspect Ratio Sphere, Cube, Plate	· CB	· Glass Beads	· $Al(OH)_3$ · $Mg(OH)_2$	· $CaCO_3$ · $BaSO_4$	· Talc · Mica	· BN · Steel
High Aspect Ratio Fiber, Flake	· CF, CNT · Natural Fibers · Synthetic Polymer Fibers	· GF			· MMN	

· CB Carbon Black
· CF Carbon Fiber
· CNT Carbon Nanotube
· MMN Montmorillonite
· BN Boron Nitrade

Table 2. List of functional fillers

Fillers	Function
$Al(OH)_3$, $Mg(OH)_2$	Flame Retardant
Boron Nitrade	Thermal conductivity
CNT, CF, CB	Electroconductivity, Mechanical Properties
Polymeric Nanofibers	Improvement of processability
Layered Silicate	Gas Barrier, Mechanical Properties, Damping Property
$SrCO_3$	Birefringence control
Starch	Biodegradability

Since the recent trend is to use eco-friendly materials, fillers from natural resources, such as natural fibers, waste wood, and silicate compounds, are preferably employed in various applications. One of the most famous eco-friendly composites is presumably plastic films having a large amount of talc and/or $CaCO_3$ to reduce the heat of energy as well as carbon dioxide at burning. Mixing starch is also employed recently to accelerate biodegradability.

Another interesting trend is the addition of nanofillers, because various properties can be modified to a great extent with a small amount. The materials obtained by this technology are often referred as nanocomposites. As exemplified in the table, nanofillers provide various functions. In particular, it has been studied intensively for a decade on the improvement of mechanical properties and gas-barrier properties for polymer composites with organically modified layered silicates (Ray and Okamoto (2003), Ray and Bousmina (2005)).

Further, nanofillers can be employed in the application of optical films, because they can avoid losing transparency by light scattering (Yano et al., (2005)). Tagaya et al. (2003)) found that $SrCO_3$ rod-like nanoparticles with 200 nm in length and 20 nm in diameter can modify the orientation birefringence without losing transparency by light scattering, because the crystalline $SrCO_3$ nanoparticles have a large value of intrinsic birefringence. This technique will be applicable for a protect film of polarizers produced by extrusion processing.

Polymeric nanofibers are a great candidate for a processing modifier of plastics, whereas conventional fibers composed of synthetic polymers are mainly used as a tire cord. Amran et al. (2009) revealed that small addition of flexible polymeric fibers can improve the melt elasticity such as strain-hardening behavior in elongational viscosity, one of the most important rheological properties for various processing operations (Yamaguchi (2004)). Although the processing has to be performed below the melting point of the fibers, it will widen the application of linear polymers such as polypropylene, polyester and polyamide. Further, Yokohara et al. (2009) reported that nanofibers of poly(butylene succinate) enhance the crystallization rate of poly(lactic acid) to a great extent because of the enlargement of the surface of PBS which acts as a nucleating agent for PLA (Yokohara and Yamaguchi (2008)). Moreover, cellulose nanofibers have attracted great attention recently to improve the mechanical toughness by a biomass-based material (Yano et al., (2005)).

CNTs and carbon fibers (CF) are also known as functional fillers as well as nanofibers of ceramics, because they act as the electrical and/or thermal conductive path when the amount is beyond a percolation threshold. In particular, increasing attention has been focused on CNTs after the discovery by Iijima (1991) because of numerous outstanding properties.

Although the interest in the material design of polymer nanocomposites grows rapidly, Brownian motion of nanofillers in a polymer liquid, especially condensed system, has not been studied in detail except for the theoretical approach by Doi and Edwards (1986) and some experimental approach (Krieger (1972), Shikata and Pearson (1994), Brady (1996)) to the best of our knowledge.

In general, a diffusion coefficient of particles by Brownian motion is given using the well-known Stokes-Einstein equation, as follows;

$$D = \frac{kT}{6\pi\eta_M r} \quad (1)$$

where k the Boltzmann constant, η_M the viscosity of medium, and r the radius of particles. The time t_D for a particle to diffuse a distance equal to its radius r is given as;

$$t_D \approx \frac{r^2}{D} = \frac{6\pi\eta_M r^3}{kT} \quad (2)$$

Moreover, Peclet number Pe, which is a kind of Weissenberg number of a suspension, is provided as;

$$Pe = \frac{\eta_M r^3 \dot{\gamma}}{kT} \quad (3)$$

When Peclet number is larger than one, not Brownian motion but hydrodynamic force becomes very important for rheological response. In other words, the contribution of Brownian motion to the shear stress is negligible at high shear rate or stress and for composite systems containing large particles. Although the equations are available only for a dilute system, the equations demonstrate that the diffusion distance and time are dependent on the size of fillers and the ambient temperature which also affects the viscosity of the medium.

The effect of Brownian motion on the rheological response in a concentrated suspension has been studied intensively by Shikata and Pearson (1994) employing hard-sphere particles. They revealed that the system shows weak elasticity because of Brownian motion, leading to redistribution to equilibrium state from shear-distorted dispersion state of particles. Further, rigid particles in condensed systems cannot move with respect to each other without intense interparticle interactions such as frictional force. Therefore, the system requires a long time for redistribution process.

Since Brownian motion leads to reorganization of filler distribution as demonstrated, it has to be understood in detail to obtain the appropriate morphology of composites. This is considerably different from conventional fillers, whose diameter is usually in the range of 10 - 30 μm.

In this chapter, the effect of Brownian motion of CNTs in a polymer liquid on the morphology and electrical conductivity is mentioned. Furthermore, a new method to localize CNTs on the polymer surface is described. In case of the conductive polymers, the cost-performance is one of the most important problems. Therefore, the method proposed in this chapter will have an impact on the industrial application. Furthermore, the imprinting technique will widen the material design of surface functional materials.

RHEOLOGICAL PROPERTIES AND ELECTRICAL CONDUCTIVITY IN EQUILIBRIUM STATE OF POLYMER COMPOSITES WITH CNTS

Prior to the discussion on the effect of Brownian motion on rheological and electrical properties, linear viscoelastic properties in a molten state of polymer composites containing CNTs are demonstrated at first. Figure 1. shows the frequency dependence of oscillatory shear modulus for polycarbonate (PC) containing various amounts of CNTs. In this study, multi-walled CNTs with a diameter of approximately 40 nm are employed. The detailed characteristics of the CNT employed were reported by Kim et al. (2005) and Chen et al. (2007). The compounds are prepared by an internal batch mixer, and then they are compressed into flat sheets by a compression-molding machine at 300 °C. As seen in the figure, neat PC shows a typical rheological behavior in the terminal zone; the slope of storage modulus *G'* is two and that of loss modulus *G"* is one. It is generally understood that the oscillatory moduli are determined by the relaxation spectrum $H(\tau)$ as follows;

$$G' = G_e + \frac{\omega^2\tau^2}{1+\omega^2\tau^2} H(\tau) d\ln\tau \tag{4}$$

$$G'' = \frac{\omega\tau}{1+\omega^2\tau^2} H(\tau) d\ln\tau \tag{5}$$

where G_e is the discrete contribution to the spectrum with $\tau = \infty$. In case of neat PC, that is a viscoelastic liquid, G_e is zero.

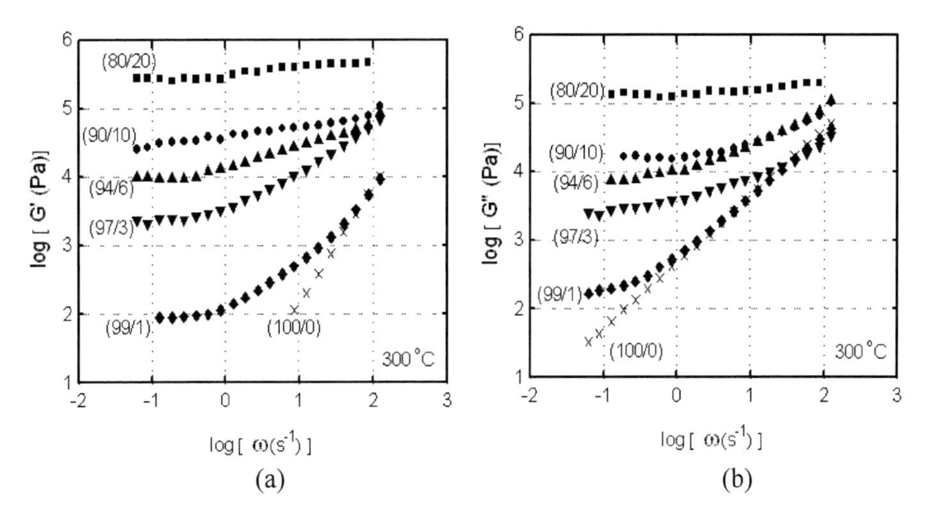

Figure 1. Frequency dependence of (a) shear storage modulus *G'* and (b) loss modulus *G"* for PC/CNT composites at 300 °C.

Both moduli increase with increasing the CNT contents, and it is pronounced in the G' curve at low frequencies. This is reasonable because G' is more sensitive to the relaxation mechanism with long relaxation time, as demonstrated in the equations. Furthermore, plateau modulus G_e seems to appear beyond 1 wt% of the CNTs, suggesting that permanent network structure composed of the CNTs will be formed in a molten PC. The magnitude of G_e, which will have a close relation with yield stress to destroy the network structure, increases with the content of the CNTs, and becomes a similar level to a rubbery plateau modulus of PC. As a result, the terminal of flow region disappears for the composites having large amounts of the CNTs. The linearity of the rheological response is also examined employing the same composites by measurements of the strain dependence of the oscillatory shear modulus at 250 °C. The obtained shear stress σ under a large strain amplitude γ_0 with an angular frequency ω is characterized by Fourier expansion as follows (Onogi et al. (1970), Matsumoto et al. (1973));

$$\sigma = G_1' \gamma_0 \sin \omega t + G_1'' \gamma_0 \cos \omega t - G_3' \gamma_0^3 \sin 3\omega t - G_3'' \gamma_0^3 \cos 3\omega t + G_5' \gamma_0^5 \sin 5\omega t + \cdots \quad (6)$$

Figure 2. shows the primary component of storage modulus, i.e., G_1' as a function of the applied shear strain at the angular frequency of 1.0 s^{-1}. The magnitude in the low strain region is a constant, whereas it decreases monotonically with increasing the strain. The decrease in the primary part indicates that a nonlinear response cannot be ignored for the contribution to the stress response. Furthermore, it is demonstrated that the linear response is restricted only in the small strain region when the amount of the CNTs increases. It is well known that simple polymer liquids show linear response in a large strain region at this frequency as shown in the neat PC.

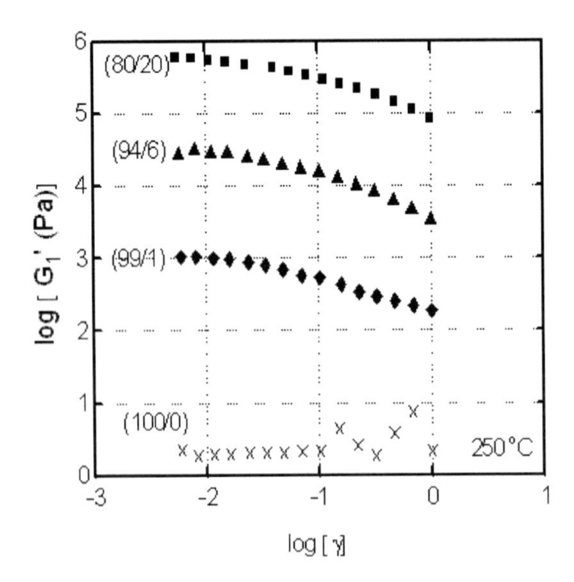

Figure 2. Strain dependence of primary component of shear storage modulus G_1' after Fourier expansion. The measurements were performed at 250 °C and at 1.0 s^{-1} employing the samples containing various amounts of CNT.

Table 3. Surface resistivity of PC/CNT composites compressed at 200 °C and 300 °C

Surface resistivity (ohm/sq.)		
CNT content (wt%)	Compression –molding temp.	
	200 oC	300 oC
1	>1015	108
3	1014	103
6	1012	103
10	109	103
20	102	102

However, the linear region becomes small when the system has a three dimensional structure, which becomes obvious with increasing the CNTs. This is reasonable, because the irrecoverable deformation of the CNT network occurs at small strains for a composite with well-developed network structure. In general, energy consumption to deform and/or destroy the structure, redistribution process of dispersed particles from distorted one, and plastic deformation always lead to a non-linear response for polymer melts containing a large amount of hard fillers. Therefore, the results in Figure 2. correspond with the oscillatory shear modulus shown in Figure 1.

Electrical property of the compressed sheet in a solid state is shown in Table 3, in which the surface resistivity is plotted against the CNT contents. Although neat PC shows a considerably high value of the surface resistivity, which is a typical phenomenon of conventional plastics, the electrical conductivity increases (the surface resistivity decreases) rapidly with the amount of the CNTs. In particular, it is apparent that the surface resistance falls off sharply around 1 wt% for the PC/CNT sheets compressed at 300 °C, suggesting that the percolation threshold of this system is approximately 1 wt%. The result shows that electrical conductive path is developed when the network structure is formed from the viewpoints of rheological response. Further, it should be noted that the sheet samples obtained by the compression-molding at 200 °C show completely different electrical conductivity as shown in the table. The mechanism is described in detail in the next section.

Dynamic Percolation of CNTs in a Molten Polymer

Electrical conductivity, as one of the most important properties of polymer composites with CNTs, is dependent on the applied processing history and post-processing annealing operation. Table 4 shows the electrical resistivity of the PC/CNT (3 wt%) composites after annealing procedure at 280 °C for various residence times. The sample sheets were obtained by the compression-molding at 200 °C. It is demonstrated that the electrical resistivity at room temperature decreases slightly with increasing the annealing time at any annealing temperature. This phenomenon obviously originates from the rearrangement of the CNTs in a molten PC during the annealing procedure.

As similar to exposure time at annealing, the annealing temperature has an important role on the electrical property. Table 5 shows the volume resistivity of PC/CNT (3 wt%) composites after annealing at various temperatures between 250 °C and 300 °C for 2 min. It is found that the electrical resistivity decreases rapidly with increasing the annealing temperature, especially above 280 °C, suggesting that CNTs reorganize through Brownian motion in a molten polymer and form a continuous conductive network. A similar result is obtained for surface resistivity (but not presented here).

These results demonstrate that the original sample obtained by compression-molding at 200 °C, prior to the annealing, is in the non-equilibrium state. This situation is directly detected by SEM observation as shown in Figure 3.

Table 4. Surface resistivity of PC/CNT (97/3) composites after annealing procedure at 280 °C for various residence times

Surface resistivity (ohm/sq.)				
Annealing temp.	Annealing time			
	2 min	5 min	10 min	30 min
280 °C	$\sim 10^{8}$	$\sim 10^{5}$	$\sim 10^{4}$	$\sim 10^{4}$

Table 5. Volume resistivity of PC/CNT (97/3) composites after annealing procedure at various temperatures from 250 to 300 °C for 2 min

Volume resistivity (ohm cm)					
Annealing temp.					
250 °C	260 °C	270 °C	280 °C	290 °C	300 °C
10^{11}	10^{11}	10^{11}	10^{7}	10^{2}	10^{2}

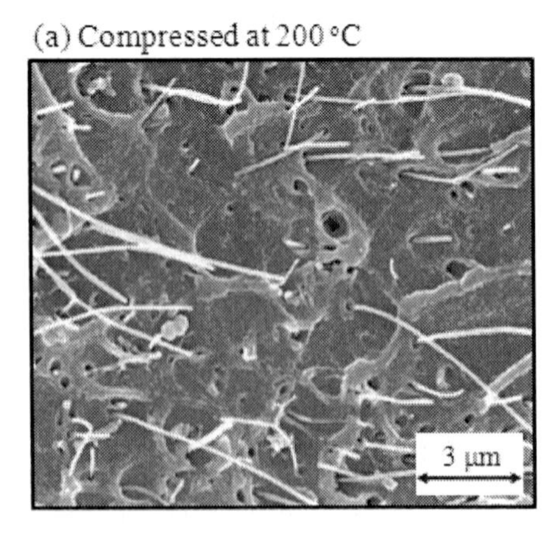

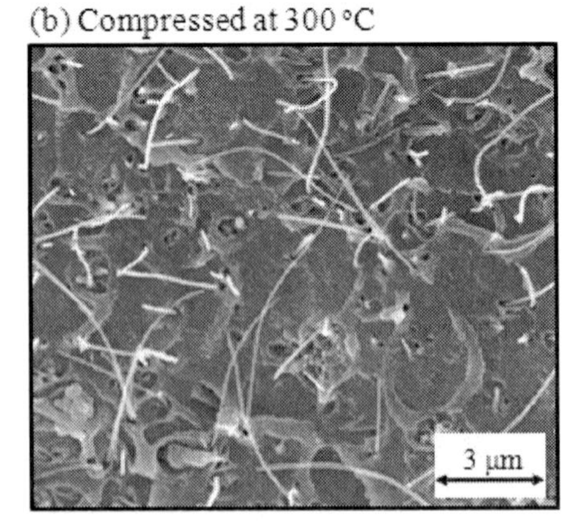

Figure 3.SEM pictures of the surface of cryogenically fractured PC/CNT (97/3) composites compressed at (a) 200 °C and (b) 300 °C. The compressive deformation was applied in the normal direction.

In the figure, compression deformation is applied to the vertical direction. Apparently, CNTs in the sample obtained at 200 °C orient perpendicular to the compression direction by applied squeeze flow during the compression-molding. On the contrary, the orientation direction of the CNTs is random. As theoretically discussed by Doi and Edwards (1986), the interparticle interaction is pronounced in a system with random orientation for rod-like particles, which is understandable by Onsager's theory on the formation of nematic liquid crystalline state. Consequently, frequent contacts between the CNTs are expected for the sample compressed at 300 °C. Further, it is suggested that the rotational diffusion time of the CNTs is shorter than 2 min at this temperature. This would be attributed to low viscosity of the matrix at high temperature, i.e., enhanced Brownian motion.

The rotational diffusion of rigid rods was theoretically discussed by Doi and Edwards (1986). The diffusion constant D_r is derived by the following equation:

$$D_r = \frac{kT}{\zeta_r} \tag{7}$$

where ζ_r is the friction constant which is provided as,

$$\zeta_r = \frac{\pi \eta_M L^3}{3\ln(L/2b)} \tag{8}$$

where b the radius of rigid rods and L the length of long axis. It should be noted in the equations that large D_r, i.e., small ζ_r, is expected in the case of fine fibers with small b, meaning that fine fibers can show Brownian motion even if the length is over 1 μm.

In this experiment, the zero-shear viscosities of the neat PC employed are found to be 17,900 [Pa s] at 200 °C and 126 [Pa s] at 300 °C. Consequently, the diffusion constant at 300 °C is found to be approximately 172 times larger than that at 200 °C. The large difference in the diffusion constant is responsible for the difference in the dispersion state of the CNTs in Figure 3. Generally speaking, the percolation theory is developed as a function of the filler shape, content, and interaction between particles (Aharoni (1972), Janzen (1975), Sumita (1992), and Natsuki et al., (1986)). The experimental results, however, demonstrate that the percolation occurs as a function of the applied annealing procedure. This phenomenon was firstly reported for polymer composites with CF with low aspect ratio and called as "dynamic percolation" (Zhang et al., (2005)). As compared with composites with CF, the dynamic percolation is considerably important for composites with CNTs. Since CNTs have enormously high aspect ratio, leading to large excluded volume for free rotation, it takes a long time to show random distribution after flow history. Moreover, it is expected that CNTs with a small diameter will show a marked reorganization process because of enhanced Brownian motion. The structure development during annealing procedure is detected also by rheological properties. Figure 4. shows the growth curves of oscillatory shear moduli at 1.0 s^{-1} for the composite with 3 wt% of the CNTs at 250 °C. The sample sheet was prepared by compression-molding at 200 °C. It is found that both moduli increase with increasing the residence time with an intense fashion in the storage modulus G'. Consequently, G' becomes almost similar level to G'' at the equilibrium condition. The rheological change suggests that network structure of the CNTs is developed in the rheometer during the measurement. This result also indicates that dispersion state of the CNTs, which affects the electrical property, would be rearranged during the applied annealing procedure in a compression-molding machine. On the contrary, it is confirmed that the values of G' and G'' measured at 300 °C are almost constant irrespective of the residence time as depicted in Figure 5. This result indicates that the rotational diffusion time to form the network structure of the CNTs is too short to measure at this temperature, which can be explained by a rotational Peclet number Pe_{rot},

$$Pe_{rot} = \frac{\omega}{D_r} \tag{9}$$

In this experiment, the effect of polymer degradation by exposure to high temperature can be ignored, which is revealed by the measurements of gel permeation chromatography (Yoon et al. (2010)).

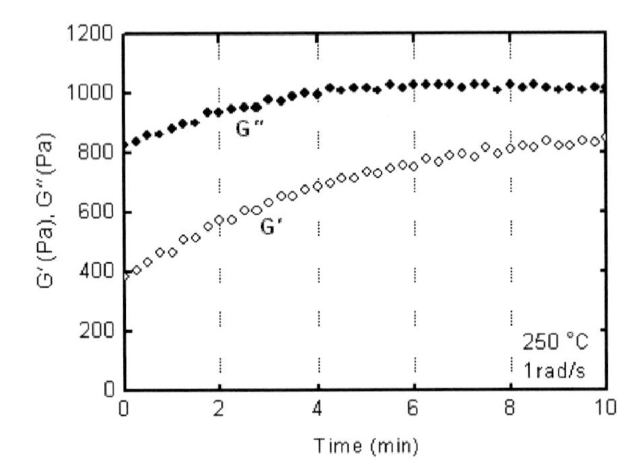

Figure 4. Growth curves of shear storage modulus G' and loss modulus G'' at 250 °C for PC/CNT (97/3) compressed at 200 °C. The angular frequency used was 1.0 s^{-1}.

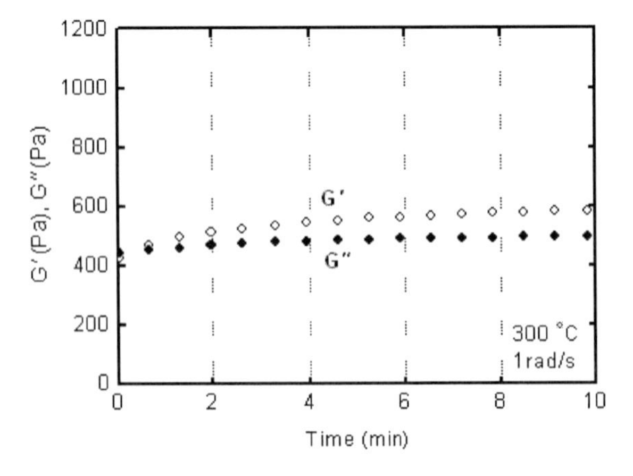

Figure 5. Growth curves of shear storage modulus G' and loss modulus G'' at 300 °C for PC/CNT (97/3) compressed at 200 °C. The angular frequency used was 1.0 s^{-1}.

The obtained results suggest that the dispersion state of CNTs in a molten PC, and thus the electrical conductivity of the composite, can be improved to a great extent by operating condition at processing, such as resin temperature, applied flow history, and post-processing annealing.

NOVEL METHOD TO LOCALIZE NANOFILLER BY BROWNIAN MOTION

In order to widen industrial application of electrical conductive materials such as polymer composites with conductive fillers, cost-performance has to be improved greatly. Therefore, the recent trend is to reduce the amount of expensive conductive fillers in composites. It has been known in rubber industries for a long time that uneven distribution of carbon blacks (CBs) often occurs in multiphase blends (Datta (1999), Duvdevani et al. (2002)), which depends on the mixing protocol and the difference in the interfacial tension with CBs. Moreover, precise control of the uneven distribution leads to conductive composites with a small amount of conductive fillers, when a conductive path is formed by the network of conductive fillers only in a continuous phase. Since the conductive fillers are not required to stay in the dispersed phase as illustrated in Figure 6.(a), the percolation threshold is reduced. This idea was proposed by Sumita et al. (1991, 1992) and called as "double percolation". Then several researchers found similar results (Gubbels et al. (1994), Zhang et al. (1997), Zhang et al. (1998), Potschke et al. (2003)).

Furthermore, Wu et al. (1999) revealed that the percolation threshold of a composite of poly(methyl methacrylate) (PMMA) with vapor grown carbon fibers (VGCFs) is greatly reduced by the addition of 1–5 wt% of polyethylene (PE). They suggested that this phenomenon is attributed to the self-assembled conductive network composed of selective adsorption of PE on VGCFs. Moreover, they clarified that the surface roughness on the carbon particles such as CBs and VGCFs decides the interaction due to van der Waals force and then concluded that the flexibility of polymer chains determines the localization of carbon fillers in a multiphase polymer blend (Wu et al. (2002)).

Localization of conductive fillers at the interphase of phase separated blends having co-continuous structure is another method to obtain a conductive composite. Mamunya (1999) clarified the relation between interfacial tension and distribution state of CBs for immiscible polymer blends and demonstrated that the morphology illustrated in Figure 6.(b) is obtained at the following conditions.

$$\Gamma_{A-C} > \Gamma_{B-C} + \Gamma_{A-B} \tag{10}$$

$$\Gamma_{B-C} > \Gamma_{A-C} + \Gamma_{A-B} \tag{11}$$

where Γ_{A-C}, Γ_{B-C}, and Γ_{A-B} are the interfacial tensions between polymer A and CB, polymer B and CBs, and polymer A and polymer B, respectively. A similar result was reported by Gubbels et al. (1994) employing a blend system composed of PE, polystyrene, and CBs.

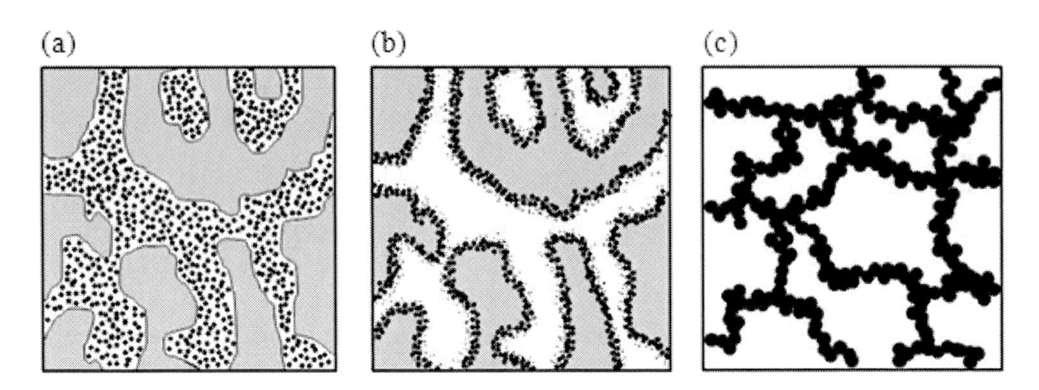

Figure 6. Uneven distribution of CBs; (a) double percolation, (b) localization at the interface, and (c) localization between the chips or particles.

Moreover, compression-molding of polymer chips or particulates coated with conductive fillers enables to provide a conductive composite (Lee, (1992)). In this technique, polymeric materials having high viscosity at processing temperature are employed because they have long diffusion time. Since CBs do not diffuse into the inside of polymer chips or particles during the processing operation, they are localized between the chips or particles as shown in Figure 6.(c). Some conductive products prepared by this method are actually employed in electrostatic dissipative floor tiles. Chan et al. (1997), Grunalan et al. (2001), Zhang et al. (2005), and Lisunova et al. (2007) also employed a similar technique using high viscous materials such as ultra-high molecular weight PE.

Finally, CNT imprinting technique using solvent was proposed recently to illustrate the conductive pattern on the material (Metil et al. (2004)). However, it is not available in industries at present, because the method requires the solvent evaporation process.

Here, we propose a novel method to localize CNTs on the surface of a polymer sheet using interphase CNT transfer from a composite with CNTs to a neat polymer, which provides a conductive sheet with a significantly small amount of CNTs. This concept will be applicable to co-extrusion process of film or sheet processing.

Figure 7. exemplifies the CNT imprinting technique employing piled sheets comprising of neat PC and polypropylene (PP) containing 20 wt% of the CNTs. The surface resistivity of the PP/CNT (80/20) is approximately 10^3 ohm/sq., which is a similar level of the PC/CNT (80/20). It is generally recognized that PP is immiscible with PC. Further, the interfacial thickness λ of the polymer pair in a molten state can be calculated by the following relation (Helfand and Tagami (1971), Yamaguchi et al. (1998)) using the value of interfacial tension Γ between PP and PC (8–18 mN/m) evaluated by Palmer and Demarquette (2005).

$$\lambda = \frac{kT}{3\xi\Gamma} \tag{12}$$

where ξ is the statistical segment step length (Flory (1969), Lohse (2005)).

The interfacial thickness is predicted to be approximately 0.1 nm, which is significantly smaller than the end-to-end distance of a single chain. Because of no/few entanglement couplings between the piled sheets, they can be separated without any difficulty after quenching. The surface resistivity of the PC sheet, which has an annealing history in the piled sheets at 300 °C for 2 min, is 10^5 ohm/sq. The result demonstrates that CNTs move from PP to PC during the applied annealing procedure. Further, the surface resistivity of the PP/CNT sheet is unchanged, i.e., 10^3 ohm/sq.

Distribution state of the CNTs on the surface of the separated PC sheet is shown in Figure 8. As seen in the edge-view picture, a number of the CNTs are detected only in the thin surface layer of the PC sheet. Moreover, the surface picture demonstrates that CNT network is clearly formed, which is responsible for the electric conductivity. Further, the surface of the separated PC sheet is characterized by infra-red spectroscopy using the attenuated total reflection mode and it reveals that PP is not detected on the surface. Therefore, it can be concluded that only CNTs move from PP sheet to PC by Brownian motion.

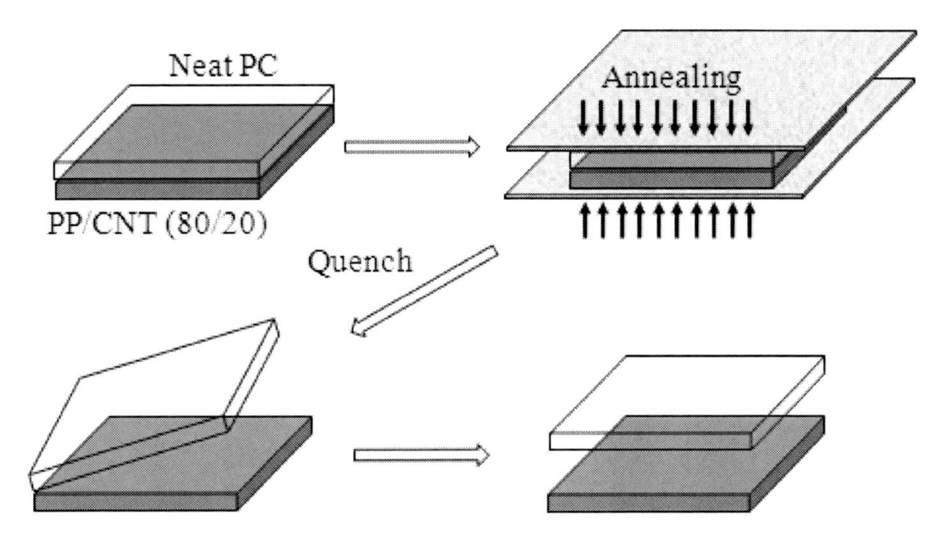

Figure 7. Experimental procedure for CNT imprinting technique; a neat PC and a PP/CNT (80/20) sheet are piled each other and heated, i.e., annealed at various conditions. The sheets are separated after quenching.

Figure 8. SEM pictures of (a) cross-section (edge-view) and (b) surface of a PC sheet separated from PP/CNT (80/20) after annealing at 300 °C for 5 min.

The diffusion distance of the CNTs into the PC sheet is dependent on the applied annealing conditions. For example, the density of the CNTs is significantly low when the annealing procedure is performed at 200 °C (Yoon et al. (2010)). Correspondingly, the surface resistivity of the PC sheet is 10^{13} ohm/sq., which is considerably higher than that annealed at 300 °C. This is reasonable because exposure to high temperature enhances Brownian motion of the CNTs, i.e., driving force of the interphase diffusion, as discussed in the previous section. The present experimental results demonstrate that CNT imprinting technique performed at high temperature enables to reduce the amount of CNTs greatly to obtain a conductive material. Figure 9. shows the diffusion distance of the CNTs plotted against the annealing time for the PC sheets separated from the PP/CNT (80/20) sheet after annealing at 300 °C. The solid line in the figure represents the following relation;

$$x = c\sqrt{(t - t_0)} \tag{13}$$

where x is the diffusion distance and c is the diffusion coefficient.

It was found that the thickness of the layer containing the CNTs increases gradually with the annealing time, which is proportional to $t^{1/2}$. The diffusion constant c is estimated from the figure to be approximately 6.5 x 10^{-9} [m^2/s]. Moreover, the figure indicates that the diffusion distance in the range $t < 1$ [min] is zero (t_0 = 1 min). Presumably, 1 min is required to raise the temperature of the sample sheets and to move the CNTs from PP to PC through the interface. The surface resistivity of the PC sheets at room temperature is plotted in Figure 10. as a function of the annealing time and temperature. It is found that the electrical resistivity decreases rapidly with the annealing time and then keeps a constant value which is the same as that of the PC/CNT (80/20) at high annealing temperature. On the contrary, the surface resistivity decreases slightly with annealing time at low annealing temperature. The resistivity would be determined by the amount of the CNTs on the surface of PC. Therefore, it is deduced that approximately 2 wt% of the CNTs are distributed in the surface layer of the PC sheet annealed for 2 min at 300 °C. Furthermore, it also suggests that the CNT content in the surface layer becomes a constant after a certain residence time.

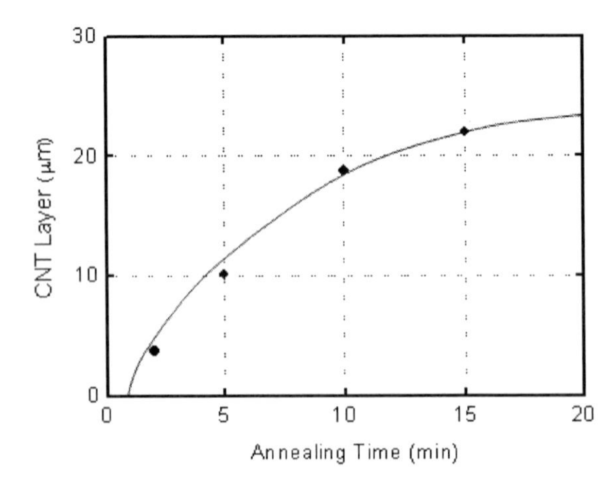

Figure 9. Thickness of CNTs-rich layer (μm) of the separated PC sheet plotted against the annealing time at 300 °C. The solid line in the figure represents the curve proportional to $t^{1/2}$.

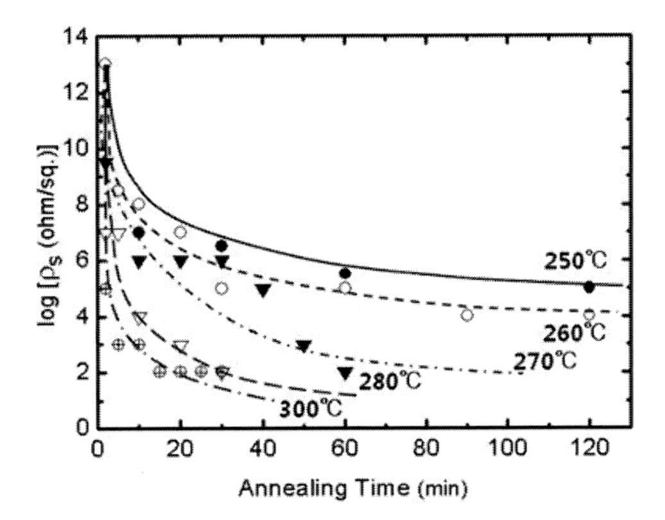

Figure 10. Surface resistivity of PC sheets after separation from PP/CNT (80/20) annealed at 250-300 °C for various residence times.

The CNT imprinting test is also performed for the piled sheets comprising of neat PP and PC/CNT (80/20) following the procedure illustrated in Figure 7. It is found that CNT transfer does not occur in this system even annealed at 300 °C for 30 min, although both polymers show a similar zero-shear viscosity at 300 °C. The surface resistivity of the separated PP sheet after the annealing is more than 10^{15} ohm/sq., i.e., the same level as that for the neat PP sheet. The compatibility between the polymers and the CNTs would be attributed to the difference in the surface tension. Dujardin et al. (1994, 1998) reported that the surface tension of CNTs is around 150 mN/m. Further, it is shown in the polymer handbook that the surface tension of PC is 32 mN/m at 200 °C and that of PP is 19 mN at 200 °C. The information suggests that the surface tension of CNTs employed would be larger than those of the polymers. Therefore, the present results indicate that the interdiffusion of CNTs occurs from the polymer with high interfacial tension to that with low interfacial tension. This idea is also supported by other experimental results employing PE, PMMA, and poly(lactide) (Yoon and Yamaguchi, (2010)).

Finally, the driving force of the current technique is Brownian motion, it will be also applicable to various nanofillers including single-walled CNTs. Recently, it is demonstrated by our group that the transfer technique is applicable to organic layered silicate (Yoon et al. (2010)).

CONCLUSION

Brownian motion of CNTs in a polymer liquid has a strong impact on the dispersion state of CNTs. As a result, the rheological properties in a molten state are sensitive to the applied annealing history as well as the processing history. Furthermore, the distribution state also has an impact on the electric conductivity, which is studied intensively for the composites with CNTs. Because of high aspect ratio, the dispersion state of the product obtained by conventional processing operations is usually in a non-equilibrium state. Therefore, the information on the redistribution process by Brownian motion will be significantly important for industrial application.

Moreover, the diffusion ability of CNTs in a polymer liquid can be applicable to the imprinting technique. CNTs will immigrate from a polymer with poor compatibility to another polymer with good compatibility during annealing procedure. Because piled polymer sheets composed of immiscible polymer pairs can be separated without difficulty, polymer sheets containing CNTs only in the surface area are obtained. The transfer phenomenon can be written by the diffusion equation. Since the amount of CNTs is considerably small to obtain the electric conductive polymer composites, this technique will be employed in various applications.

ACKNOWLEDGMENT

The authors would like to express their gratitude to Hodogaya Chemicals Co. Ltd. for the kind supply of the samples employed in this study.

REFERENCES

Aharoni, S. M. *J. Appl. Phys*. 1972, 43, 2463-2465.

Amran, M. B. A.; Okamoto, K.; Koshirai, A.; Kasai, T.; Yamaguchi, *M. J. Polym. Sci. Polym. Phys. Ed.* 2009, 47, 2008-2014.

Bin, Y.; Xu, C.; Zhu, D.; Matsuo, *M. Carbon* 2002, 40, 195-199.

Brady, J. F. *Curr. Opin. Colloid Interface Sci*. 1996, 1, 472- 480.

Chan, C.; Cheng, G.; Yuen, *M. M. F. Polym. Eng. Sci*. 1997, 37, 1127-1136.

Chen, J.; Shan, J. Y.; Tsukada, T.; Munekane, F.; Kuno, A.; Matsuo, M.; Hayashi, T.; Kim, Y. A.; Endo, *M. Carbon* 2007, 45, 274-280.

Datta, S. *Polymer Blends*; *Elastomer Blends*; Wiley: New York, NY, 1999; Vol. 2, pp 477-515.

Doi, M.; Edwards, S. F. *The Theory of Polymer Dynamics*; Oxford University Press: Oxford, UK, 1986; pp 289-323.

Dujardin, E.; Ebbesen, T. W.; Hiura, H.; Tanigaki, K. *Science* 1994, 265, 1850-1852.

Dujardin, E.; Ebbesen, T. W.; Krishnan, A.; Trecy, M. M. *Advanced Materials* 1998, 10, 1472-1475.

Duvdevani, I.; Tsou, A.; Yamaguchi, M.; Gogos, C. G. 161st Technical Meeting of Rub. Div. 2002, American Chemical Society, Georgia.

Flory, P. J. Statistical Mechanics of Chain Molecules; Interscience: New York, NY, 1969.

Gubbels, F.; Jerome, R.; Teyssie, P.; Vanlathem, E.; Deltour, R.; Calderone, A.; Parente, V.; Bredas, J. L. *Macromolecules* 1994, 27, 1972-1974.

Grunlan, J. C.; Gerberich, W. W.; Francis, *L. F. J. Appl. Polym. Sci.* 2001, 80, 692-705.

Helfand, E.; Tagami, *Y. J. Chem. Phys*. 1971, 56, 3592-3601.

Iijima, S. *Nature* 1991, 354, 56-58.

Janzen, J. *J. Appl. Phys*. 1975, 46, 966-970.

Kim, Y. A.; Hayashi, T.; Endo, M.; Kaburagi, Y.; Tsukada, T.; Shan, J.; Osato, K.; Tsuruoka, S. *Carbon* 2005, 43, 2243-2250.

Krieger, *I. M. Adv. Colloid Interface Sci.* 1972, 3, 111-136.

Lisunova, M. O.; Mamunya, Y. P.; Lebovka, N. I.; Melezhyk, A. V. *Eur. Polym. J.* 2007, 43, 949-958.

Lohse, D. J. *Polymer Review* 2005, 45, 289-308.

Natsuki, T.; Endo, E.; Takahashi, *T. Phys. Stat. Mech. Appl.* 2005, 352, 498-508.

Palmer, G.; Demarquette, *N. R. Polymer* 2005, 46, 8169-8177.

Pötschke, P.; Bhattacharyya, A. R.; Janke, *A. Polymer* 2003, 44, 8061-8069.

Ray, S. S.; Okamoto, *M. Prog. Polym. Sci.* 2003, 28, 1539-1641.

Ray, S. S.; Bousmina, *M. Prog. Materials Sci.* 2005, 50, 962-1079.

Sumita, M.; Asai, S.; Miyadera, M.; Jojima, E.; Miyasaka K. *Colloid Polym. Sci.* 1986, 264, 212-217.

Tagaya, A.; Ohkita, M.; Mukoh, R.; Sakaguchi, R.; Koike, Y. *Science* 2003, 301, 812-814.

Lechkov, M.; Prandzheva, S. *Encyclopedia of Polymer Composites: Properties, Performance and Applications*; Nova Science Publishers: New York, NY, 2009.

Lee, B. *Polym. Eng. Sci.* 1992, 36-42.

Mamunya, *Y. P. J. Macromol Sci. Phys.* 1999, B38, 615-622.

Matsumoto, T.; Segawa, Y.; Masuda, T.; Onogi, *S. Trans. Soc. Rheol.* 1973, 17, 47-62.

Metil, M. A.; Zhou, Y.; Gaur, A.; Jeon, S.; Usrey, M. L.; Strano, *M. S. Nano Lett.* 2004, 4, 2467-2471.

Onogi, S.; Masuda, T.; Matsumoto, *T. Trans. Soc. Rheol.* 1970, 14, 275-294.

Palmer, G.; Demarquette, *N. R. Polymer* 2005, 46, 8169-8177.

Shikata, T.; Pearson, *D. S. J. Rheol.* 1994, 38, 601-616.

Sumita, M.; Sakata, K.; Hayakawa, Y.; Asai, S.; Miyasaka, K.; Tanemura, K. *Colloid Polym. Sci.* 1992, 270, 134-139.

Wu, G.; Asai, S.; *Sumita. M. Macromolecules* 1999, 32, 3534-3536.

Wu, G.; Asai, S.; Zhang, C.; Miura, T.; Sumita, *M. J. Appl. Phys.* 2000, 88, 1480-1487.

Wu, G.; Asai, S.; Sumita, M.; Yui, *H. Macromolecule*s 2002, 35, 945-951.

Xanthos, M. *Functional Fillers for Plastics*; Wiley-VCH: New York, NY, 2010.

Yamaguchi, *M. J. Appl. Polym. Sci.* 1998, 70, 457-463.

Yamaguchi, M. *Melt Elasticity of Polyolefins; Impact of Elastic Properties on Foam Processing in Polymeric Foam, Mechanisms and Materials*; Lee, S. T.; Ramesh, N. S.; Ed.; Chap. 2; CRC Press: New York, NY, 2004.

Yano, H.; Sugiyama, J.; Nakagaito, A.; Nogi, M.; Matsuura, T.; Hikita, M.; Handa, K. *Advanced Materials* 2005, 17, 153-155.

Yokohara, T.; Okamoto, K.; Yamaguchi, *M. J. Appl. Polym. Sci.* 2010, 117, 2226-2232.

Yokohara, T.; Yamaguchi, *M. Eur. Polym. J.* 2008, 44, 677-685.

Yoon, H.; Okamoto, K.; Yamaguchi, *M. Carbon* 2009, 47, 2840-2846.

Yoon, H.; Okamoto, K.; Umishita, K.; Yamaguchi, *M. Polym. Composite*, in press.

Yoon, H.; Ha, J. U.; Xanthos, M.; Yamaguchi, M. *Proceeding of Annual Meeting of Polymer Processing Society* 2010, PPS26, Banff, Canada.

Zhang, C.; Yi, X. S.; Yui, S.; Asai, S.; Sumita, *M. Mater. Lett.* 1998, 36, 186-190.

Zhang, C.; Ma, C.; Wang, P.; Sumita, *M. Carbon* 2005, 43, 2544-2553.

Zhang, C.; Sheng, J.; Ma, C.; Sumita, *M. Mater. Lett.* 2005, 59, 3648-3651.

Zhang, C.; Wang, P.; Ma, C.; Wu, G.; Sumita, *M. Polymer* 2006, 47, 466-473.

In: Brownian Motion: Theory, Modelling and Applications ISBN: 978-1-61209-537-0
Editors: R.C. Earnshaw and E.M. Riley

Chapter 7

BROWNIAN DYNAMICS SIMULATION OF SURFACTANT MICELLAR MICROSTRUCTURE AND RHEOLOGY

Jinjia Wei[1], Yasuo Kawaguchi[2] and Chenwei Zhang[1]
[1]State Key Laboratory of Multiphase Flow in Power Engineering, Xi'an Jiaotong University, Xi'an, China
[2]Department of Mechanical Engineering, Faculty of Science and Technology, Tokyo University of Science, Noda, Chiba, Japan

NOMENCLATURE

F Force (N)
$[\mathbf{I}]_k$ the momentum of inertia of rod k.
K Boltzmann's constant
L computational cell length (m)
m_k mass of rod k
N micelle number
n bead number in a model rodlike micelle
$\vec{n}_{klij}$ unit vector pointing from bead i in rod k to bead j in rod l.
Pe Peclet number
$\vec{r}_{kc}$ position of the center of mass of rod k
$\vec{r}_{ki}$ position of bead i in rod k
$\vec{r}_{klij}$ vector pointing from bead i in rod k to bead j in rod l, Figure 2
T temperature (^{o}C or K)
t time (s)
U potential (J)
V volume of the computational cell (m^3)
v velocity (m/s), Figure 2

x Cartesian coordinate
y Cartesian coordinate
z Cartesian coordinate

Greek Symbol

α_v volume fraction of micelles
$\Delta\eta$ additive shear viscosity
$\Delta\sigma_{\alpha\beta}$ additive shear stress
ε energy parameter in potentials (J)
ϕ rotation angle of rodlike micelle (deg), Figure 2
$\dot{\gamma}$ shear rate (1/s)
η shear viscosity (Pa s)
η_{0n} shear viscosity at zero shear strain rate in the absence of potential (Pa s)
η_E elongational viscosity (Pa s)
$[\boldsymbol{K}]$ velocity gradient tensor of the macroscopic fluid flow
ξ Stokes' friction constant (kg/s)
ψ_1 first normal stress coefficient
$\psi_{1,n0}$ first normal stress coefficient at zero shear strain rate in the absence of potential
λ relaxation time (s)
ν kinematic viscosity (m^2/s)
ρ density (kg/m^3)
σ diameter of bead (m)
θ rotation angle of rod k (deg), Figure 2

Subscript

α Cartesian component (x, y, z)
β Cartesian component (x, y, z)
h hydrodynamic
p potential
r Brownian
s solvent

ABSTRACT

Three-dimensional Brownian dynamics simulation was conducted for dilute micellar surfactant solution under a steady shear flow and a steady uniaxial elongational flow. The rodlike micelle in the surfactant solution was assumed as a rigid rod made up of beads that were lined up. Lennard-Jones potential and soft-sphere potential were employed as the inter-bead potentials for end-end beads and interior-interior beads, respectively. The motion of the rodlike micelles was determined by solving the translational and rotational equations for each rod under hydrodynamic drag force, Brownian force and inter-rod potential force. Velocity Verlet algorithm was used in the simulation. The micellar microstructures and the rhelogical properties of the surfactant solution at different shear rates and elongation rates were obtained. The micellar network structure was formed at low shear or elongation rates and was destructed by high shear or elongation rates. The computed shear and elongational viscosities and the first normal stress coefficient showed shear thinning characteristics. The relationship between the rheology and the microstructure of the surfactant solution was revealed. The effect of surfactant solution concentration on the micellar structures and rheological properties was also investigated.

1. Introduction

Surfactant drag reduction is a well-known flow phenomenon in which the turbulent friction drag can be greatly reduced by adding small amounts of surfactant into a carrier fluid at the same flow rate [1,2]. In the Reynolds number region of industrial interest, the amount of turbulent friction drag reduction (DR) sometimes approaches 90% [3]. Polymer solutions also possess similar effect. Compared to polymer solutions, however, surfactant solutions are easy recovery after mechanical degradation because of self-assembling nature of their micelles. Therefore, they can recover quickly from destruction caused by pump in recirculation systems [4]. Recently, surfactants have been widely accepted to be the most practical drag-reducing additives in district heating and cooling systems (DHC) for reducing pumping power because they are rather stable and show no mechanical degradation [5]. The surfactant monomer in the surfactant solution has a hydrophilic (water-loving) head and a long hydrophobic (oil-loving) alkyl chain tail. The hydrophobic tails concentrate in the center to avoid contacting with solvent while the hydrophilic heads lie on the surface in contact with solvent molecules. This process is named micellization. Thus the monomers aggregate to form spherical micelles with hydrophilic head outside and hydrophobic tail inside. With further increasing concentration, the rodlike micelles can be further formed and the micellar network structure is constructed by the combination of the rodlike micelles. Rheological measurements indicated the great change of rheological properties of the surfactant solution compared to the solvent. The surfactant solutions always show strong viscoelastisity and high elongational viscosity. It is generally believed that the formation of rodlike or threadlike micellar network structures changes the rheological characteristics of surfactant solution greatly and thus causes the drag reduction occurrence. Evidence to support the existence of rodlike micelles has been provided by a number of researchers using various light scattering techniques [6-10], nuclear magnetic resonance (NMR)[7,11] and small angle neutron scattering (SANS) methods [9,12]. Recent developments in Cryo-TEM (transmission electron microscopy) have made it probable to get direct images of micellar structure without altering the structures in sample preparation [13]. Rodlike or threadlike micellar network structures have been found in various surfactant/salt systems. The cross-sectional diameter of rodlike micelles is usually approximately 5 nm. On a length scale greater than the persistence length, a micelle can be thought of as a flexible structure, whereas at lesser lengths, the micelle appears to be linear. Therefore, the micellar network structure can be treated as the connection of rigid rodlike micelles at the ends. Since all these micelles with dimensions of nanometer order-in-magnitude are embedded in a solvent, it does not accord with continuum hypothesis. Their dynamics is of Langevin rather than of Newtonian type. A complete time scale separation between solvent and micelle relaxation justifies the picture of the motion of one micelle described by Brownian dynamics. The rod experiences a Brownian force arising from the random thermal bombardment on the rod by the surrounding random motion of the solvent molecules, a hydrodynamic force due to the velocity difference between the rod and the flow field, and an inter-rod potential interaction force. A suitable inter-cluster potential should be employed to make the rods combine at their ends to form a micellar network structure. The rheological properties can be obtained from the corresponding micellar structure by use of Brownian dynamics theory and statistical mechanics. Some Brownian dynamics studies of hard spherocylinders have been carried out by Lowen [14], Branka and Heyes [15], and Mori et al. [16,17]. In these studies, the potentials employed for describing the interactions between rods are the Yukawa potential, the repulsive Lennard-Jonnes potential or the Gay-Berne potential. All these potentials were unable to get network structures existing in the surfactant solutions. The length-to-width ratios of the rods in these studies were restricted to a low value up to 6 due to the limit of the applicable range of the rod translational and rotational self-diffusion formula. On the other hand, Doi et al. ([16] investigated a simple model of infinitely thin needles. Both these two cases are not suitable for describing the self-diffusion of the rod-like micelles which have a length-to-width ratio of 10 or larger. In this chapter, we introduce a new inter-cluster potential to describe the inter-micelle interaction for the micellar network formation. A rigid osculating multibead rod model is employed in which the rod translational and rotational self-diffusions are sums of those of the element beads in the rods. The properties of surfactant solutions can be obtained from simulations of different time- and length-scales from the atomistic to the macroscopic. Molecular dynamics simulation (MDS) can be used to study the dynamics of surfactant and solvent molecules at chosen infinite molecule numbers. The forming process of spherical or even cylindrical micelles can be computed by MDS. However, MDS is limited to relative low number of molecules because of power of computers. Brownian dynamics simulation (BDS) is a mesoscopic simulation. In this method, the effect of solvent is simplified as random Brownian force (caused by fluctuation of solvent molecules) and hydrodynamics force (caused by relative velocity between coarse-grained particles and solvent molecules). Therefore, finite power of computers can be used to compute the dynamics of coarse-grained particles (e.g. micelles in surfactant solution). It is practical to study the developing process of microstructure and origin of rheological characteristics. The direct numerical simulation (DNS) based on the macroscopic rheological properties has been used as a powerful tool for analyzing the turbulent characteristics of the DR flow such as the enhancement of streamwise velocity fluctuations at large Reynolds numbers and the attenuation of the Reynolds stress. Such information can help us to attain a better understanding of the mechanisms of DR. It is of great importance to study the microstructure of surfactant solution from a microscopic viewpoint since it is the origin of the DR. The relationship between the macroscopic rheological properties of dilute surfactant solution and the underlying microscopic processes is an interesting but complicated subject. BDS can be used to develop a new model for describing the microstructure of the surfactant solution and to relate microstructure to chemical structure, rheological behavior and drag reduction effectiveness. The final purpose of this chapter is to provide useful drag reduction surfactant additives. In the present study, we use BDS to examine the effects of shear rate and solution concentration on micellar network formation in dilute surfactant solution and the corresponding rheological properties, and thus to reveal the relationship between the DR, rheological characteristics and surfactant microstructures.

2. PROCESSING OF BROWNIAN DYNAMICS SIMULATION ON SURFACTANT SOLUTIONS

In this chapter, the surfactant solutions owing threadlike micelles are discussed. Threadlike micelles are flexible structure. However, there is a short length named persistence length that the micelle exhibits rigid property when threadlike structure is shorter than it. Then a piece of micelle which is equal to persistence length can be simplified as the original particle. We consider the surfactant solution as a system of N rigid rodlike micelles suspended in an incompressible continuum fluid medium of viscosity of η_s. The rodlike micelles interact through both inter-rod forces and hydrodynamic forces mediated via the continuum fluid. The micelles also receive fluctuating Brownian forces arising from the apparently random thermal bombardment by surrounding solvent molecules. The rigid osculating multibead rod model is employed for modeling surfactant solutions of rodlike micelles. The micelle is assumed to be made up of n evenly spaced beads of diameter σ, which are linearly connected as shown in Figure 1. It represents a cylindrical rod of length $n\sigma$, diameter σ and an aspect ratio of n. Different from a sphere case, the rod is multiscale and the BD simulation of it is a challenge. The positions of the beads are given in terms of their distances from the center of mass along the rod axis. The rods interact via the bead-bead potential interaction. Thus the interaction energy between two rigid linear rods is a sum of pairwise contributions from distinct bead i in rod k and bead j in rod l. Figure 1. shows the inter-rod interaction. As mentioned above, the micellar network structure is considered as the connection of rodlike micelles at the micelle ends.

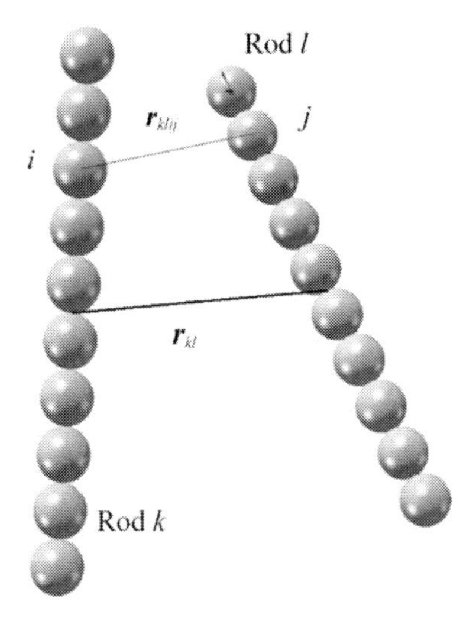

Figure 1. Schematic of rod model made up of n beads of diameter σ The equations of the translational and rotational motions of a rod read.

$$m_k \frac{d^2\vec{r}_{kc}}{dt^2} = \sum_{i=1}^{n} (\vec{F}_{ki}^h + \vec{F}_{ki}^r + \sum_{l=1;l\neq k}^{N} \sum_{j=1}^{n} \vec{F}_{klij}^p) \quad (1)$$

and

$$[\mathbf{I}]_k \frac{d^2\theta_k}{dt^2} = \sum_{i=1}^{n} \left[(\vec{r}_{ki} - \vec{r}_{kc}) \times (\vec{F}_{ki}^h + \vec{F}_{ki}^r + \sum_{l=1,l\neq k}^{N} \sum_{j=1}^{n} \vec{F}_{klij}^p) \right] \quad (2)$$

where m_k and $[\mathbf{I}]_k$ are respectively the mass and momentum of inertia of rod *k*, and

$$[\mathbf{I}]_k = \sum_i m_{ki}[(r_{ki} - r_{kc})^2[\mathbf{1}] - (\vec{r}_{ki} - \vec{r}_{kc})(\vec{r}_{ki} - \vec{r}_{kc})] \quad (3)$$

where $\vec{r}_{kc}$ and θ_k are respectively the position of center of mass and rotation angle of rod *k*. The definition of orientation angles of the rigid rodlike micelle in a 3-D Cartesian coordinate system is shown in Figure 2. Here, $\vec{r}_{ki}$ is the position of bead *i* in rod *k*, and [**1**] is the unit tensor. In order to simulate the formation of the network structure, for end-end beads potential between two rods, Leonard-Jones potential is employed:

$$U_{klij}(r_{klij}) = 4\varepsilon[(\sigma/r_{klij})^{12} - (\sigma/r_{klij})^6] \quad (4)$$

where $\vec{r}_{klij}$ is the vector pointing from the center of bead *i* in rod *k* to bead *j* in rod *l*; and ε represents the strength of the interaction. To prevent overlap among micelles, for interior-interior beads potential between two rods, a repulsive soft-sphere potential is assumed:

$$U_{klij}(r_{klij}) = 4\varepsilon(\sigma/r_{klij})^6 \quad (5)$$

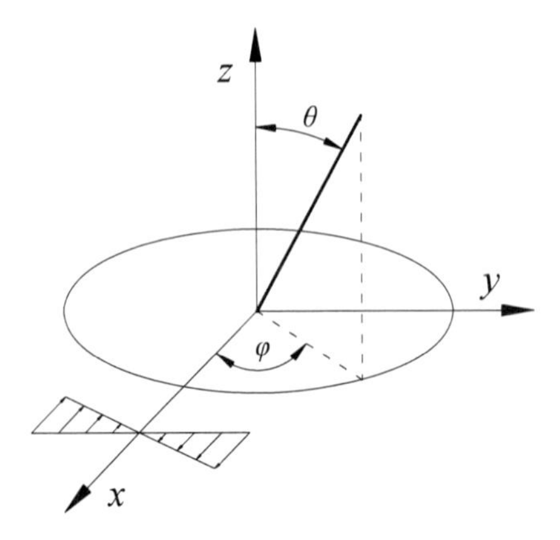

Figure 2. Orientation angles for a rigid rodlike micelle in Cartesian coordinate system.

We call this kind of combined interrod potentials between micelles as the WK (Wei-Kawaguchi) potential. This is different from the conventional BD simulation in which only one potential was used for interrod interaction. Here ε is the depth of potential. The selection of ε is a problem. If it is too large, the micellar structure will become very rigid and if it is too small, the network structure cannot be formed. Both of the two cases are not consistent with the TEM image and rheological measurements. Here we selected ε = 10 KT for the simulations, where K is the Boltzmann constant. $\vec{F}^{p}_{klij}$ is the potential force on bead i in rod k arising from bead j in rod l.

$$\vec{F}^{p}_{klij} = \frac{dU_{klij}}{d\vec{r}_{ki}} = \frac{dU_{klij}}{dr_{klij}} \vec{n}_{klij} \tag{6}$$

where $\vec{n}_{klij}$ is the unit vector pointing from the center of bead i in rod k to bead j in rod l.

$$\vec{n}_{klij} = \frac{\vec{r}_{klij}}{r_{klij}} \tag{7}$$

$\vec{F}^{h}_{ki}$ is the hydrodynamic force on bead i in rod k and is assumed to be proportional to the relative velocity of the sphere with respect to the macroscopic fluid flow.

$$\vec{F}^{h}_{ki} = -\xi(\vec{v}_{ki} - [\mathbf{K}]\vec{r}_{ki}) \tag{8}$$

where ξ is the Stokes' friction constant, $\xi = 3\pi\eta_s\sigma$; $[\mathbf{K}]$ is the velocity gradient tensor of the macroscopic fluid flow. $\vec{F}^{r}_{ki}$ is the random Brownian force on bead i in rod k due to the thermal motion of the solvent molecules. The short time steps are required to handle the fast motion of the solvent molecules, and the very long runs are needed to allow the evolution of the slower mode of the rodlike micelles, making the simulation very time consuming. Here, we are not concerned with the motion of the solvent molecules, and the random force caused by the solvent molecules is assumed to be white noise, i.e. has a correlation time shorter than any process of interest. The amplitude of the random force is then given by the fluctuation-dissipation theorem [19]

$$< \vec{F}^{r}_{ki}(t)\vec{F}^{r}_{ki}(t') >= 2kT\xi[\mathbf{1}]\delta(t-t') \tag{9}$$

Here $<$ denotes an average over the probability space on which $\vec{F}^{r}_{ki}$ is defined, and $\delta(t-t')$ is the delta function.

The additive shear stress $\Delta\sigma_{\alpha\beta}$ due to the existence of rodlike micelle may be given by

$$\Delta\sigma_{\alpha\beta} = \frac{1}{2V}\left[\sum_{k=1}^{N}\left(\sum_{i=1}^{n}(F_{ki\alpha}^{h}(r_{ki\beta} - r_{kc\beta})) + \sum_{l=k+1}^{N}(F_{kl\alpha}^{p} r_{kl\beta})\right)\right] \tag{10.a}$$

where

$$\vec{F}_{kl}^{p} = \sum_{i=1}^{n}\sum_{j=1,}^{n} \vec{F}_{klij}^{p} \tag{10.b}$$

is the force on the center of mass of rod k arising from interactions with rod l; $\vec{r}_{kl}$ is the vector pointing from the center of mass of rod k to that of rod l; V is the volume of the computational cell; α and β are the Cartesian components.

For a 3-D shear flow in the x-direction, we have

$$v_x = \dot{\gamma} r_y; \quad v_y = 0; \quad v_z = 0 \tag{11}$$

where v is velocity and $\dot{\gamma}$ shear rate. For this flow, the additive viscosity and two normal stress coefficients can be expressed as

$$\Delta\eta = \Delta\sigma_{xy} / \dot{\gamma} \tag{12}$$

$$\psi_1 = (\sigma_{xx} - \sigma_{yy}) / \dot{\gamma}^2 \tag{13}$$

$$\psi_2 = (\sigma_{yy} - \sigma_{zz}) / \dot{\gamma}^2 \tag{14}$$

For a 3-D steady elongational flow in the x-direction, we have

$$v_x = \dot{\varepsilon} r_x; \quad v_y = -0.5\dot{\varepsilon} r_y; \quad v_z = -0.5\dot{\varepsilon} r_z \tag{15}$$

where $\dot{\varepsilon}$ is elongation rate. In this flow only one normal stress difference can be measured:

$$\sigma_{xx} - \sigma_{yy} = (\Delta\sigma_{xx} - \Delta\sigma_{yy}) + 3\eta_s\dot{\varepsilon} = \eta_E\dot{\varepsilon} \tag{16}$$

The flow field can also be represent by a tensor.

In this chapter, dimensionless quantities are used where length (σ), energy (kT) and friction factor ξ of beads are set to unity. It follows that time is reduced by $\xi\sigma^2 / kT$ and shear rate or elongation rate $kT / \xi\sigma^2$. The dimensionless shear rate or elongation rate is called bead Peclet number, $\mathrm{Pe} = \dot{\gamma}\xi\sigma^2 / kT$ or $\mathrm{Pe} = \dot{\varepsilon}\xi\sigma^2 / kT$, which is the ratio of the time for a bead to freely diffuse a distance σ to the flow scale $1/\dot{\gamma}$ or $1/\dot{\varepsilon}$. At large Pe, the inverse shear rate or elongation rate is less than $\xi\sigma^2 / kT$ and the structure rearrangement due to the shear or elongation will dominate. Conversely, at small Pe, the shear or elongation induced structure will be a perturbation of the Brownian structure.

For the steady shear flow and uniaxial elongational flow, the Lee-Edwards' sliding periodic image boundary conditions were used to maintain velocity continuity at the simulation cell boundaries. By periodic boundary conditions, we represent an infinite suspension as a spatially periodic array of identical cubic cells with cell length of L. The replicas of the central cell move with the shear flow to ensure continuity in the fluid velocity across the boundaries of the unit cell. A bead leaves the central cell through a side $x = \pm L/2$, reenters through the opposite side. A bead which leaves through an edge $y = \pm L/2$ or $z = \pm L/2$ will re-appear on the opposite edge but with a displacement in the x-direction according to the instantaneous position of the surrounding replicas.

For the steady elongational flow, periodic boundary conditions were also used. The fluid should extend along one dimension (x here) and contracts in the remaining orthogonal directions so as to maintain a constant density. In the most general treatment of this problem one must follow the time evolution of a fluid element which in simulations is represented by a cubic box containing the micelles. All points on the boundaries of the box change at every time step according to the following equations during the simulation period.

$$\begin{aligned} r_x(t) &= r_x(t=0)\exp(\dot{\varepsilon}t) \\ r_y(t) &= r_y(t=0)\exp(-0.5\dot{\varepsilon}t) \\ r_z(t) &= r_z(t=0)\exp(-0.5\dot{\varepsilon}t) \end{aligned} \tag{17}$$

These equations are obtained by integrating Eq. (15). They ensure that the evolution of cell boundaries is compatible with the particle dynamics and that the system volume remains a constant of the motion. Velocity Verlet algorithm was used in the simulation. This is a Verlet-equivalent algorithm which stores positions, velocities and accelerations all at the same time and which minimizes round-off error. The velocity Verlet algorithm can be described as follows

Step 1. Give initial values of translational and angular positions, translational and rotational velocities and accelerations and etc for rods and beads.

Step 2. Update the initial fields.

Translational position of the center of mass of a rod at $t + \delta t$

$$r_c(t+\delta t) = r_c(t) + \delta t v_c(t) + 1/2\delta t^2 a_c(t) \tag{18}$$

Velocity of the center of mass of a rod at $t + \frac{1}{2}\delta t$

$$v_c(t + 1/2\delta t) = v_c(t) + \frac{1}{2}\delta t a_c(t) \tag{19}$$

Rotational position of a rod at $t+\delta t$

$$\phi(t+\delta t)=\phi(t)+\dot{\phi}(t)\delta t \quad (20)$$

$$\theta(t+\delta t)=\theta(t)+\dot{\theta}(t)\delta t \quad (21)$$

where $\dot{\phi}(t)=\omega_z(t)-(\omega_x(t)\sin\phi(t)-\omega_y(t)\cos\phi(t))/\tan\theta(t)$

$$\dot{\theta}(t)=\omega_x(t)\cos\phi(t)+\omega_y(t)\sin\phi(t)$$

Angular velocities at $t+\frac{1}{2}\delta t$

$$\omega_i(t+1/2\delta t)=\omega_i(t)+\frac{1}{2}\delta t\dot{\omega}_i(t) \quad (22)$$

$i=x, y, z$, Cartesian coordinates.

From $r_c(t+\delta t)$, $\phi(t+\delta t)$, $\theta(t+\delta t)$ and the relative position of a bead r_{cj} (r_j-r_c)on a rod, one can get the position of a beads easily as follows

$$r_j(t+\delta t)=f(r_c,\phi,\theta,r_{cj}) \quad (23)$$

j=1, 2, …, n n is beads number on a rod.

Step 3. Calculate force and torque acting on a rod

From $v_c(t+1/2\delta t)$, $\omega_i(t+1/2\delta t)$ and the relative position vector of a bead r_{cj} on a rod, one can get the velocity of a bead easily as follows

$$v_j(t+\delta t)=f(v_c,\omega_i,r_{cj}) \quad (24)$$

j=1, 2, …, n n is beads number on a rod.

With $r_j(t+\delta t)$ and $v_j(t+1/2\delta t)$ of a bead obtained, one can compute the hydrodynamic friction force F_j^h, Brownian force F_j^r and interactive potential force F_j^p acting on the bead.

$$F_j=F_j^h+F_j^r+F_j^p \quad (25)$$

Total force acting on a rod

$$F_c = \sum_{j=1}^{n} F_j \tag{26}$$

Total torque acting on a rod

$$M = \sum_{j=1}^{n} F_j \times r_{cj} \tag{27}$$

Step 4. Calculate translational and rotational accelerations $t + \delta t$

$$a_c(t + \delta t) = F_c(t + \delta t) / m_c \tag{28}$$

$$\dot{\omega}_i(t + \delta t) = M_i(t + \delta t) / I_i \tag{29}$$

$i = x, y, z$, Cartesian coordinates,
where I_i are three principal moments of inertia.
Step 5. Get translational and rotational velocities at $t + \delta t$

$$v_c(t + \delta t) = v_c(t + \delta t) + \frac{1}{2}\delta t a_c(t + \delta t) \tag{30}$$

$$\omega_i(t + \delta t) = \omega_i(t + \delta t) + \frac{1}{2}\delta t \dot{\omega}_i(t + \delta t) \quad i = 1, 2, 3 \tag{31}$$

Step 6. Sample data for statistical analysis at a given interval.
Step 7. Go to step 2 for next cycle until defined conditions are met.

In the simulation, the beads number in a rod was 10. The cutoff distance in the potential interaction was 5σ above which the potential was set to zero. Three different micelle volume concentrations, 0.126%, 0.189% and 0.453%, were simulated. They are very dilute surfactant solutions. The computational domain is a 3-D cubic cell with side length 50σ.

At the start of a simulation, the rods were placed in the primary cell in a α-f.c.c. (face-centered cubic) structure, which has been the starting configuration for many simulations [20]. The ensemble of rods is allowed to equilibrate in the absence of shear flow for 10^7-10^8 time steps to make the interested quantities cease to show a systematic drift and start to oscillate about steady mean values. The shear flow field was then turned on at $t = 0$, and the statistics of the rod trajectories were taken at regular intervals. The non-dimensional time step size is 10^{-4} in the simulation.

3. RESULTS AND DISCUSSION

3.1. Micellar Structure

Figure 3. shows the micellar structures at Peclet number (dimensionless shear rate) Pe of 0.01, 0.1, 1.0 and 10.0 under simple shear flow, respectively. Micellar network structure formed at low Pe of 0.01 and is destroyed completely at high Pe of 10.0 by shear stress. The rodlike micelles become more and more parallel to the shear flow direction with increasing Pe. This leads to a nonuniform orientation angular distribution of micelle directions, in other words, a preferred alignment angle. This phenomenon can also be seen from orientation of rodlike micelles. Figure 4.(a) and (b) show the simulated distributions of orientation angle θ and ϕ of rodlike micelles with Pe as a parameter, respectively. The distribution is defined as probability of orientation angle during the statistical period.

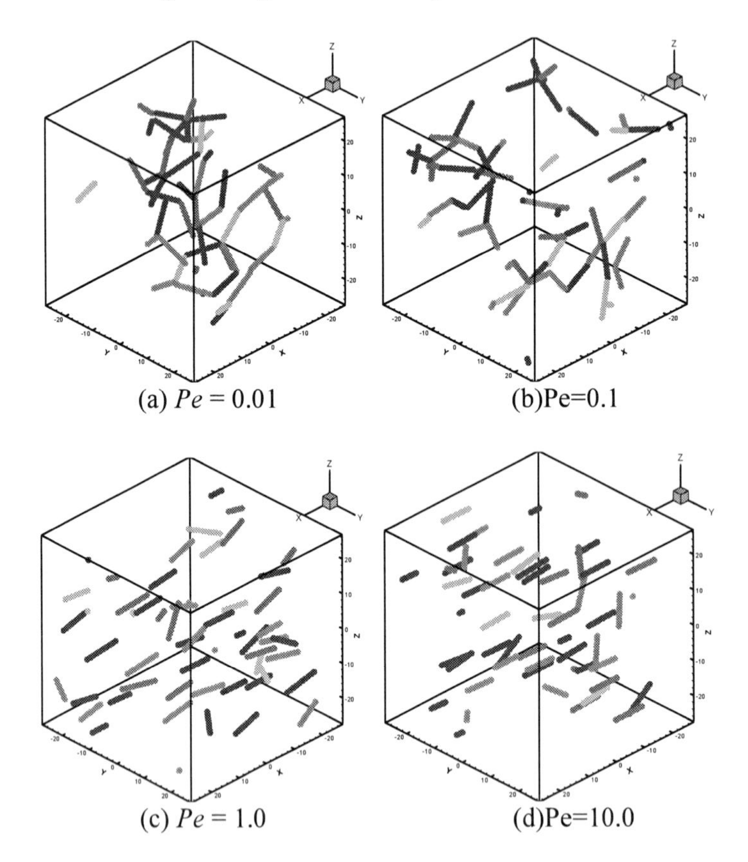

Figure 3. Micellar structures at different Peclet numbers.

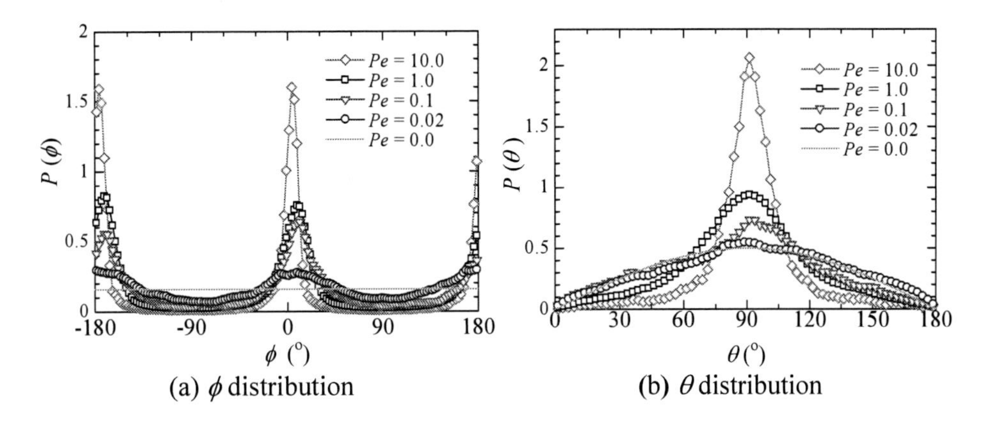

Figure 4. simulated distributions of orientation angle θ and ϕ of rodlike micelles.

The peaks of the angle distribution increase in amplitude as Pe is increased. The position of the maximum for θ distribution is fixed at 90°, whereas the peak position moves towards 0° for ϕ distribution with increasing Pe, indicating that the rods are becoming more and more aligned with the shear flow. At the largest Pe of 10, the distributions of θ and ϕ are sharply peaked, indicating that the combination of micelles is destroyed by the strong shear stress and the rods spend a large percent of time to align with the flow direction. Shear induced birefringence is a well-known important rheological phenomenon. The position of the maximum value of ϕ is closely related to the extinction angle χ in the birefringence measurements. We can see that this position is at 45° at the smallest Pe of 0.01 and moves towards 0° with increasing Pe. This is consistent with the measured result that the extinction angle χ is 45° at the equilibrium state and decreases with increasing shear rate [21].

Pair distribution function $g(r)$ describes the structures of rodlike micelles. It gives the probability of a pair of micelles at a center-center distance r apart. $g(r)$ is defined as[20]:

$$g(r) = \frac{V}{N^2}\left\langle \sum_{k}\sum_{l \neq k} \delta\left(\boldsymbol{r} - \boldsymbol{r}_{kl}\right)\right\rangle \tag{32}$$

Where V is the volume, N is the number of rod, $\boldsymbol{r}_{kl}$ is the center-to center distance vector between rod k and l. Figure 5. shows the pair distribution functions of the center-of-mass coordinates of the rods vary with Pe numbers. The $g(r)$ at Pe = 0.01 shows the most number of peaks, exhibiting complex structures. Therefore, micelles will connect at adjacent endcaps and further form network. With increasing Pe, the peaks decrease in amplitude and become broad. The curve of $g(r)$ becomes flat, indicating the transition of structure and destruction of network gradually. Finally, for Pe exceeding 5.0, network was completely destroyed, which validating the transition shown in Figure 3.

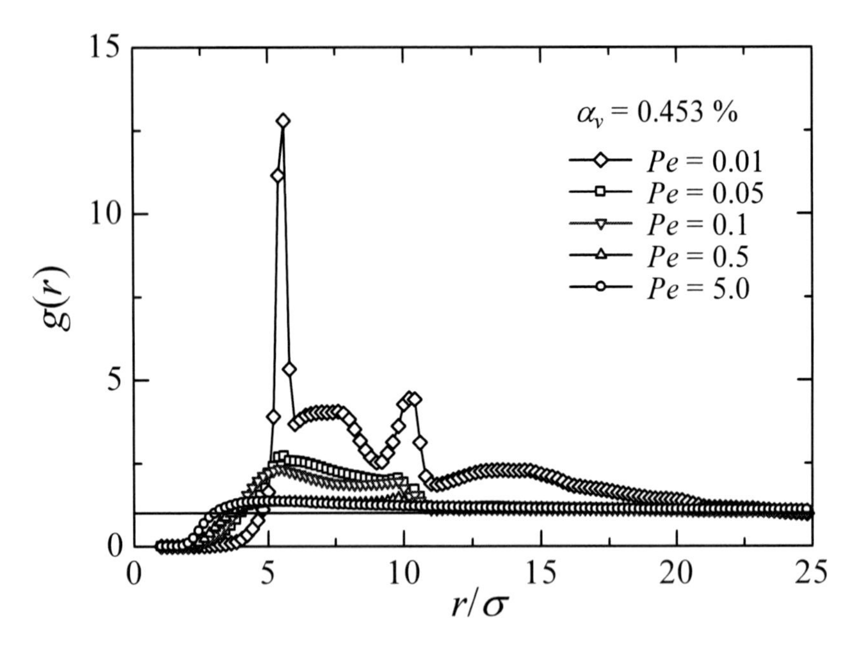

Figure 5. Pair distribution functions of the center-of-mass.

3.2. Dynamics

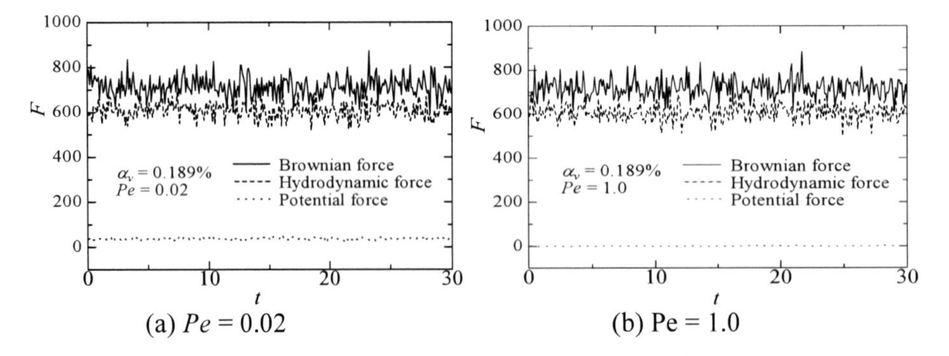

Figure 6. Transient change of root-mean-square of Brownian force, hydrodynamics and potential force.

The variation of shear rate results the change of strength of flow field, and consequently the dynamics of rods. Figure 6. shows the transient change of root-mean-square of Brownian force, hydrodynamics force and potential force at Pe = 0.01 and 1.0, respectively. The time and force are both dimensionless. It can be obtained that the root-mean-square of potential force is much smaller than that of Brownian force and hydrodynamics force. It does not mean, however, that the absolute value of potential force is smaller than that of Brownian and hydrodynamic force correspondingly. Potential force is function of distance between particles. If the distance is small enough, the potential force will be extreme large comparing to Brownian and hydrodynamics force. However, most of particle-particle distances exceed 5, as shown in Figure 5, resulting in small values of root-mean-square of potential force, which is averaged over all particle pairs. For given particles, Brownian and hydrodynamics force only depend on thermal condition and relative flow field, respectively. Therefore, the average root-mean-square of potential force may be much smaller than that of Brownian and hydrodynamic force. With increasing *Pe*, the root-mean-square of potential force decreases. This can be explained based on the fact that high shear rates prevent the rods from contacting each other and induce rods aligning with flow direction and weaken interactions. The thermal condition has no changes and Brownian force varies slightly. Hydrodynamics force also show no obvious changes, indicating that relative velocity between rods and shear flow has no obvious changes, although the shear rate increases. Brownian force is a little larger than hydrodynamics force in amplitude. Figure 7. shows the kinetic energy and potential energy of the system at different Pe. Here we defined the kinetic energy based on the velocities of rod and therefore the kinetic energy has no direct relation with thermal state. Generally speaking, the system keeps a constant energy at a given *Pe*. The negative value of the potential energy at lower *Pe* is due to the micellar network structure formation in which the attractive force between the rod ends plays an important role. With increasing *Pe*, the absolute value of potential energy decreases and the kinetic energy increases. For the Peclet number above 1.0, the potential energy becomes almost zero, indicating that the micellar network structure has almost been completely destroyed by the shear rate, which agrees with the micellar structures, rod orientation, radial pair distribution and rheological curves shown later. The energy for the movement of rods is mainly obtained from shear flow through friction force caused by relative velocity between rod and shear flow. Therefore, the kinetic energy is small at low shear rates and becomes larger when shear rate increases.

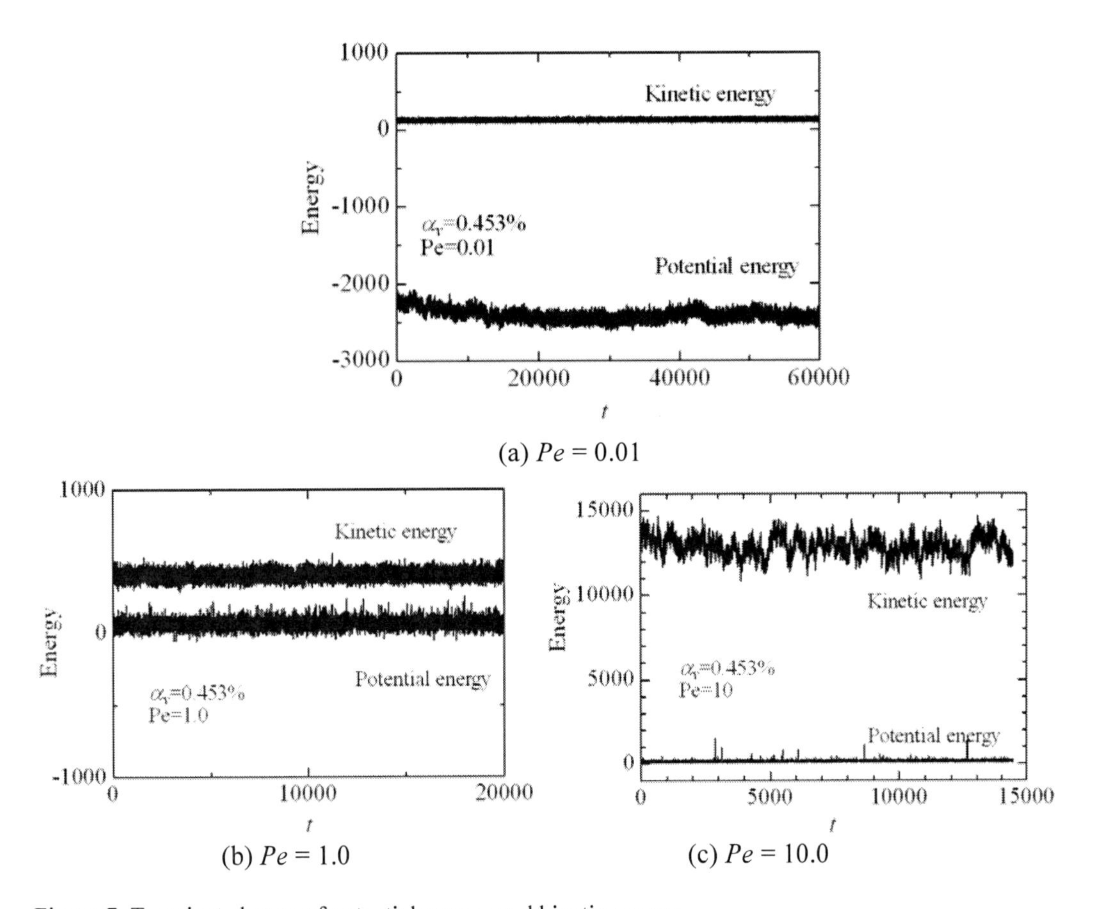

(a) $Pe = 0.01$

(b) $Pe = 1.0$

(c) $Pe = 10.0$

Figure 7. Transient change of potential energy and kinetic energy.

3.3. Rheology

Surfactant solutions are distinguished from Newtonian fluid in that they possess the complex rheological characteristics. Surfactant solutions usually exhibit high viscosity and dramatic normal force, which is caused by internal microstructure and dynamics.

3.3.1. Shear Flow

Figure 8. shows the calculated dimensionless shear viscosity $(\eta-\eta_s)/(\eta_{0n}-\eta_s)$ versus dimensionless shear rate $n(n^2-1)/72\,\mathrm{Pe}$ with surfactant volume fraction α_v as a parameter. η_{0n} is the zero shear viscosity in the absence of potential. It should be noted that it is difficult to probe the rheology at very small Peclet numbers due to the deteriorating signal response-to-noise ratio and the relatively slow reorientational time scale of the rod. The difficulty in probing rheology at low strain rates have also been found in the NEMD simulations of molecular fluids [22] and BD simulations of molecular colloid liquids [23,24]. It can be seen that $(\eta-\eta_s)/(\eta_{0n}-\eta_s)$ increases with increasing surfactant concentration and decreases with increasing shear rate, showing shear thinning characteristics. The shear thinning characteristics is related to the shear aligning of rods shown in Figures 3. and 4. The analytical [25] and numerical results for the case of α_v= 0.126% in the absence of potential are also shown. At shear rates less than 10 for the case of α_v= 0.126%, $(\eta-\eta_s)/(\eta_{0n}-\eta_s)$ shows a significant increase in the presence of WK potential interaction between rodlike micelles than those in the absence of potential interaction. When only soft-sphere potential (Eq.(5)) is considered, the effect of soft potential between beads of different rods is not as large as the L-J potential. In view of pure repulsive force of soft-sphere potential, it is concluded that the shear viscosity is mainly caused by *attractive force* ????? at dilute surfactant solutions. It can also be seen that there exists a critical shear rate above which the shear viscosity curves become parallel to the analytical curve for no potential case. This indicates that the micellar network structure caused by the WK potential interaction is completely destroyed and the contribution of WK potential to the rheology disappears. The drop from the viscosity curve to the parallel dotted line is the increase of viscosity due to the WK potential interaction.

Figure 9. shows the calculated dimensionless first normal stress coefficient ψ_1/ψ_{10n} versus dimensionless shear rate $n(n^2-1)/72\,\mathrm{Pe}$ with surfactant volume fraction α_v as a parameter. $\psi_{1,0n}$ is the first normal stress coefficient in the absence of potential. The first normal stress coefficient shows a shear thinning characteristic and increases with increasing surfactant concentration. The shear thinning is related to the shear aligning of rods shown in Figures 3 and 4. The analytical [25] and numerical results for the case of α_v= 0.126% in the absence of potential are also shown. The effect of soft-sphere potential is also much smaller than WK potential. At shear rates less than 10 for the case of α_v= 0.126%, ψ_1/ψ_{10n} with the WK potential interaction between rodlike micelles is much higher than that without potential interaction.

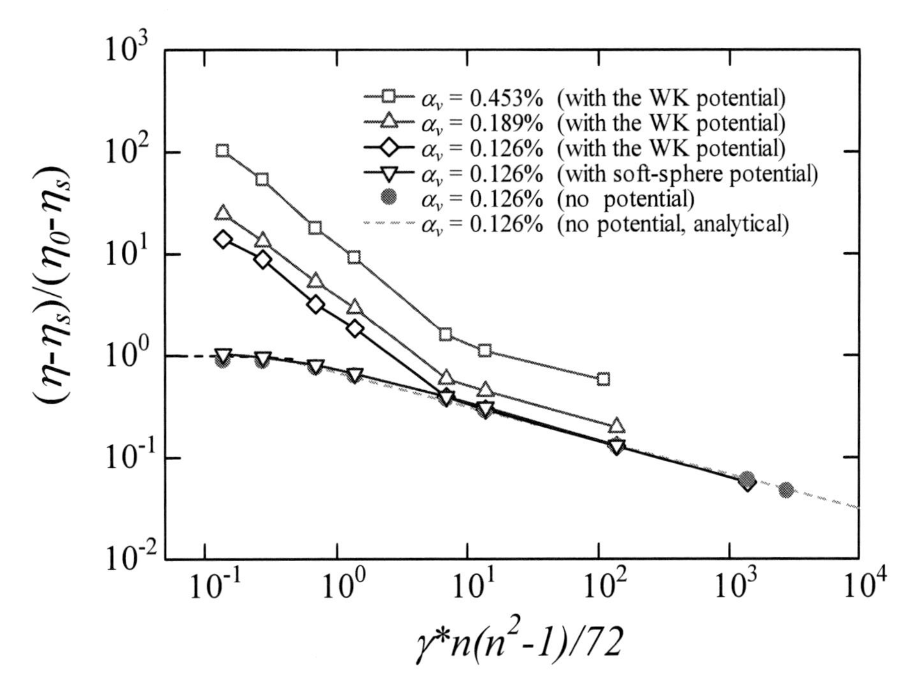

Figure 8. Dimensionless shear viscosity versus dimensionless shear rate.

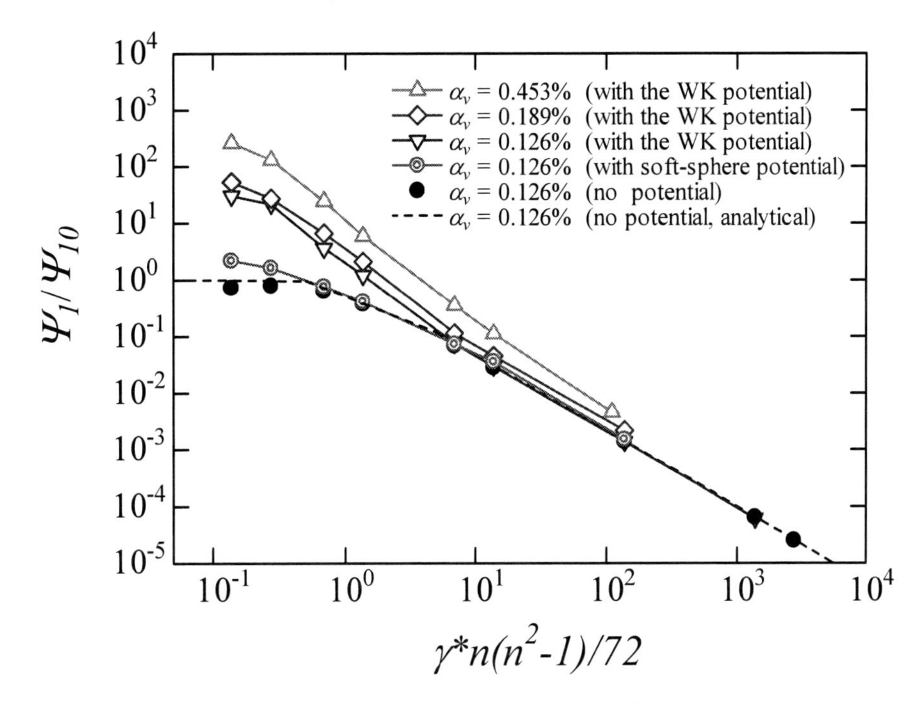

Figure 9. Dimensionless first normal stress coefficient versus dimensionless shear rate.

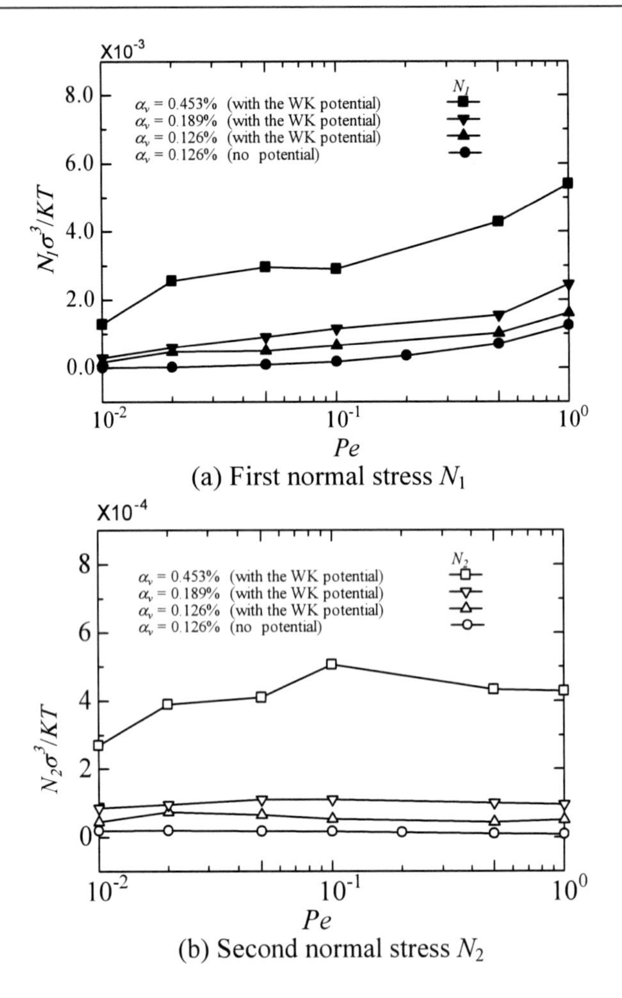

Figure 10. First and second normal stresses versus *Pe*.

Figure 10. shows the first and second normal stresses. We can see the absolute values of N1 and N2 are larger with the WK potential than that without the WK potential. N1 and N2 increase with increasing concentration. N1 is much larger than N2 in amplitude.

3.3.2. Uniaxial Elongational Flow

The properties of surfactant solutions with imposed uniaxial elongational flow are also considered. Figure 11. shows the calculated dimensionless elongational viscosity $(\eta_E - 3\eta_s)/3(\eta_{0n} - \eta_s)$ versus dimensionless elongation rate $n(n^2 - 1)/72\,\mathrm{Pe}$ with surfactant volume fraction α_v as a parameter. η_{0n} is the zero shear viscosity in the absence of potential.

The analytical [25] and numerical results for the case of 0.126% in the absence of potential are also shown. It can be clearly seen that there exists a transition from a low elongation viscosity plateau ($(\eta_E - 3\eta_s)/3(\eta_{0n} - \eta_s)$ = 1) at low elongation rates to a high one ($(\eta_E - 3\eta_s)/3(\eta_{0n} - \eta_s)$ = 2) at high elongation rates for no potential case. For the WK potential case, $(\eta_E - 3\eta_s)/3(\eta_{0n} - \eta_s)$ decreases with elongation rates, leveling out at high elongation rates. The strain thickening characteristics in the absence of potential is due to the increasing alignment of rods with elongation. The elongation viscosity may be evaluated using viscous dissipation argument.

The rate of viscous dissipation due to the presence of the rods increases as the alignment of rods increases, and takes maximum value $0.5N\xi(n-1)^2\sigma^2\dot{\varepsilon}^2/V$ when the rods completely align with flow direction. Here V is the volume of computational box. The strain thinning behavior in the presence of potential is related to the increasing destruction of micellar network structure. The rods in solution interact each other more during weak elongation than during strong elongation, causing the elongational viscosities to be higher at low elongation rates and to drop as the rods become increasingly aligned in the flow direction at higher elongation rates. This strain-thinning phenomenon has also been found by Cathey and Fuller [26] in their uniaxial and biaxial elongational viscosity measurements of semi-dilute solutions of rigid rod polymers and well explained by Doi and Edwards [27]. At elongation rates less than 0.1 for the case of 0.126%, $(\eta_E-3\eta_s)/3(\eta_{0n}-\eta_s)$ shows a significant increase in the presence of WK potential interaction between rodlike micelles than those in the absence of potential interaction. We can also see that there exists a critical elongation rate above which the elongational viscosity curves become parallel to the analytical curve for no potential case. This indicates that the micellar network structure due to the WK potential is completely destroyed and the contribution of WK potential to the rheology disappears.

The drop from the viscosity curve to the parallel dotted line is the increase of elongational viscosity due to the WK potential interaction. The elongational viscosities increase with increasing surfactant concentration. The increase of elongational viscosities by the formation of micellar network structure in the surfactant solution can suppress the vortex stretching, thus resulting in the reduction of turbulence energy production and friction drag in turbulent flow.

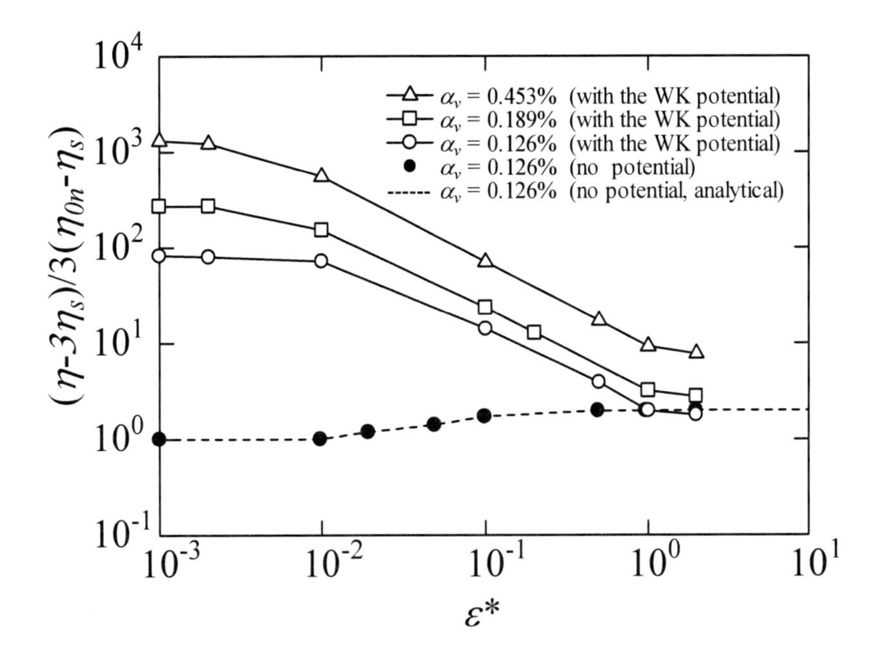

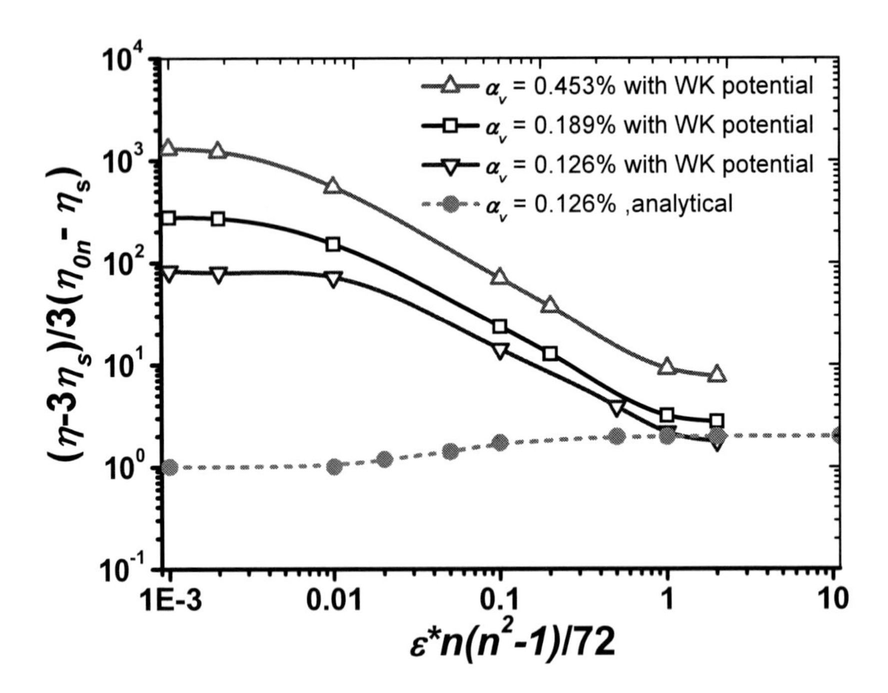

Figure 11. Dimensionless elongational viscosity versus dimensionless elongationg rate.

Conclusion

By using a new WK potential in Brownian dynamics simulation, the surfactant micellar network structure was obtained at low shear or elongation rates and was destroyed by shear or elongational flow at high shear or elongation rates. The alignment of model rod-like micelles increased with increasing strain rate. The shear and elongational viscosities and first normal stress coefficient showed shear thinning characteristics and they increased significantly by introducing the WK potential interaction between model rod-like micelles, and the attractive force plays a main role. The strain thinning characteristics was related to the shear or elongation aligning of the micelles with the flow direction. The shear and elongational viscosities and first normal stress coefficient increased with increasing surfactant concentration. The large viscoelasticity and elongational viscosity due to the formation of micellar network structure were considered to cause turbulent drag reduction.

Acknowledgment

We gratefully acknowledge the financial support from the NSFC Fund (No. 51076124), the open project of Institute of Rheological Mechanics & Material Engineering (Central South University of Forestry and Technology), the program for new century excellent talents in university (NCET-07- 0680), and the Fundamental Research Funds for the Central Universities.

REFERENCES

[1] Mysels, K. J., 1949, *Flow of Thickened Fluids*, U.S. Patent 2,492,173.

[2] White, A., 1967, *Nature*, 214, pp. 585-586.

[3] Zakin, J.L.; Zhang, Y., Ge, W. *Giant micelles: Properties and Applications*, CRC Press Taylor and Francis Group: New York, 2007, 473-492.

[4] Gyr, A.; Bewersdorff. H. W. *Drag Reduction of Turbulent Flows by Additives*, Kluwer Academic Publishers, The Netherlands, 1995.

[5] Lu, B.; Li, X.; Zakin, J. L.; Talmon, *Y. J. Non-Newton. Fluid Mech.* 1997, 71, pp. 59-72.

[6] Debye, P.; Anacker, E. W. *J. Phys. Colloid Chem.* 1951, 55, pp. 644-655.

[7] Porte, G.; Appell J.; Poggi, Y., *J. Phys. Chem.* 1980, 84, pp. 3105-3110.

[8] Young, C. Y.; Missel, P. J.; Mazer, N. A.; Benedec, G. B.; Carey, *M. C. J. Phys. Chem.* 1978, 82, pp. 1375-1378.

[9] Ikeda, S.; Hayashi, S.; Imae, *T. J. Phys. Chem.* 1981, 85, pp. 106-112.

[10] Linder, P.; Bewersdoff, H. W.; Heen, R.; Sittart, P.; Thiel, H.; Langowski, J.; Oberthur, *R. Progr. Colloid Polym. Sci.* 1990, 81, pp. 107-112.

[11] Olsson, U.; Soderman, O; Guering, *P. J. Physical Chem.* 1986, 90, pp. 5223-5232.

[12] Qi, Y.Y.; Littrell, K.; Thiyagarajan, P.; Talmon, Y.; Schmidt, J.; Lin, Z.Q.; Zakin, J.L. *J. Colloid Interface Sci.* 2009, 37, 218-226.

[13] Clausen, T. M.; Vinson, P. K.; Minter, J. R.; Davis, H. T.; Talmon, Y.; Miller, *W. G. J. Physical Chem.* 1992, 96, pp. 474-484.

[14] Lowen, *H. Phys. Rev.* E 1994, vol. 50, pp. 1232-1242.

[15] Branka, AC.; Heyes, *D.M. J. Chem. Phys.* 1998, vol. 109, pp. 312-317.

[16] Mori, N.; Kumagae, M.; Nakamura K. *Rheol. Acta* 1998, vol. 37, pp. 151-157.

[17] Mori, N.; Fujioka, H.; Semura, R.; Nakamura, K. *Rheol. Acta* 2002, vol. 42, pp. 102-109.

[18] Doi, M; Yanamoto, I.; Kano, *F. J. Phys. Soc, Jpn*, 1984, vol. 53, p. 3000.

[19] Liu, T.W. *J. Chem. Phys.* 1989, vol. 90, pp. 5826-5842.

[20] Allen, M.P.; Tildsley, D.J. *Computer Simulation of Liquids*, Oxford Science, Oxford, 1987.

[21] Lu, B. *Characterization of Drag-Reducing Surfactant Systems by Rheology and Flow Birefringence Measurements*, Ph. D Dissertation, The Ohio State University, 1997.

[22] Kubo, R.; Toda, M.; Hasitsume, N. *Statistical Physics*, Springer, 1985.

[23] Hounkonnou, M.N.; Pierieoni, C.; Ryckaert, *J. P. J. Chem. Phys.*1992, 97, pp. 9335-9344.

[24] Branka, A.C.; Heyes, *D.M. J. Chem. Phys.* 1998, 109, pp. 312-317.

[25] Bird R.B.; Curtiss, C. F.; Armstrong, R. C.; Hassager, O. *Dynamics of Polymeric Liquids: Vol. 2, Kinetic Theory*, John Wiley &Sons, 1987.

[26] Cathey C. A.; Fuller, *G. G. J. Non-Newton. Fluid Mech.* 1988, 30, pp. 303-316.

[27] Doi, M.; Edwards, S. F. *The Theory of Polymer Dynamics*, Oxford Science Publications, 1986.

In: Brownian Motion: Theory, Modelling and Applications ISBN: 978-1-61209-537-0
Editors: R.C. Earnshaw and E.M. Riley

Chapter 8

Development of the Fractional Brownian Motion and Its Applications on Coastal Dispersion Modelling

Bo Qu and Paul S. Addison
Department of Civil and Transportation Engineering, Napier University,
Merchiston Campus, Edinburgh, UK, and others

Abstract

The methods for generating a Brownian and simple random walks are introduced. An improved fractional Brownian motion(fBm) model has developed and compared with the Mandelbrot fBm model.

A new particle tracking technique is introduced where non-Fickian diffusion is generated by employing fBm. The modelling of pollutant dispersion in an idealised coastal bay is carried out and the differences between a traditional Brownian particle tracking model and the new fBm particle tracking model within the coastal bay are compared in detail.

The results show that the fBm particle tracking model gives more flexibility in controlling the spreading of the diffusing cloud and, moreover, it is closer to reality (in that non-Fickian diffusion may be modelled).

Simulation of real observed data from Numthumbrian coastal waters is then carried out and an improved accelerated fBm method is developed for application in large water bodies such as coastal waters and the ocean surface.

Our simulation results and the HR Wallingford simulation results are compared and tested with the observed data sets. The novel and practical fBm particle tracking model may become a useful engineering tool for the prediction of contaminant spread in environmental fluid flows.

Keywords: Brownian motion, fractional Brownian motion, Fickian, non-Fickian, Costal Dispersion, Diffusion

1. Background

1.1. Introduction

In recent years, pollutant dispersion in rivers, estuaries, coastal and ocean waters has become an important environmental issue for water engineers. The need to predict the transport of pollutants has resulted in a rapid rise in the use of numerical models. There has been much research and many attempts to produce models which explain and predict the dispersion of pollutant in fluids. Computer modelling of the dispersion of pollutants in fluids has traditionally utilised standard Advection-Diffusion equations. However, these traditional methods do not usually give successful predictions of the distributions of pollutants actually measured in fluids due to the complicated nature of environmental flows. Particle tracking models have received more and more attention in recent years whereby the random paths of particles can be used as a simple way to solve diffusion equations using high-speed computers. To date, particle tracking models presented in the literature employ random Brownian motion to simulate turbulent diffusion. These traditional models generally assume that particle tracks are neutrally persistent where the particle executes a simple random walk without memory. However, the main drawback of these models is that they consist of statistically independent steps and cannot account for the correlation which exists in the flow field (Feder, 1988). It is known that particle movements in turbulent fluids are persistent, where the memory of the particle plays an important role in indicating the future direction of the particle. To simulate persistent motion, one may resort to the emerging field of fractal statistics.

1.2. Fractals and Fractional Brownian Motion

Fractal geometry provides both a description and a mathematical model for many of the seemingly complex forms found in nature. Shapes such as coastlines, mountains and clouds are not easily described by traditional Euclidean geometry. However, they often possess a remarkable simplifying invariance under changes of magnification. This self-similarity is the essential quality of fractals in nature. Although, fractal geometry has only a twenty year history, it has proved an important tool with which to both characterise and model natural objects and processes. One of the most useful mathematical models for random fractals found in nature has been fractional Brownian motion (fBm). It is an extension of the central concept of Brownian motion which can simulate both persistent and anti-persistent random motions. This paper details a newly developed model for the numerical synthesis of fractional Brownian motion and incorporates this within a fBm particle tracking model which allows for great flexibility in the modelling of pollutant dispersion in fluids.

1.3. Statistical Theory of Diffusion: Brownian Motion

Discovered by Robert Brown in 1826, Brownian motion is the sustained irregular motion performed by small particles in a fluid and it is observable with the aid of a microscope.

Later, in 1905, Einstein showed convincingly that this 'Brownian motion' is maintained by collisions with molecules of the surrounding fluid. The random wanderings of many Brownian particles result in their mean dispersion.

Figure 1. shows Brownian motion generation for one particle. Each particle randomly walks about (see Figure 1, left) taking steps from a Gaussian distribution. Brownian motion in one dimension may be generated as follows:

$$B(t_i) = \sum_{j=1}^{i} R(t_j) \tag{1.1}$$

Where $B(t_i)$ is the observed Brownian location at discrete time t_i and $R(t)$ is a random number taken from a Gaussian distribution. Each particle path is then constructed by simply adding together a series of random numbers. Figure 1 (right) is a Brownian motion trajectory in two dimensions.

If a number of particles are allowed to take random steps through time, then the diffusion of a contaminant cloud can be simulated. This is shown schematically in Figure 2.

Figure 2. (a) shows ten, one-dimensional Brownian motion traces all beginning at the origin. The overall diffusion of the particles can be observed from the figure. In practice, many thousands of particles are used to simulate diffusive clouds. Using a large number of particles results in diffusive processes which have a standard deviation which grows with time as (see Figure 2 (b)):

$$\sigma_c = \sqrt{2Dt} \tag{1.2}$$

i.e. in a Fickian manner. This behaviour of a large number of Brownian particles forms the basis of traditional particle tracking methods.

The statistical theory of Brownian motion provides a useful background for an attack on the more difficult problem of turbulent diffusion, although the reader will see below that in reality turbulent diffusion is non-Fickian.

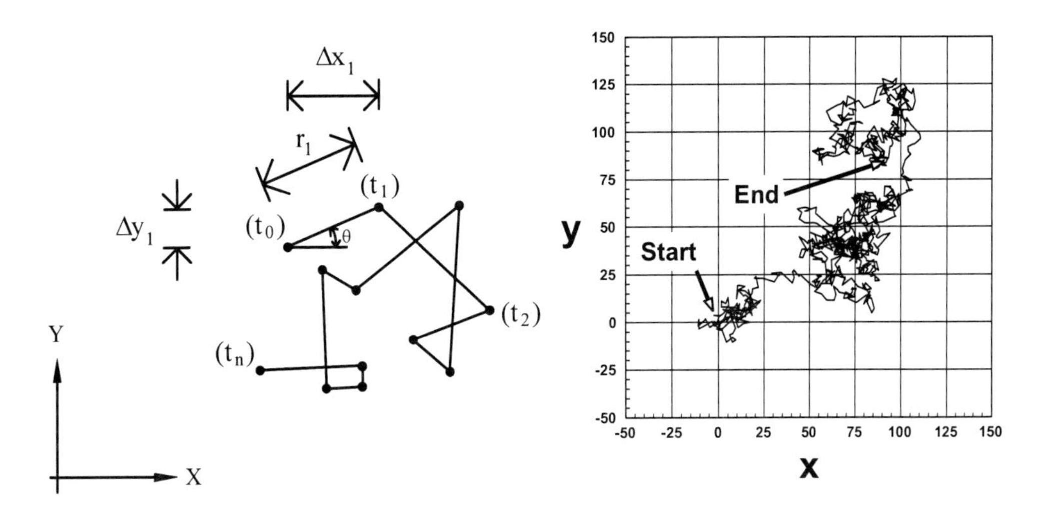

Figure 1. Brownian Motion Generation for One Particle.

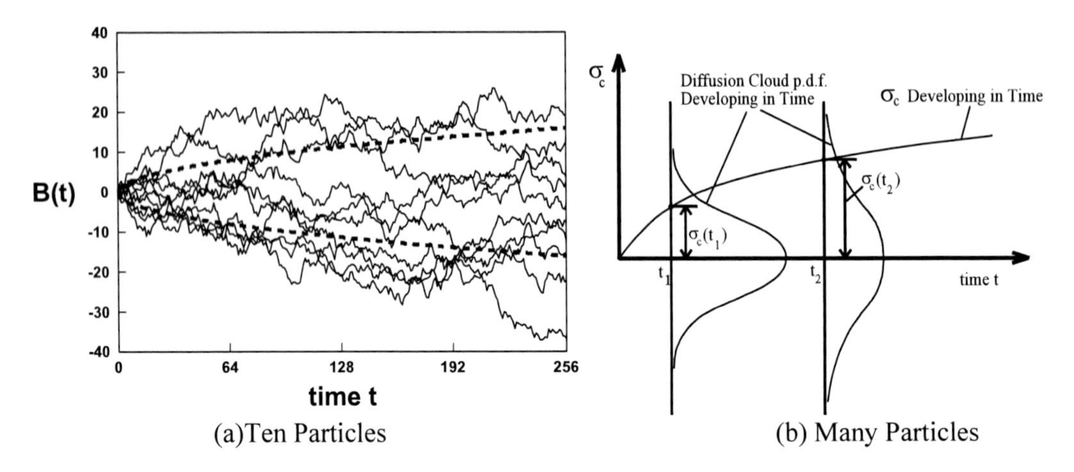

(a)Ten Particles (b) Many Particles

Figure 2. The Diffusion Through Time of a Cloud of Brownian Particles.

2. Brownian Motion

2.1. Brownian Motion Definition

Small particles of solid matter suspended in a liquid can be seen under a microscope to move about in an irregular and erratic way. This movement, known as Brownian motion, was observed by the Scottish botanist Robert Brown around 1827. The modelling of this movement is the one of the great topics of statistical mechanics. Brownian motion is widely used within particle tracking models to predict pollutant dispersion in fluids, e.g. river flows, atmospheric flows and oceanic flows. What kind of motion can be called Brownian motion? Hastings and Sugihara (1993, p25) give the following definition:

Brownian Motion Definition: A continuous process $\{B(t)\}$ is called a Brownian process, if, for any step Δt, the increments $\Delta B(t) = B(t + \Delta t) - B(t)$ are

(1) Gaussian,
(2) of mean 0, and
(3) of a variance proportional to Δt.
Here (2) and (3) are equivalent to
(4) successive increments $\Delta B(t)$ and $\Delta B(t + \Delta t)$ are uncorrelated.

Brownian motion $B(t)$ can be expressed as a continuous-time random function, which is the integral of a Gaussian white noise $WT(s)$,

$$B(t) = \int_{-\infty}^{t} WT(s)ds. \tag{2.1}$$

The random variables $WT(s)$ are uncorrelated and have the same Gaussian, or normal distribution with zero mean and unit standard deviation, denoted N(0,1). Because successive increments of the Brownian trace, $B(t)$-$B(t- \Delta t)$, have a Gaussian distribution, it can be seen that a discretised approximation to a Brownian motion trace, *B*(*i*), may be produced at discrete times, $t_i = i \cdot \Delta t$, (where *i* is an integer), by summing up a series of independent random steps taken from a Gaussian distribution,

$$WT(i) = B(i) - B(i-1),$$

we have

$$B(i) = \sum_{j=0}^{i} WT(j). \tag{2.2}$$

The finite resolution of the Brownian motion trace generated is governed by the choice of the time increment Δt. The reader should note that in equation (2.2), *B*(*i*) is used to represent the synthesised Brownian motion at time t_i, whereas in equation (2.1), *B*(*t*) is used to represent the continuous Brownian motion at time *t*.

2.2. Brownian Motion Generation

It is known that Gaussian white noises, *WT*(*i*), are uncorrelated and have a Gaussian (i.e. normal) distribution N(0,1). Brownian motion is the integral of *WT*(*i*). To generate Brownain motion, we need first to produce Gaussian white noise *WT*(*i*).

As Gaussian white noises are independent of each other, the Gaussian distribution of the pulses *WT*(*i*) is given by the probability density function

$$\Phi(x) = \frac{1}{\sqrt{2\pi}\sigma_{step}} \exp(-\frac{(x-\mu)^2}{2\sigma_{step}^2}) \tag{2.3}$$

Therefore, the standard deviation of each step, σ_{step}, in Brownian motion needs to be calculated first. The FTN77 random variables $x = R(i)$ are all within [0,1] with mean $\mu = 0.5$. Hence, for each step, the standard deviation σ_{step} can be calculated by

$$\sigma_{step}^2 = \int_0^1 (x-\mu)^2 dx = \frac{1}{12}$$

Hence,

$$\sigma_{step} = \frac{1}{2\sqrt{3}} \tag{2.4}$$

We use two methods to generate Gaussian random numbers. One is the Central Limit Theorem, whilst the other is the Box-Muller method.

2.3. Central Limit Theorem Method

According to the Central Limit Theorem, if Z_n is the standardised sum of any n identically distributed random variables $R(i)$, the probability distribution of Z_n tends to the normal distribution as $n \rightarrow \infty$. For the random number generator within the FTN77 software, we have:

$$E(R(i)) = 1/2, \qquad E(\sum_{i=1}^{n} R(i)) = n/2.$$

Where E is the expectation operator. Hence, to produce Gaussian random variables Z_n with normal distribution N(0,1), from the constant p.d.f. $R(i)$, we require:

$$\begin{aligned} Z_n &= ((\sum_{i=1}^{n} R(i)) - E(\sum_{i=1}^{n} R(i)))/\sigma_{nstep} \\ &= ((\sum_{i=1}^{n} R(i)) - n/2)/(\sigma_{step} \times n^{1/2}) \\ &= 2\sqrt{3}\,(\sum_{i=1}^{n} R(i)) - n/2)/n^{1/2} \\ &= 3.4641 \times (\sum_{i=1}^{n} R(i)) - n/2)/n^{1/2}. \end{aligned} \tag{2.5}$$

Hence, for each i, Gaussian white noise is

$$WT(i) = \lim_{n \to \infty} Z_n$$

Brownian motion may then be generated by summing up a series of discrete white noise pulses. The discrete Gaussian pulses in Figure 3 were generated using the method just described using a finite 'n' and the Brownian motion shown in Figure 3 was generated by summing up these Gaussian pulses. This method for generating Gaussian random numbers requires the modeller to choose the number of random numbers n, which may cause significant errors if the number n is not large enough. The Box-Muller method can avoid this problem and leads to more accurate results.

2.4. The Box-Muller Method

In this method, Gaussian random numbers are obtained exactly by means of a one-to-one transformation of two random variables with mean of zero and a unit standard deviation. For example, if $R(1)$and $R(2)$ are two independent random variables, then

$$WT(1) = (-2\log_e R(1))^{1/2} \cos(2\pi R(2)),$$

and $$WT(2) = (-2\log_e R(1))^{1/2}\sin(2\pi R(2)) \qquad (2.6)$$

have independent Gaussian distributions with each zero mean and unit standard deviation.

Press et al (1986, 1992, p279-280) illustrates the use of the Box-Muller method for generating Gaussian random numbers in FORTRAN programs.

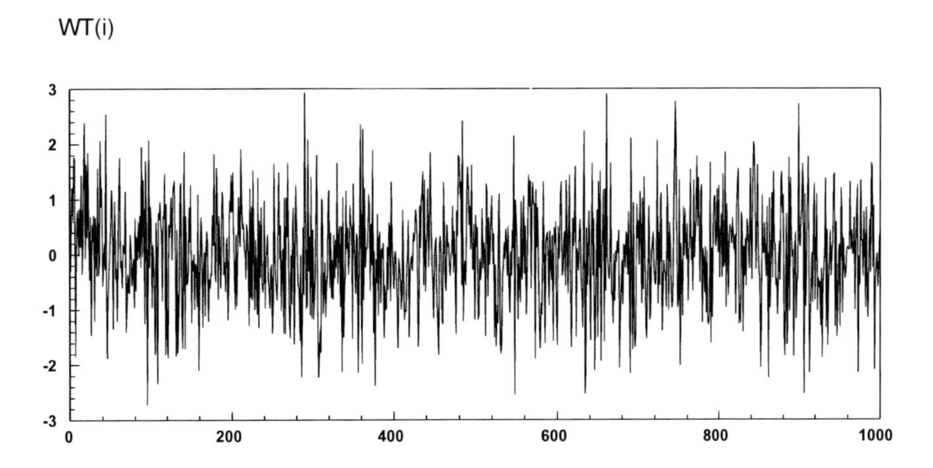

Figure 3. Gaussian Pulses with 1000 Steps and D = 1, Generated Using the Central Limit Theorem Method.

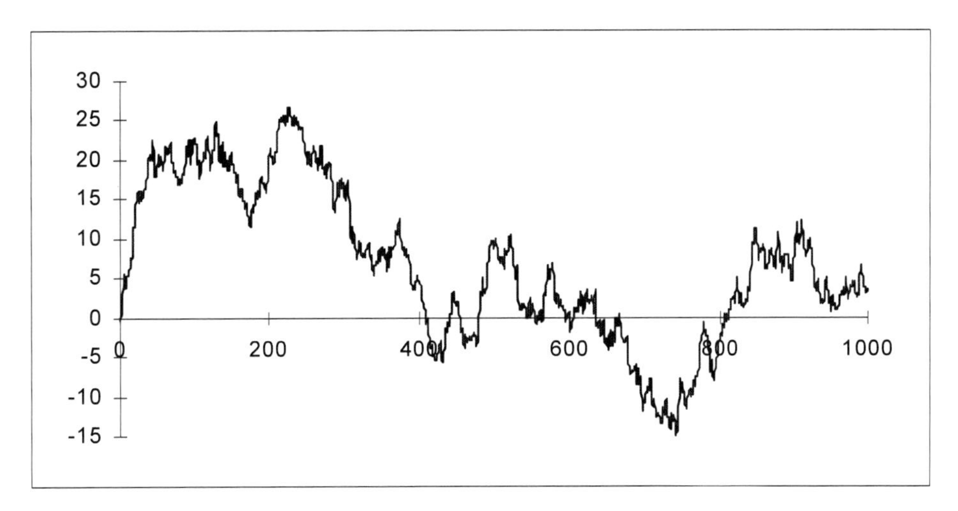

Figure 4. A Sample of Brownian Motion with 1000 Steps. Generated from the Integration of the Gaussian Pulses in Figure 3.

2.5. The Properties of a One-Dimensional Brownian Motion Time Trace

Apart from Brownian motion *B*(*t*) has the property of self-affinity, it also has following properties:

(1) It follows from (2) that for a large number of Brownian particles released at time $t = 0$, diffusing away from the origin through time, *the standard deviation of the diffusion cloud,* $\sigma_c(t)$*, is proportional to the square root of time*.

$$\sigma_c(t) \propto t^{\frac{1}{2}} \tag{2.7}$$

The scaling relationship is

$$\sigma_c(t) = \sqrt{2Dt} \tag{2.8}$$

where D is the diffusion coefficient.

Equation 2.7 shows that the variance of the diffusing cloud is proportional to time. Hence D can be obtained from a plot of variance against time.

$$D = \frac{Var(t)}{2t} \tag{2.9}$$

where $Var(t) = \sigma_c^2(t)$ is the variance of the diffusion cloud at time t.

Figure 5. shows the variances of clouds of diffusing particles undergoing Brownian motion against time. The gradient of each is actually twice the diffusion coefficient D. Here, for simplicity of illustration, the diffusion coefficient is set to $D = 0.5$, which gives from (2.9),

$$\text{Var}(t) = \sigma_c^2(t) = 2Dt = t\,.$$

The number of steps taken by each particle is denoted by t (i.e. unit time step) and number of particles by P. The variances at each step are numerically calculated for each diffusing cloud and the results are shown in Figure 5.

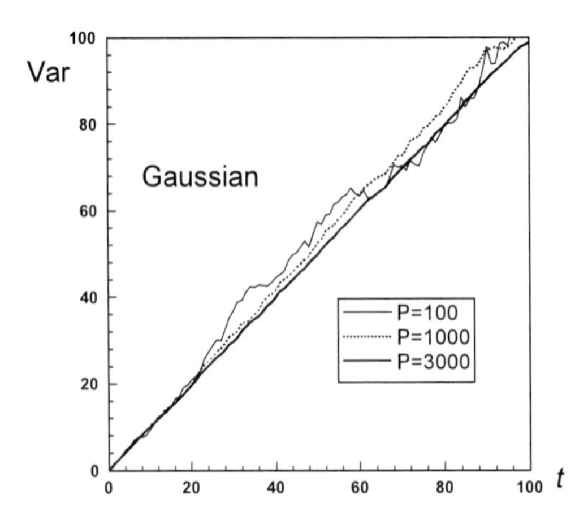

Figure 5. Variances versus Time Steps for Brownian motion ($D = 0.5$. P is the Number of Particles Used in the Simulation).

Figure 5. confirms that the author's program is reasonably accurate for large numbers of particles. This is the basic method used to check if the generated Brownian motion is correct. It was observed that with increasing numbers of particles, the results become more accurate, as expected.

3. Fractional Brownian Motion

3.1. Fractional Brownian Motion: A Generalisation of Brownian Motion

It is known that for Brownian motion, the standard deviation of a diffusing cloud of particles $\sigma_c(t)$ is proportional to the square root of time $t^{\frac{1}{2}}$, e.g.

$$\sigma_c(t) \propto t^{\frac{1}{2}} \tag{3.1}$$

Mandelbrot and Van Ness (1968) introduced the concept of fractional Brownian motion, $B_H(t)$, as a generalisation of regular Brownian motion by adding an additional parameter, the 'Hurst exponent' H ($0 < H < 1$), replacing (3.1) by

$$\sigma_c(t) \propto t^H \tag{3.2}$$

Actually, the Hurst exponent H was first introduced by Edwin Hurst (1950) when he tried to find a solution of the problem of determining the reservoir storage required on a given stream. H is a parameter which can control the 'smoothness' of the trace and is related to the fractal dimension of the trace D_f as:

$$H = 2 - D_f \tag{3.3}$$

The cases where $H \neq \frac{1}{2}$ are proper fractional Brownian motions; where $H = \frac{1}{2}$ is the special case of independent increments valid for Brownian motion.

It has been observed that, in many cases, the diffusion generated by Brownian motion based models does not reproduce the physical reality. However, fractional Brownain motion (fBm) can produce a wider range of diffusive phenomena which meets the requirement of a variety of modelling purposes.

There are many applications about pollutant dispersion modelling in fluids using either the Advection-Diffusion equation model or the traditional particle tracking model (where Brownian motion is used for diffusion modelling). The author has not found any article concerning the use of fBm to predict pollutant dispersion in fluids.

A novel fBm particle tracking model will be developed by the author which can be used in a wide range of pollutant dispersion applications, such as river, coastal regions and open ocean surface.

3.2. The Definition of Fractional Brownian Motion

The generation of fBm is not as simple as generating Brownian motion, because an fBm trace does not take statistically independent steps (as Brownian motion does), but rather each point on a fBm trace depends upon the whole of the history of the fBm previous to that point. In other words, an fBm has a long-term memory associated with it. Mandelbrot and Van Ness (1968) defined the random function $B_H(t)$ with zero mean roughly as a moving average of $dB_H(t)$, in which past increments of $B_H(t)$ are weighted by the kernel $(t\text{-}s)^{H-\frac{1}{2}}$, as

$$B_H(t) = \frac{1}{\Gamma(H+\frac{1}{2})} \int_{-\infty}^{t} (t-s)^{H-\frac{1}{2}} dB(s) \tag{3.4}$$

Here $\Gamma(x)$ is the gamma function, and H is the Hurst exponent of the trace. This definition states that the value of the random function at time t depends on all previous increments $dB(s)$ at time $s < t$ of a Gaussian random process $B(t)$ with average zero and unit variance.

Mandelbrot and Van Ness (1968) replaced the above definition (equation 3.4) with another which forces the function through the origin. Given the value $B_H(t=0)$, hence

$$B_H(t) - B_H(0) = \frac{1}{\Gamma(H+\frac{1}{2})} \left(\int_{-\infty}^{t} (t-s)^{H-\frac{1}{2}} dB(s) - \int_{-\infty}^{0} (0-s)^{H-\frac{1}{2}} dB(s) \right)$$

$$= \frac{1}{\Gamma(H+\frac{1}{2})} \left[\int_{-\infty}^{0} ((t-s)^{H-\frac{1}{2}} - (-s)^{H-\frac{1}{2}}) dB(s) + \int_{0}^{t} (t-s)^{H-\frac{1}{2}} dB(s) \right] \tag{3.5}$$

Defining

$$K(t-s) = \begin{cases} (t-s)^{H-\frac{1}{2}}, & 0 \le s \le t \\ (t-s)^{H-\frac{1}{2}} - (-s)^{H-\frac{1}{2}}, & s < 0 \end{cases} \tag{3.6}$$

then equation (3.5) can be written as:

$$B_H(t) - B_H(0) = \frac{1}{\Gamma(H+\frac{1}{2})} \int_{-\infty}^{t} K(t-s) dB(s) \tag{3.7}$$

This kernel vanishes quickly enough as $s \to -\infty$.

Letting H be such that $0 < H < 1$, the following random function $B_H(t)$ is called fractional Brownian motion with parameter H. For $t > 0$, $B_H(t)$ is defined by

$$B_H(t) - B_H(0)$$
$$= \frac{1}{\Gamma(H+\frac{1}{2})}\left\{\int_{-\infty}^{0}[(t-s)^{H-\frac{1}{2}} - (-s)^{H-\frac{1}{2}}]dB(s) + \int_{0}^{t}(t-s)^{H-\frac{1}{2}}dB(s)\right\} \quad (3.8)$$

3.3. Properties of Fractional Brownian Motion

Fractional Brownian motions have the following properties:

(1) They are Gaussian processes and their increments $B_H(t) - B_H(t-\Delta t)$ constitute a stationary random process.
(2) They are self affine processes, in the sense that, if time is changed in the ratio b, and the function $B_H(t) - B_H(0)$ is changed in the ratio b^H.

It is easy to prove this from (2.19). By changing the time scale a factor b to obtain

$$B_H(bt) - B_H(0) = \frac{1}{\Gamma(H+\frac{1}{2})}\int_{-\infty}^{bt} K(bt-s)dB(s) \quad (3.9)$$

Here, a new integration variable $s = b\hat{s}$ is introduced (see Feder, 1988, p173) and the following result for an independent Gaussian process is used

$$dB(s = b\hat{s}) = b^{\frac{1}{2}}dB(\hat{s}).$$

Using the relation $K(bt-b\hat{s}) = b^{H-\frac{1}{2}}K(t-\hat{s})$, (3.21) becomes

$$B_H(bt) - B_H(0) = \frac{1}{\Gamma(H+\frac{1}{2})}\int_{-\infty}^{bt} K(bt-s)dB(s)$$

$$= \frac{1}{\Gamma(H+\frac{1}{2})}\int_{-\infty}^{t} b^{H-\frac{1}{2}}K(t-\hat{s})\cdot b^{\frac{1}{2}}dB(\hat{s})$$

$$= b^H\left(\frac{1}{\Gamma(H+\frac{1}{2})}\int_{-\infty}^{t} K(t-\hat{s})dB(\hat{s})\right)$$

$$= b^H \{B_H(t) - B_H(0)\}$$

In particular, $t = 1$ and $\Delta t = bt$ may be chosen and so that the increment of the fractional Brownian 'particle' position is given by

$$B_H(\Delta t) - B_H(0) = |\Delta t|^H \{B_H(1) - B_H(0)\} \propto |\Delta t|^H .$$

(3) As H changes, the correlation between past and future will change accordingly. When H > 0.5, fractional noises are persistent, such persistence increasing with H; When H < 0.5, fractional noises are anti-persistent, and will decrease with H; When H = 0.5, Brownian motion is obtained where the fractional noises are independent.

Suppose in one dimension, an incremental correlation function $R_u(t)$ of the future increments $B_H(t)$ with the past increments $B_H(-t)$ is defined as

$$R_u(t) = \frac{\overline{(-B_H(-t)B_H(t))}}{\overline{B_H(t)^2}}$$

(3.10)

where the over-bar denotes the ensemble mean.

Then one finds that

$$R_u(t) = 2(2^{2H-1} - 1) \tag{3.11}$$

which only depends on H and is independent of t.

From (3.11), if $H > 1/2$, $R_u(t) > 0$, the fBm is persistent. That means that if the distance from the time axis has been increasing for a period of time, it is expected to continue to increase in the future. Conversely, if the distance from the t-axis is observed to decrease for a period of time, it is expected to continue to decrease in the future.

If $H < 1/2$, $R_u(t) < 0$, the fBm is anti-persistent. Anti-persistent tends to show a decrease in values following previous increases, and shows increases following previous decreases. The record of an anti-persistent process appears very 'noisy'.

If $H = 1/2$, $R_u(t) = 0$. The increments of random walk are uncorrelated. This is the special case of fBm which is regular Brownian motion.

3.4. Methods for the Generation of Fractional Brownian Motion

There are many methods which can produce discrete approximations to fractional Brownian motion given by equation (3.8) (Voss (1985); Peitgen et al (1992)). However, many of the popular methods (i.e. midpoint displacement, successive random additions, spectral methods, etc.) are not practical for incorporation within a particle tracking technique as they require either a predetermined sequence length for the fBm or a knowledge of both the start and end point of the fBm before the intermediate points can be calculated (see also Wheatcraft and Tyler (1988)).

In this paper, two methods were developed by the author: Firstly the FBM method, and then an improved FBMINC method. Both methods allow each successive step to be added to the end of the fBm trace, in a similar manner to the generation of regular Brownian motion discussed in section 1.

3.5. FBM Model

The discrete version of $B_H(t)$ given in equation (3.5) has to be modified with the proper kernel in order to make the sum convergent.

However, any calculation of B_H must use a finite number of terms and the sums can only cover a range M in the integer time t. Herein, M is called the memory. The definition of fBm (3.20) is discretised to give the following:

$$B_H(t_i) - B_H(0)$$
$$= \frac{1}{\Gamma(H+\frac{1}{2})}\left[\sum_{j=i-M}^{0}[(i-j)^{H-\frac{1}{2}} - (-j)^{H-\frac{1}{2}}]R(t_j) + \sum_{j=1}^{i}(i-j)^{H-\frac{1}{2}}R(t_j)\right]$$
$$= \frac{1}{\Gamma(H+\frac{1}{2})}\left[\sum_{j=i-M}^{-1}[(i-j)^{H-\frac{1}{2}} - (-j)^{H-\frac{1}{2}}]R(t_j) + \sum_{j=1}^{i-1}(i-j)^{H-\frac{1}{2}}R(t_j)\right] \tag{3.12}$$

Equation (3.12) is the FBM model developed by the author (see Qu and Addison, 2009). Here the memory M can be any positive integer, $0 \le i \le$ NSTEP. It is easy to see that (3.12) is only valid when $M \ge i$. The following formula may be applied to the case of $M < i$:

$$B_H(t_i) - B_H(0) = \frac{1}{\Gamma(H+\frac{1}{2})}\left[\sum_{j=0}^{i-1}(i-j)^{H-\frac{1}{2}}R(t_j)\right] \tag{3.13}$$

In general, the work of the thesis mainly considers the case where $M \ge i$. To generate NSTEP steps of an fBm using (3.12) (here NSTEP is the number of steps in the computational model), NSTEP+M random steps are needed from a Gaussian distribution. The larger the memory, M, used, the better the approximation to the real fBm. The reason why M should be at least as large as NSTEP will be discussed later.

One thing to notice is that the above definition of an fBm trace passes through the origin, $B_H(t_0) = 0$. Also, when $H = 0.5$, the fBm trace reduces to regular Brownian motion. Figure 6 contains plots of synthesised fBm time traces produced using the FBM model with $H = 0.2$, 0.5, 0.8, where the same random number sequences are used.

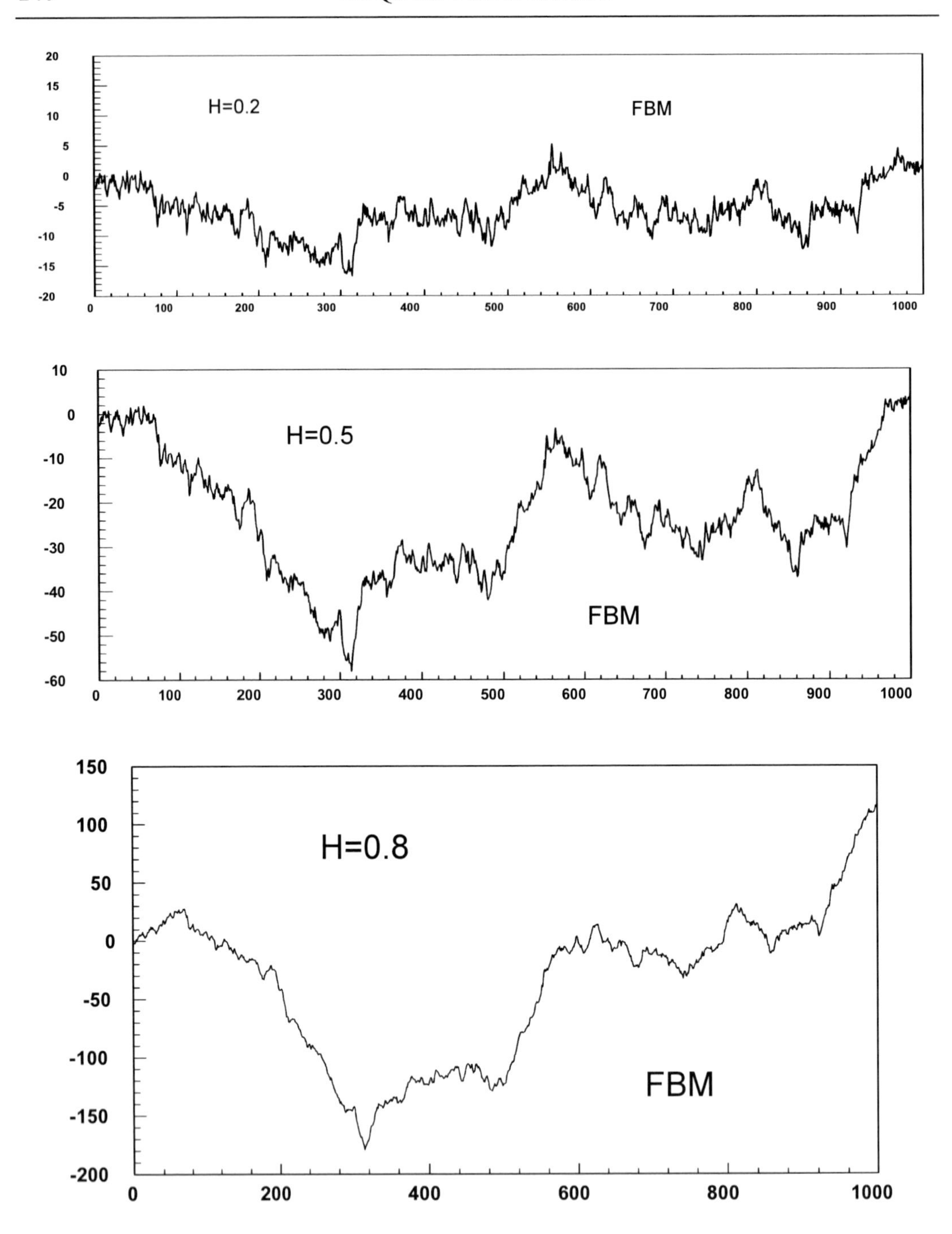

Figure 6. FBM Model (Gaussian): fBm in One Dimension with 1000 NSTEP, M = 5000 (Note: Different Scales Used for Vertical Axes).

3.6. FBMINC Model

The FBM model is a discrete version of fBm which has simplified the generation of fBm, however, it does not give a good approximation when $M <$ NSTEP. It is also found that the standard deviation of each step jump $\sigma(B_H(t_i) - B_H(t_{i-1}))$ always increases with the time step i. An improved FBMINC model was developed which overcame these two faults and gave a better approximation of fractional Brownian motion. This model is called the FBMINC model.

The sequence of increments of $B_H(t)$, namely the sequence of values of $\Delta B_H(t) = B_H(t) - B_H(t-1)$, with t an integer, is called 'discrete-time fractional noises'. It can be deduced from a Brownian motion $B(s)$ by the formula

$$\Delta B_H(t) = B_H(t) - B_H(t-1) = \frac{1}{\Gamma(H+\frac{1}{2})} \int_{-\infty}^{t} K_H(t-s)dB(s). \tag{3.14}$$

with the 'kernel function'

$$K_H(u) = \begin{cases} u^{H-\frac{1}{2}}, & 0 < u < 1; \\ u^{H-\frac{1}{2}} - (u-1)^{H-\frac{1}{2}}, & u > 1. \end{cases} \tag{3.15.a}$$

(3.15a) can be deduced from (3.6). The process as follow: Because (3.6) is the kernel for $B_H(t)$ - $B_H(0)$ (see Figure 7).

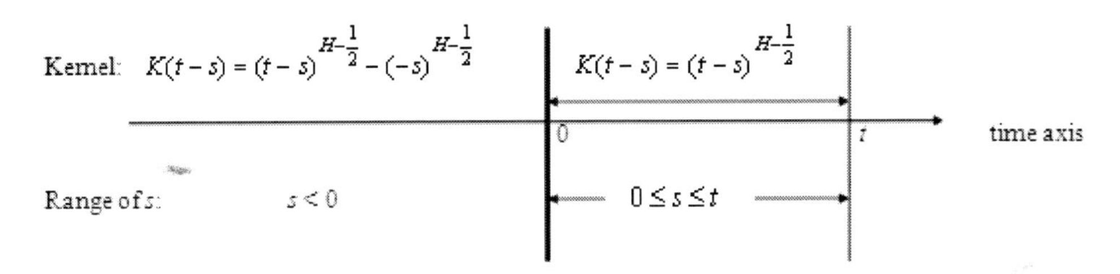

Figure 7. Kernel for $B_H(t) - B_H(0)$ (Refer Back to Equation (3.6)).

A kernel for $B_H(t) - B_H(t-1)$ can be obtained by modifying the kernel of $B_H(t) - B_H(0)$. Figure 8 visualises the new kernel K_H for $B_H(t) - B_H(t-1)$.

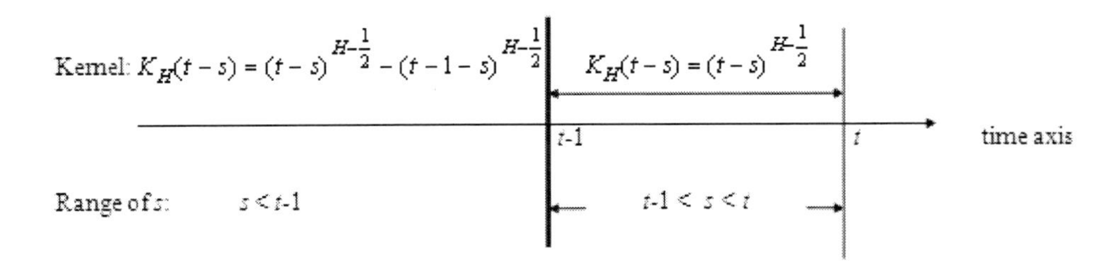

Figure 8. Kernel for $B_H(t) - B_H(t-1)$ (Compare to Figure 7).

Hence, the Kernel, K_H, can be written as

$$K_H(t-s)=\begin{cases}(t-s)^{H-1/2} & t-1\le s\le t\\ (t-s)^{H-1/2}-(t-s-1)^{H-1/2} & s<t-1\end{cases} \tag{3.15.b}$$

Let $u = t - s$. When $s \in (t\text{-}1, t)$, $u \in (0,1)$;
When $s < t\text{-}1$, $u > 1$.

Hence, (3.15a) is obtained. From (3.14) and (3.15a), replace u with $i - j$, the discrete version of $\Delta B_H(i)$ is as follow:

$$\Delta B_H(i)=\frac{1}{\Gamma(H+\frac{1}{2})}\sum_{j=i-M}^{i}K_H(i-j)R(j)$$

$$=\frac{1}{\Gamma(H+\frac{1}{2})}[\sum_{j=i-M}^{i-1}[(i-j)^{H-\frac{1}{2}}-(i-j-1)^{H-\frac{1}{2}}]R(j)+\sum_{j=i-1}^{i}(i-j)^{H-\frac{1}{2}}R(j)]$$

$$=\frac{1}{\Gamma(H+\frac{1}{2})}[\sum_{j=i-M}^{i-2}[(i-j)^{H-\frac{1}{2}}-(i-j-1)^{H-\frac{1}{2}}]R(j)+R(i-1)]\cdot$$

So,

$$B_H(i)-B_H(i-1)$$
$$=\frac{1}{\Gamma(H+\frac{1}{2})}[\sum_{j=i-M}^{i-2}[(i-j)^{H-\frac{1}{2}}-(i-j-1)^{H-\frac{1}{2}}]R(j)+R(i-1)] \tag{3.16}$$

For $H = \frac{1}{2}$, $\Delta B_H(t)$ is reduced to a discrete time Gaussian white noise. Clearly, the following relationship is valid:

$$B_H(t)-B_H(0)=\sum_{i=1}^{t}[B_H(i)-B_H(i-1)] \tag{3.17}$$

Equation (3.17) and (3.18) form the two step FBMINC model developed by the author [9-12].

3.7. The Comparison of the FBM and FBMINC Models

Figure 9. shows six fBm traces generated by both the FBM and FBMINC models. Three H values are used: $H = 0.2, 0.5, 0.8$, and the diffusion coefficient is set to $D = 1$.

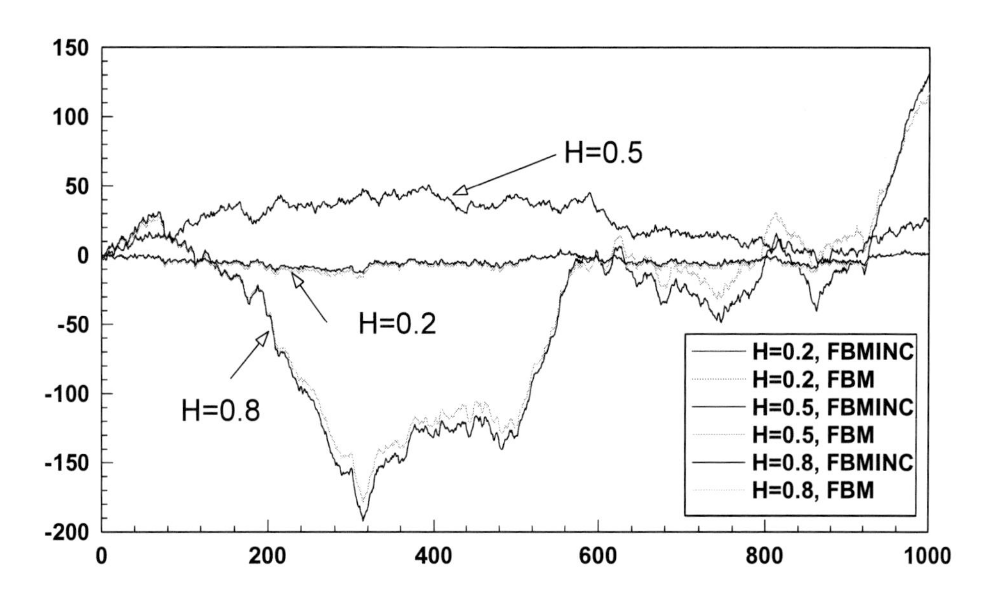

Figure 9. Fractional Brownian Motion Produced by the FBM and FBMINC Models. Where $M = 5000$, $D = 1$, $\Delta t = 1$.

The time interval is $\Delta t = 1$, memory $M = 5000$ and the same random seed is used in each case to begin the simulation. The plot shows that there is no significant difference between the two models, especially for $H = 0.2$ and $H = 0.5$. Both models generate traces almost on top of each other for the same H. However, the $H = 0.8$ traces exhibit noticeably divergent behaviour compared to the other two.

Although the FBM and FBMINC models synthesise fBm using different methods, they produce similar fBm traces. In the following, the more accurate model for synthesising fBm traces is determined. Both models are compared from three aspects: the standard deviation of each step, the small memory (when M<NSTEP) and the effect of the number of particles in the cloud.

(1) Consideration of the Standard Deviation of Each Step: $\sigma(B_H(t) - B_H(t-1))$. Theoretically, $\sigma(B_H(t) - B_H(t-1))$ should be a constant for all time t. From Figure 10, it can be noticed that for the FBM model, $\sigma(B_H(t) - B_H(t-1))$ increases with the number of time steps, although this effect is reduced as the memory used is increased. However, for the FBMINC model, $\sigma(B_H(t) - B_H(t-1))$ is always constant, regardless of the number of steps and memory used (see Figure 11). This is the principal advantage of the FBMINC model.

(2) Consideration of M < NSTEP. Another advantage of the FBMINC model is that it can be used while M < NSTEP. However, according to its definition, the FBM model is not suitable for this case. We need to redefine the FBM model for the case when M < NSTEP.

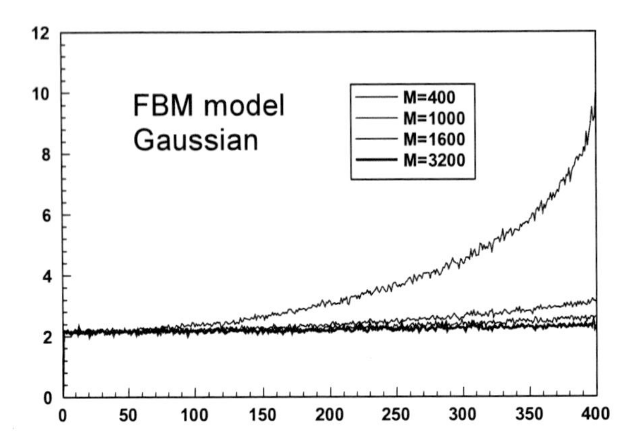

Figure 10. FBM Model: $\sigma(B_H(t) - B_H(t-1))$ versus Time Step. Where $P = 1000$, H = 0.8, D = 1. The Memory, M, Used for Each Run is Given in the Figures.

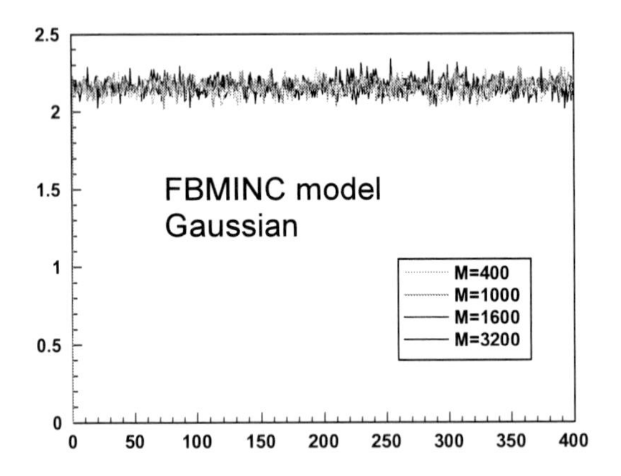

Figure 11. FBMINC Model: $\sigma(B_H(t) - B_H(t-1))$ versus Time Step. Where P = 1000, H = 0.8, D = 1.

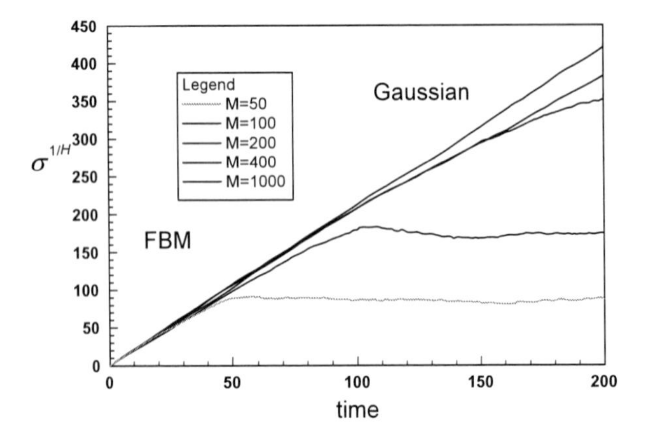

Figure 12. FBM Model. $\sigma^{\frac{1}{H}}$ versus Time Step with 400 Particles, D = 1, H = 0.8.

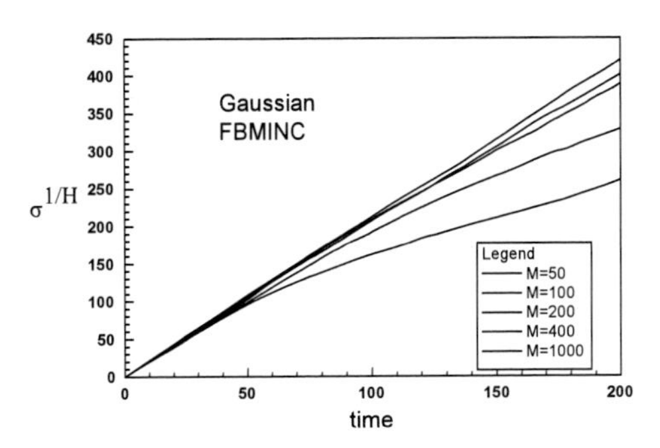

Figure 13. FBMINC Model: $\sigma^{\frac{1}{H}}$ versus Time Step with 400 Particles, $D = 1$, $H = 0.8$.

Figure 12. and Figure 13. contain plots of $\sigma^{\frac{1}{H}}(t)$ versus time step *t*, for the FBM and FBMINC models respectively. Figure 13 show an obvious cut-off in the plots for $M <$ NSTEP of the FBM model. However, Figure 12, 13 show that the FBMINC model can produce better results than the FBM model, especially when $M <$ NSTEP. In the figures , *D* is set to unity, hence we expect $\sigma^{\frac{1}{H}}(t) = 2Dt = 2t$.

(3) The Effect of the Number of Particles in the Cloud. Figure 15 and 16 show that both the FBM and FBMINC models all give good prediction of fBm traces as long as the number of particles and memory used are large enough. For a memory only several times the value of NSTEP, the FBMINC model gives a better prediction than the FBM model.

From comparison tests (1), (2) and (3), we find that the FBMINC model is the better model as it produces more accurate statistics and hence it will be used in syntheses of fBm. However, many of the examples of fBm used to illustrate this paper use the original FBM model which was developed first by the authors.

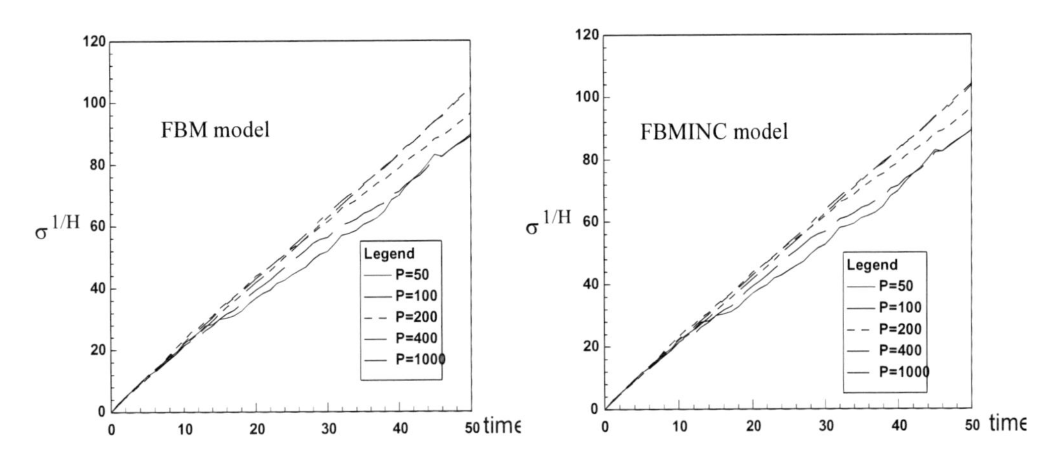

Figure 14. The Effect of the Number of Particles in a Cloud for Both the FBM and FBMINC Models. Here $H = 0.8$, $D = 1$, NSTEP = 50, $M = 400$.

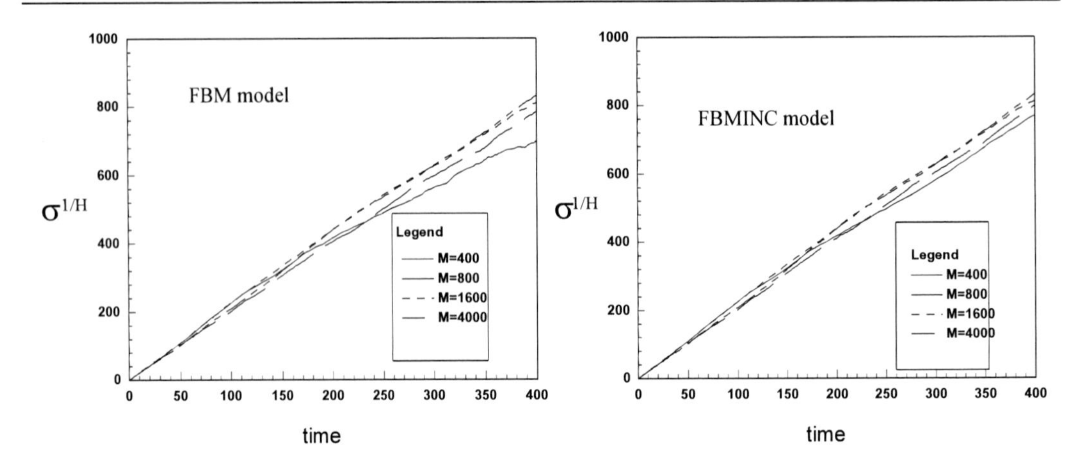

Figure 15. The Effect of Memory for the FBM and FBMINC Models. Here $H = 0.8$, $D = 1$, NSTEP = 400, $P = 400$.

3.8. Fractional Brownian Motion in Two Dimensions

To produce two-dimensional fBm particle trajectories, two independent fBm traces are used in both the x and y directions. Figures 16. contain fBm trajectories in two dimensions with $H = 0.1 \sim 0.9$. We can see the trajectories becoming less densely packed as H increases.

Figure 17 contains two plots of the location of a particle cloud at four distinct times for two values of H: $H = 0.5$, corresponding to regular Brownian motion of traditional particle tracking techniques; and $H = 0.8$, corresponding to a super-diffusing particle cloud. The expected standard deviation of the particle cloud is superimposed on the lower plot. The clouds are advected simply by a constant velocity for ease of visualisation. One obvious consequence of non-Fickian diffusive behaviour is the long term prediction of the concentrations of contaminant which will be lower at the centre (and correspondingly higher at the edge) of the super-diffusing cloud ($H > 0.5$) than would be predicted by the traditional model.

CONCLUSION

The method for generating a Brownian motion and simple random walks were illustrated in detail in the first section. Fractional Brownian motion (fBm) was shown to be a useful tool for generating non-Fickian diffusion. The definition of fBm shows how each point on an fBm trace depends upon the whole history of the fBm previous to that point. Two models for synthesising fractional Brownian motion (FBM model and FBMINC model) were developed by the authors. The differences of the two models are compared from the three aspects: the standard deviation of each step, the small memory and the effect of the number of particles in the cloud. The results show the FBMINC model is a better model as it produces more accurate statistics. The effect of the memory, number of steps and number of particles on the statistics of synthesised fBm was evaluated. FBm in two dimensions were generated.

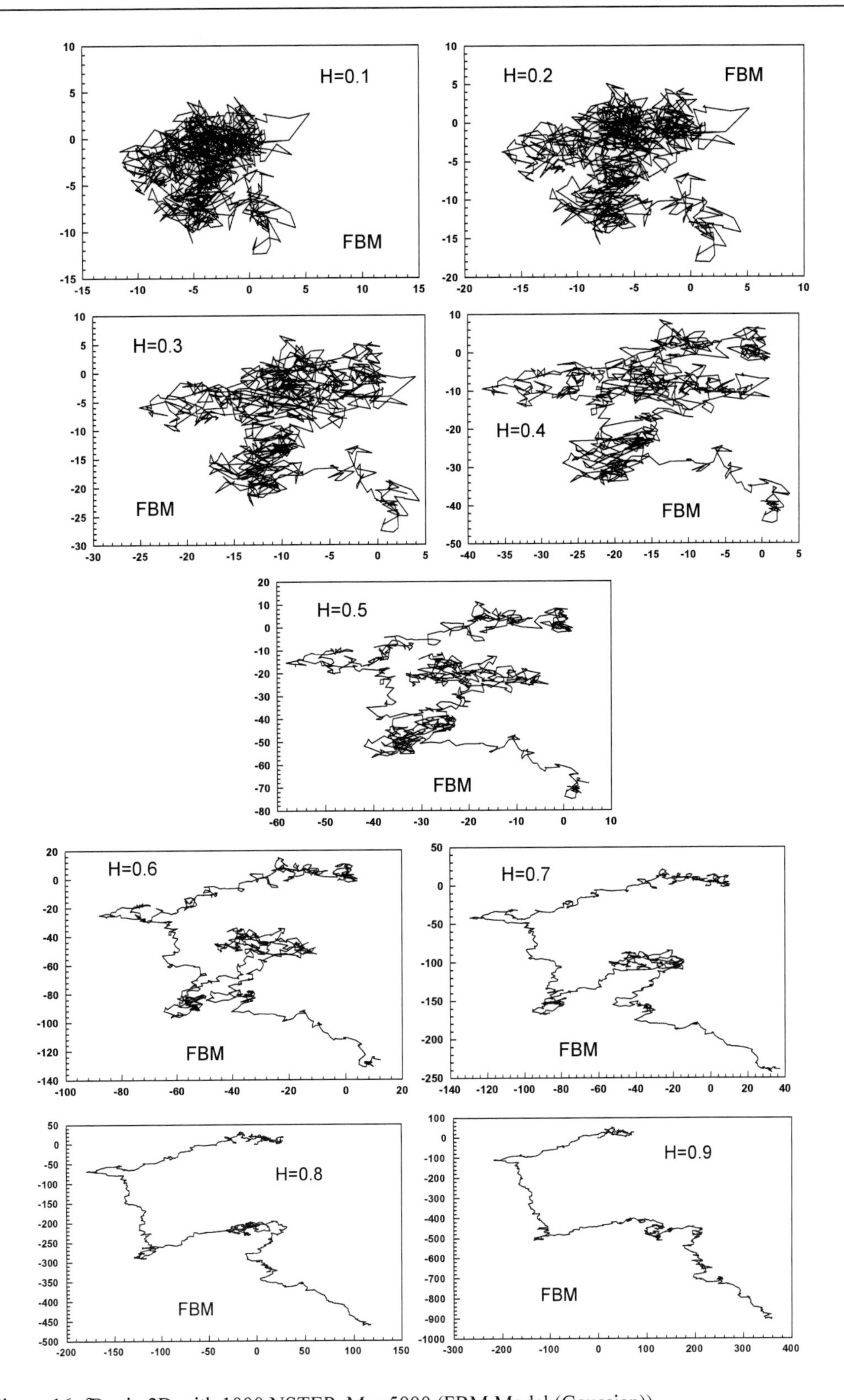

Figure 16. fBm in 2D with 1000 NSTEP, M = 5000 (FBM Model (Gaussian)).

Next section will detail the fBm particle tracking model used in coastal bay to mimic the flow trajectories by controlling Hurst exponent. The traditional Brownian motion particle tracking and fBm particle tracking results will be compared. It will show that fBm particle tracking will be more flexible and closer to reality. The fBm particle tracking model will be used on modeling pollutant dispersion in open sea surface. An accelerated fBm particle tracking model will be obtained to simulate the real observed data from Numthumbrian coastal waters ([13]). The most common and useful methods for determining H value from physical data will also introduced. Our simulation results and the HR Wallingford simulation results will be compared and tested with the observed data sets. The novel and practical fBm particle tracking model can become an easy and useful engineering tool for the prediction of pollutant dispersion in both small and large area of fluids.

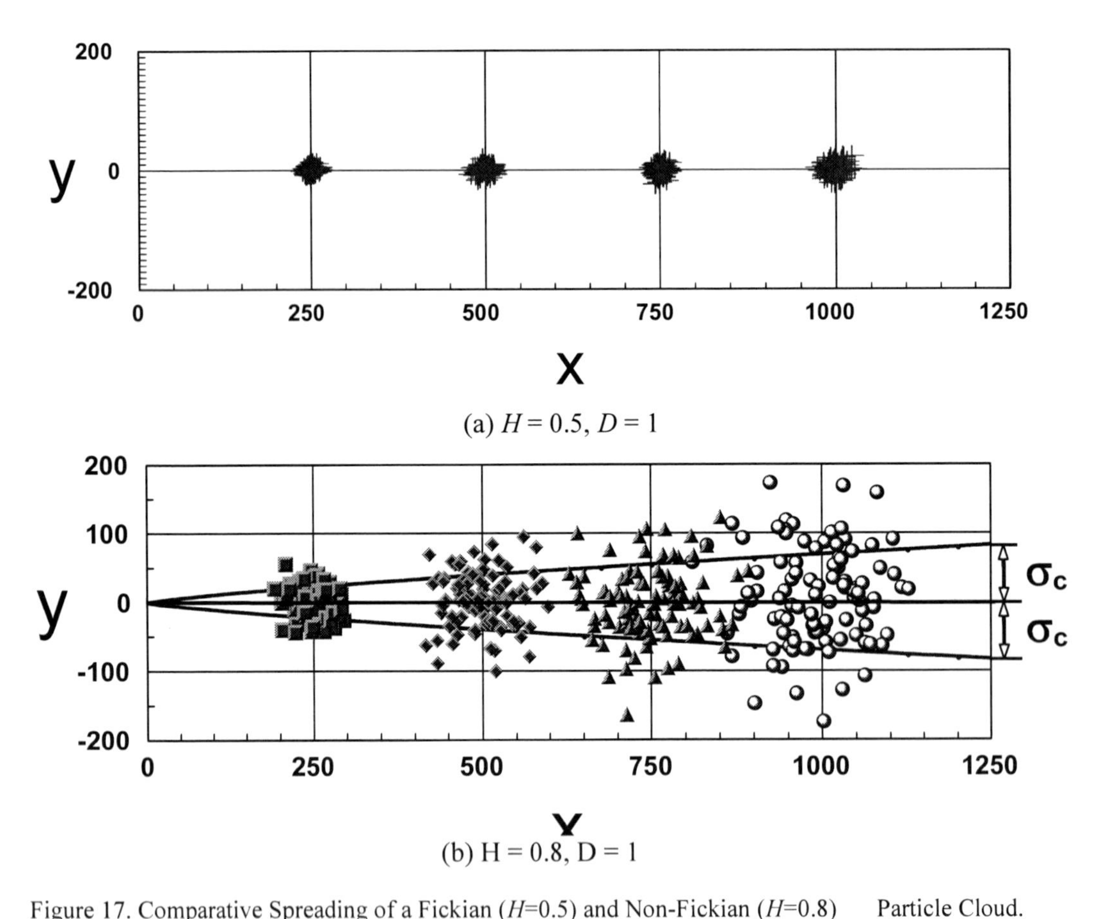

Figure 17. Comparative Spreading of a Fickian (H=0.5) and Non-Fickian (H=0.8) Particle Cloud.

REFERENCES

[1] Feder J. (1988), *Fractals*. Plenum Press, New York and London.

[2] Hastings H. M. and Sugihara.G. (1993), *Fractals: A User's Guide for The Natural Sciences*. Oxford University Press, Oxford.

[3] Press W. H., Flannery B. P., Teukolsky S. A., Vetterling W. T. (1986, 1992), *Numerical Recipes in Fortran. The Art of Scientific Computing.* Cambridge University Press, Cambridge.

[4] Mandelbrot B. B. and Van Ness J. W. (1968), *Fractional Brownian Motions, Fractal Noises and Applications*. SIAM review, Vol. 10, No 4, 422-437.

[5] Hurst H. E. (1950), *Long-Term Storage Capacity of reservoirs.* American Society of Civil Engineers, Vol. 116, 770-808.

[6] Voss R.F. (1985), *Random Fractal Forgeries, in Fundamental Algorithms for Computer Graphics.* R. A. Earnshaw, ed., NATO AS1 Series, V. F17, Springer-Verlag. Berlin.

[7] Peitgen H-O, Jurgens H., Saupe D. (1988, 1992), *Chaos and Fractals, New Frontiers of Science*. Springer-Verlag New York, Inc.

[8] Wheatcraft S. W. and Tyler S. W (1988), *An Explanation of Scale-Dependent Dispersivity in Heterogeneous Aquifers Using Concepts of Fractal Geometry.* Water Resour. Res., Vol. 24, No. 4, 566-578.

[9] Paul S Addison, Bo Qu, Alistair Nisbet and Gareth Pender (1997) 'Modelling Contaminant Spread on the Ocean Surface and within Soils using fBm's: Two Civil Engineering Applications', *Fractals in Engineering*, Springer, 375-382.

[10] Paul S Addison, Bo Qu, Alistair Nisbet and Gareth Pender (1997) 'A Non-Fickian, Particle-Tracking Diffusion Model Based on Fractional Brownian Motion', *International Journal for Numerical Methods in Fluids*, Vol. 25(12), 1373-1384.

[11] Paul S Addison, Bo Qu, A. S. Ndumu and I. C. Pyrah (1998) 'A Particle Tracking Model for Non-Fickian Subsurface Diffusion', *Mathematical Geology*, Vol.30, No.6, 695-716.

[12] Bo QU and Paul S Addision (2009) Development of FBMINC model for particle diffusion in fluids. *International Journal of Sediment Research,* 24(4), 439-454.

[13] Bo Qu, Paul S. ADDISION and Christopher T. MEAD (2003). *Coastal Dispersion Modelling Using an Accelerated fBm Particle Tracking Method.* Coastal Engineering Journal, Vol. 45, No. 1, pp 139-158.

In: Brownian Motion: Theory, Modelling and Applications
Editors: R.C. Earnshaw and E.M. Riley
ISBN: 978-1-61209-537-0

Chapter 9

DIFFERENTIAL GAMES AND EQUILIBRIUMS OF FUSION REACTORS

Danilo Rastovic
Control Systems Group, Nehajska, Zagreb, Croatia

ABSTRACT

In this paper the authors consider plasma confinement in tokamaks and stellarators as the problem of differential game that will solve the Grad-Shafranov equilibriums. Turbulence can show deterministic and stochastic regimes.

Keywords: tokamak, differential game, fuzzy scaling.

PACS numbers :02.30. Jr ; 05.40.-a; 28.52.Cx ; 52.65.Kj

1. INTRODUCTION

Magnetically confined fusion plasmas are thermodynamically nonequilibrium systems where particles and energy are injected deep in the plasma, providing heat which flows towards much colder edge region. The best-known example is Low to High (L-H) confinement transition, where the transport barrier forms at the edge of the plasma, but internal transport barriers have been found as well. We want come to the equilibrium case. The equilibrium is described by the Grad-Shafranov equation, which depends on two arbitrary functions. It can be considered as differential game that is described by Vlasov-Poisson-Fokker-Planck (VPFP) equations with external noises. Trajectories that start close to a chaotic saddle will behave chaotically for a while before setting into another (possibly periodic) attractor. But in our case of tokamak and stellarator, in H mode we must work with multivariable control of several chaotic attractors in Cantori sets. Thus , we must work with a different scenarios. If we apply fuzzy scaling, then with the IF-AND-THEN rules we can describe the conflict situations between different differential games in each fuzzy Cantori.

Since the system is not isolated, the entropy is not required to be minimum, but it can be assumed to be stationary even locally, expressing the local balance between the entropy injected externally and the entropy produced in plasma [1]. In the case of electrostatic processes the thermodynamic treatment points to the so called „reactive" instabilities are the most relevant from physical point of view. The probability P can now be determined from the requirement that the entropy be stationary with respect to arbitrary variations of P under constraints. We have seen that the entropy of the total system formed by the collective equilibrium plus background is invariant under the adiabatic variations so that no heat can be emitted or the absorbed by the total system in a reversible transformation as it is described by the Grad –Shafranov equation. There is the connection between the electrostatic entropy and the Lagrangian description of particle motion. The natural field of application of the stationary magnetic model concerns relaxed states in which the dissipation processes are counterbalanced by external sources (Ohmic or auxiliary) such that the magnetic entropy and the plasma-state are constant in time, at least approximately. On the basis of the data collected from different machines, we are able to analyse with our method different heating scenarios.

2. Grad -Shafranov Equation

The Grad-Shafranov equation is the equilibrium equation in ideal magnetohydrodynamics (MHD) for a two dimensional plasma, for example the axisymmetric toroidal plasma in a tokamak. Interestingly the flux function ψ is both a dependent and an independent variable in this equation

$$\Delta^* \psi = -\mu_0 r^2 \frac{dp}{d\psi} - F\frac{dF}{d\psi} \tag{1}$$

where μ_0 is the magnetic permeability, $p(\psi)$ is the pressure, and $F(\psi) = rB_\phi$, for $\vec{B}$ is the magnetic field.

The eliptic operator Δ^* is given by

$$\Delta^* \psi = r \cdot \frac{\partial}{\partial r}\left(\frac{1}{r}\frac{\partial \psi}{\partial r}\right) + \frac{\partial^2 \psi}{\partial z^2} \tag{2}$$

The nature of the equilibrium is largely determined by the choices of two functions F(ψ) and p(ψ) as well as the boundary conditions [2], [3].

Two dimensional, stationary, magnetic structures are described by the balance of pressure forces and magnetic forces, i.e. :

$$\nabla p = \vec{j} \times \vec{B} \tag{3}$$

where p is the plasma pressure and $\vec{j}$ is the electric current .

The equilibrium of the Grad-Shafranov equation in advanced tokamaks corresponds to the saddle points of corresponding cooperative differential game of finitelly many players for appropriate Vlasov-Poisson-Fokker-Planck equation. Some players (the designer, ion cyclotron range of frequency) want to obtain a minimal electric field and maximal temperature using as goal function the maximum energy and the some players (the device, neutral beam injection) want to obtain a maximal pressure using as goal function the maximum energy. The minimax strategy is an optimum strategy for all players. An interpretation for such case could be obtained in the context of the entropy function. Namely, the stationary entropy in this case will be the minimax entropy according to the result of [4] .

3. Vlasov-Poisson-Fokker-Planck Equation

Let us consider the nonlinear transport equation $f' = H(f, p_1)$ where H denotes a nonlinear operator that is differentiable on a dense domain of a separable Banach space $X \times U$ and p means control vector of parameters. The main problem encountered are the availability of a good model of the process that can be sufficiently linearized around any operating point. After linearization in some equilibrium neighbourhood the following equations hold true

$$df' = A_1' \, df + B_1' \, dp_1 \qquad A_1' = \frac{\partial H}{\partial f}, \quad B_1' = \frac{\partial H}{\partial p_1} \tag{4}$$

An equilibrium state of an appropriate semigroup evolution $T_1'(t)$ with the infitesimal generator A_1' is a state g such that $T_1'(t)g = g$ for all $t \geq 0$. In the case of linear transport phenomena the equilibrium distribution will be a Maxwell-Boltzmann distribution. After linear transformation of coordinates in some zero equilibrium neighbourhood , we can write $f' = A_1 f + B_1 u$ [5]. We consider VPFP equation and the properties of weak solutions

$$\frac{\partial f}{\partial t} + (v \cdot \nabla_x) f + div_v((E_\sigma - \beta v) f) - \sigma \, \delta_v f = 0 \tag{5}$$

We consider the singular neutron transport equation as a simulation orbit of appropriate singular neutron transport equation [6]. This approach is good for design of neutron emission spectrometer for the stabilization of plasma ions. We can write VPFP equation as semigroup problem $f'(t) = (A + BF)f + Kw(t), \; w(t) = F_1 f + w_1$

where the notations are

$$A = -(v \cdot \nabla_x), \quad BF = -\nabla_v (E_\sigma - \beta v), \quad BF_1 = \sigma \nabla_v \cdot \nabla_v \tag{6}$$

The model of collisonless plasmas, specially in controlled fusion or laser-plasma fusion is too idealized and collisional effects need to be incorporated.

The development on integrated , simultaneous, real-time controls of plasma shape, current, pressure, temperature, edge localized modes (ELMs) are now seen to be essential for further developments of quasi-steady state conditions with feedback, or the stabilization of transient phenomena with event-driven actions.

The EFDA-JET Real Time Project has developed a set of real-time plasma measurements and experimental control. The actuators include toroidal, poloidal and divertor coils,gas and pellet fuelling, neutral beam injection, radio frequency waves (ICRH) . The EFDA Real-Time-Project is essential groundwork for future reactors such as ITER.

The plasma diagnostics used for real-time experiments are for infra red (IR) interferometry, polarimetry, visible, UV and X-ray spectroscopy, bolometry, neutron and magnetics. Further analysis systems produce integrate results, such as temperature profiles on geometry derived from MHD solutions.

The status of the plasma control system is one of the keys to the successful operation of ITER. Current experiments increasingly use feedback control to improve their performance. This system will orchestrate all parts of the tokamak during a pulse to make sure that the discharge is successful. It relies on information from ITER diagnostics and will send to various actuators- for instance heating and current drive systems, fuelling, magnetic coils- to ensure that the plasma behaves the way we want it to.

The progresive development of superfast computers adds an interesting touch to this topic. Access to high-power computing makes the incorporation of simulation tools into real-time plasma control possible. This enable much better control of the plasma regimes than can be achieved now. ITER will be able to have very long pulses and may run several experiments during one pulse. Proper feedback control is absolutely essential be able to use this opportunity.

4. Linearization Methods

For the ageing device, we need not worry as long as the nominal equilibrium remains stable. This naturally raises the question of what happens if the equilibrium $f^*(\lambda)$ undergoes a bifurcation at $\lambda = \lambda_0$, which may have a catastrofic consequence for the device. To avoid such a problem one may try to control the system $f' = F(f,\lambda) + Bu(f,\lambda)$, $\lambda =$ const, stable when $\lambda = \lambda_0$.

We have to modify the linearization A, $A = \partial F / \partial f$ of F at the bifurcation point [7] .

We obtain as a consequences the tree of the evolution. In each branch of the tree (or Cantori) the evolution is nonsingular, but also there are singular points somewhere at the points of branching . In each case of scenarios we must take such feedback that we obtain a saddle point if we consider a Vlasov-Poisson-Fokker-Planck equation with external disturbances, as a differential games problem. It can be applied to the optimization of tokamak design.

Let us consider as approximation the finite dimensional case. We shall refer to the following linear time-varying system:

$$f'(t) = A(t)f(t) + B(t)u(t) + G(t)w(t)$$
$$z(t) = C(t)f(t) + D(t)u(t), \quad f(t_0) = 0 \qquad (7)$$

where $f(t) \in R^n$ is the system state, $u(t) \in R^m$ is the control input, $w(t) \in R^l$ is an exogenous disturbance input and $z(t) \in R^q$ is the controlled variable.

Full information (FI) problem admits a solution if and only if the Riccatti diferential equation admits a positive-semidefinite solution if D is full column rank (FCR). State feedback(SF) problem admits a solution if and only if the FI problem admits a solution . Now assume $B = (B_1 0)$ with B_1 FCR and $D = (D_1 D_2)$ for D_2 be FCR. Consider the simplified state feedback system (SSF)

$$f' = Af + B_1 u_1 + Gw$$
$$z = (I - D_2 D_2^*)Cf + (I - D_2 D_2^*)D_1 u_1 \qquad (8)$$

In this case the SF problem admits a solution if and only if SSF problem admits a solution [8],[9].

Approach uses a state transformation for the singular optimal control context. We can use this approach when we consider the finite dimensional approximations of the VPFP equation in each Cantorus.

Toroidal and poloidal momentum confinement in neoclassical quasilinear theory in tokamaks and stellarators is under developing. The theory is applied to explain the changing of the toroidal and poloidal flow direction after low mode to high mode (H-Mode) transition observed in some experiments. Toroidal momentum confinement in tokamaks is most likely to be anomalous. This implies that a quantitative theory for the toroidal flow is difficult if not impossible. Fortunately for H-mode the radial electric field is predominately determined by the poloidal momentum equation or the parallel momentum equation. Poloidal momentum equation can be described fairly by the neoclassical theory [10]. In the anomalous transport theory the crucial role should be given by recurrent unstable behaviour of plasma with the evolution toward stabilizable Grad-Shafranov equation. It open the new ways toward interpretation of the Kolmogorov -Arnold-Moser theorem.

The selection of neutral beam injectors (NBIs) in the actual experiment is important to reduce to the loss of fast ions and its ratio to the total injection power. Destabilization of neoclassical tearing modes (NTMs) is avoided by two different methods. One is optimizing the current profile and pressure profile so that a step pressure gradient is not located at the mode rational surfaces. The other is the NTM stabilization using electron cyclotron current drive [11].

To keep trace of the hot spots during plasma discharges, TC , visible cameras and IR cameras should be installed. While the lower hybrid (LH) system is to major tool to sustain the plasma current on hunderts of seconds long pulses, the ion cyclotron range of frequencies (ICRF) system is the mean heating tool.

The size of the Cantor manifolds is not uniform. We have no Kolmogorov-Arnold-Moser (KAM) theory yet for infinite-dimensional tori of such Hamiltonians. We shall see what is the role of recurrence for obtaining of good machine behaviour.

5. Approach with Differential Games

The equations of motion for continuous stochastic differential games for two players are a set of operator stochastic differential equations :

$$df = F(f,t,u_1(f,t),u_2(f,t))dt + B(f,u_1(f,t)u_2(f,t))dw \tag{9}$$

where $f \in R^n$, w is an m-dimensional Wiener process; $u_i(f,t)$ the strategy of player , i, i=1,2 and B is $n \times m$ matrix function.

The payoffs to each player depend on the strategies employed, and for player i is :

$$I^i(f_0,t_0,u_1,u_2) = E_{f_0,t_0}\left[\int_0^T L^i(\xi,t,u_1,u_2)dt + F^i(\xi(\tau),\tau)\right] \tag{10}$$

where i=1,2, ξ is the solution of ( ).

The problem consists in finding strategies u_1, u_2 that will form a saddle point for the functional I:

$$I(f_0,t_0,u_1,u_2^*) \leq I(f_0,t_0,u_1^*,u_2^*) \leq I(f_0,t_0,u_1^*,u_2) \tag{11}$$

Minimax Principle is satisfied if the appropriate PDEs hold [12].

Stabilizing modes that limit plasma beta is a key goal in fusion reactor design. Future burning plasma experiments, such as ITER, will be dealing with a plasma consisting of a core thermal component characterized by some base temperature (e.g. 15-25 keV in ITER) , an energetic alpha particle component with much higher energies and energetic ions produced by neutral beam injection (NBI) and /or by ion cyclotron heating. Such plasmas often give rise to nonlinear phenomena that are based on both MHD fluid properties and intrinsic kinetic particle properties.

Numerical simulation based on the gyrokinetic formalism for drift wave turbulence, such as the ion temperature gradient (ITG) mode, have been extensively performed with the aim of understanding anomalous transport mechanism in a core region of magnetically confined plasmas.

Tokamak disruption occur when plasma thermal or magnetic energy are lost very rapidly as a result of some instability. It is thought that the transport of magnetically confined plasmas is mainly caused by plasma turbulence.

Linear quadratic differential games have many applications. Since the presence of uncertainities is a matter of fact in any realistic applications, we assume that the players do not know exactly the state equation. Due to the presence of the uncertainities, the players cannot evaluate, even if existing, their optimal saddle point strategies; there the aim of the paper [13] is to introduce a class of suboptimal strategies which garantee to each player a given value for the objective functional. Such situation arrises in the Grad-Shafranov equilibrium problem where magnetic field $\vec{B}$ and pressure p must be controlled by two different players. In the VPFP equations it is obtained via electric control field and disturbances as players strategies.

We have the equation as simplified model

$$f'(t) = A(t)f(t) + B(t)u(t) + G(t)w(t), \quad f(t_0) = f_0 \tag{12}$$

and by the objective functional

$$I(f_0, u, w) = f^T(T)Lf(T) + \int_0^T \left[f^T(t)Q(t)f(t) + u^T(t)R(t)u(t) - w^T(t)S(t)w(t)\right]dt \tag{13}$$

The objective functional has to be minimized by Player 1 and maximized by Player 2, by using the control vectors $u_{[t_0,T]}$ and $w_{[t_0,T]}$ respectively. A pair (μ^*, ν^*) is said to be a feedback saddle point solution of the game if it is satisfies the inequality

$$I(f_0, \mu^*, \nu) \leq I(f_0, \mu^*, \nu^*) \leq I(f_0, \mu, \nu^*) \tag{14}$$

We can assume that Player i (i=1,2) describes the dynamics of the game with the following uncertain state equation

$$f'(t) = (A_i(t) + F_i(t)\Delta_i(t)E_i(t))f(t) + B(t)u(t) + H(t)w(t) \tag{15}$$

The control law $\mu^0 = -R^{-1}B^T Xf$ is a quadratic garanteeing cost strategy for player 1. If the solution of another one the Riccatti differential equation exists, then we obtain a quadratic garanteeing cost strategy for Player 2. It can be obtained by solving the approprate finite dimensional restriction of VPFP equations with disturbances . Infinite-dimensional Riccati equations are very hard to resolve; therefore some effort has been spent to discuss particular problems which can be solved via finite-dimensional equations. The problem is solved by showing that original problem has a solution if and only if certain differential game with time delay has [14].

The original game is transformed to the game without delay, which can be solved via standard techniques. The resulting control law is given by the sum of two contributes: a traditional memoryless plus a distributed delay state feedback. Consider the linear Vlasov-Poisson-Fokker-Planck time-varying system with lumped and distributed time delays in the form

$$f'(t)=A(t)f(t)+\sum_{k=1}^{r}F_K(t)f(t-\tau_K)+\int_{-\tau}^{0}H(t,\alpha)f(t+\alpha)d\alpha+B(t)u(t)+G(t)w(t), \tag{16}$$
$$f(s)=0,\quad t\in[0,T],\quad s\in[-\tau,0]$$

u(t) is the control input and w(t) is the disturbance. We focus on the study of the differential game with time delay consisting of state equation and cost functional

$$\min\max\ I(u,w)$$

$$I(u,w)=\left(\|f(T),u\|^2_{Q_f,R}-\wp^2\cdot\|w\|_2^2\right)\quad for\ some\ \wp,\ \|G\|\langle\ \wp \tag{17}$$

Q_f and $R(t)$ being positive-definite matrices,

$$\|(f,u)\|_{Q_f,R}=\left[\|f\|^2+\|u\|^2\right]^{1/2} \tag{18}$$

It follows that we can consider the VPFP confinement system as several differential games, that are defined in each fuzzy Cantorus. Consider the differential game consisting of the state equation

$$y'(t)=\hat{B}(t)u(t)+\hat{G}(t)w(t),\quad y(0)=y_0 \tag{19}$$

and the cost functional

$$C(y_0,u,w)=\|(y(T),u)\|_{Q_f,R}-\wp^2\cdot\|w\|_2^2 \tag{20}$$

Then

a). there exists a saddle point feedback solution if and only if the Riccati differential equation

$$-P'(t)=P(t)\left(\frac{1}{\wp^2}\hat{G}(t)\hat{G}^T(t)-\hat{B}(t)R^{-1}(t)\hat{B}^T(t)\right)P(t)$$
$$P(t)=Q_f \tag{21}$$

admits a positive semidefinite solution. In this case the optimal strategies are given by

$$u_f(t)=-R^{-1}(t)\hat{B}^T(t)P(t)y(t)$$
$$w_f(t)=\frac{1}{\wp^2}\hat{G}^T(t)P(t)y(t) \tag{22}$$

Problem

$$\begin{aligned} f'(t) = A(t)f(t) + F(t)f(t-\tau_1) + \int_{-\omega_1}^{0} H(t,\alpha)f(t+\alpha)d\alpha + \\ B(t)u(t-\tau_2) + \int_{-\omega_2}^{0} L(t,\alpha)u(t+\alpha)d\alpha + G(t)w(t) \end{aligned} \tag{23}$$

is solvable if and only if the appropriate Riccati differential equation admits a positive semidefinite solution P. In this case we have appropriate control law . For proof of these assertions see [15].

We have general formalism of optimal control in the presence of noise. Optimal feedback design based on deterministic model is minimization of total fluctuation energy of the instabilities, as well as minimization of control power

$$I(t_f) = \int_0^{t_f} \left[f(t)Qf(t) + u^T(t)Ru(t) \right] dt \tag{24}$$

For Lagrangian

$$L(f, f') = f^T(t)Qf(t) + u^T(t)Ru(t) + \lambda^T(t)\left[Af(t) + Bu(t) - f'(t)\right] \tag{25}$$

It holds that $\frac{d}{dt}\left(\frac{\partial L}{\partial f'}\right) - \frac{\partial L}{\partial f} = 0$, with optimal control $u(t) = -R^{-1} \cdot B^T \cdot \lambda(t)$.

The appropriate reduced model is $f' = Af + Bu + D\psi$ where $D\psi$ stochastic noise is. The solution under optimal feedback is $f'(t) = (A - B\,K_C)f(t) + D\psi(t)$

and $f(t) = \phi(t, t_0)f(t) + \int_0 \phi(t,\tau)D\psi(\tau)d\tau.$ (26)

Optimal control of the resistive wall mode in the tokamak is solved as an optimum control problem by minimization (via a procedure of variational calculus [16]) . The aim is to applied a minimization of some functional for obtaining a saddle points as solution of Grad-Shafranov equation for the generalization of the optimal control theory on two players.

The control of a plasma shape is realized by using more independent coil circuits. For long-range correlation analysis of plasma turbulence for self-organized and critical system the characteristic is global scaling involving space scales that exceed diffusive scale. It might lead to novel ways of controlling turbulence. But, if we have a fuzzy scaling, then we can describe the problem as differential game over n-th fuzzy Cantorus. Cantor set is ideal fractal description with normal scaling, but in tokamak the most natural approach is to use fuzzy scaling. The paper [17] develops a new approach that characterizes directly Markov perfect Nash equilibrium in stochastic differential games as solution of a system of semilinear partial differential equations (PDEs), that can be seen as a set of generalized Euler equations. We shall consider the special case of differential game, i.e. the game for Vlasov-Poisson-Fokker-Planck equation. We consider an N-person differential game over a fixed and bounded time interval $[0,T]$ with $0 \langle T \leq \infty$. The state process $f \in R^n$ satisfies the system of controlled stochastic differential equations (SDEs)

$$df(s) = F(s, f(s), u(s))ds + B(s, f(s))dw(s), \quad t \leq s \leq T \qquad (27)$$

A strategic profile $\{u_1(t), u_2(t), \ldots u_N(t)\}$ is called admissible if

i) For every (t, f) the system of SDEs admits a pathwise unique strong solution ;
ii) For each i=1,2.. N there exists some function $\phi^i : [0,T] \times R^n \to U^i$ of class $C^{1,2}$

such that u^i is in relative feedback to ϕ^i, i.e. $u^i(s) = \phi^i(s, f(s))$
for every $s \in [0,T]$.

Given initial conditions $(t, f) \in [0,T] \times R^n$ and an admissible strategic profile u , the payoff of each player must to be maximized . In a non-cooperative setting the aim of the players is to maximize their individual payoff I^i. Since this aspiration depends on the strategies selected by other players also, it is generally impossible to attain. An adequate concept of solutions is Nash equilibrium, which prevents unilateral deviations of the players from its recommendation of play. The standard approach to determine Nash equilibrium is to solve Hamilton-Jacobi-Bellman system of PDEs. The deterministic Hamiltonian function of the ith player is

$$H^i(s, f, u, p^i) = L^i(s, f, u) + (p^i)^T g(s, f, u) \qquad (28)$$

6. Fuzzy Scaling

We consider in ideal case the control systems of the VPFP equations

$$f' = Af + Bu + Gw, \quad u = Ff + w_1, \quad w_1 = F_1 f + w_2 \qquad (29)$$

Where A is direct sum of operators

$$BFf = \nabla_v((E_\sigma - \beta v) f) \qquad (30)$$

and $BF_1 f = \sigma \cdot \nabla_v \cdot \nabla_v f$, vector f is direct sum of vectors $f_i, i = 1,2,\ldots m$

E(t,f) is the force field acting on particles. The strictly positive parameters σ and β model a certain type of interaction between particles.

Let us take after first bifurcation the new time-scale $\tau = \varepsilon \cdot t$

$$f'(\tau)=\frac{1}{\varepsilon}Af(\tau)+\frac{1}{\varepsilon}Bu(\tau)+\frac{1}{\varepsilon}Gw(\tau),\quad 0\le\tau\le T_1 \tag{31}$$

We obtain the system $f'(\tau)=\frac{1}{\varepsilon^2}Af(\tau)+\frac{1}{\varepsilon^2}Bu(\tau)+\frac{1}{\varepsilon^2}Gw(\tau)$ (32)

for new time $\tau_2=\varepsilon^2\cdot t+T_1$

For next bifurcations and the times we have got a new systems. Finally,

$$f'(\tau)=\frac{1}{\varepsilon^m}Af(\tau)+\frac{1}{\varepsilon^m}Bu(\tau)+\frac{1}{\varepsilon^m}Gw(\tau)\qquad T_{m-1}\le\tau\le T_m \tag{33}$$

and in this way the desired control system is obtained.

If in function of control $w_1=F_2f+w_2$ we take the additional input w_2 it can be interpreted as a neutral beam injection for heating and fuelling of fusion plasma.

The concept of inertial manifold system is exactly described by a low-order ordinary differential equations system. From this the synthesis of linear optimal controllers will follow.

Similarities in transport between stellarators and tokamaks are as follows : i). The radial heat transport in anomalous higher than neoclassical theory in the edge to the intermediate region, ii). The global energy confinement time shows a similar parametric dependence, and the power degradation of confinement is observed in both devices. It means that the applications of fuzzy scaling and differential games are natural for such kind of problems[18].

In the case of tokamaks, the appropriate Kolmogorov-Arnold-Moser theory via Cantori is so far idealized theory. We must take in considering the fuzzy Cantori in the real case because of the nature of tokamak machine and so called Edge Localized Modes (ELMs) effects. It means that we consider the following situation. Consider the following IF-THEN rules in the case of VPFP equation:

$$\begin{aligned}&IF\ m(1)\ is\quad M(i1)\ and\ \ldots\ IF\ m(r)\ is\quad M(ir)\\&THEN\quad (df/dt)=(A(i)+E(i))f(t)+A(di)f(t-h)+Bu(t)+Gw(t)\\&i=1,2,\ldots k\end{aligned} \tag{34}$$

where $M(ij)$ are fuzzy sets and $m(1),\ldots m(r)$ are given premise variables, and $E(i)$ are the uncertain matrices. The fuzzy system is hence given by the sum of equations

$$(df/dt)=a(i)[(A(i)+E(i))f(t)+A(di)f(t-h)+Bu(t)+Gw(t)] \tag{35}$$

where $a_i(t)$ are the fuzzy basis functions [19].

In the new concept it all should be given in the context of differential games theory.

The paper [20] considers dynamically stable cooperative solutions in randomly furcating stochastic differential games. Analytically tractable payoff distribution procedures contigent upon specific random realizations of the state and payoff structure are derived. Pontryagin solved differential games in open-loop solution in terms of the maximum principle. In the presence of stochastic elements the optimality principle must remain optimal throughout the game and a subgame consisting is required for a dynamically stable cooperative solution.

Changes in preference, technology, legal arrangements and the physical environments are examples of factors which constitute the change in payoff structures. The state dynamics of the game is characterized by the vector-valued stochastic differential equations

$$df(t) = F(t, f(t), u_1(t), u_2(t))dt + B(t, f(t))dw(t), \quad f(t_0) = f_0 \tag{36}$$

To obtain a Nash equilibrium solution of the game we first consider the solution for the subgame. In this case there exists continuously differentiable functions which satisfy some partial differential equations. We assume that a particular noncooperative Nash equilibrium is adopted in the entire subgame. In order to formulate the subgame in the second to the last time interval $[t_{m-1}, t_m)$, it is necessary identify the expected terminal payoffs at time. We obtain player i's payoffs at the time t_m. The expected terminal payoff of player i , i $\in \{1,2\}$ in the subgame can be computed under the assumption that a particular Nash equilibrium is adopted in each of the possible subgame scenarios.

We can consider the case when the players want to cooperate and agree to act and allocate the cooperative payoff according to a set of agreed upon optimality principles. The solution optimality principle for the cooperative game includes

i) An agreement on a set of cooperative strategies /controls and
ii) A mechanism to distribute total payoff between players.

Both group rationality and individual rationality are required in a cooperative plan. Hence under cooperation the players will adopt the cooperative strategy. It yields the dynamics of cooperative state trajectory. In the individual rationality they will compare their cooperative payoff to their noncooperative payoff at same time. The players agree to maximize the sum of their payoffs and equally divide the excess of the cooperative payoff over the noncooperative payoff.

CONCLUSION

An open loop two-person linear quadratic (LQ for short) stochastic differential game is obtained. A forward-backward stochastic differential equations and a generalized differential Riccati equation are introduced, whose solvability leads to the existence of the open-loop saddle points for the corresponding two-person LQ stochastic differential game, under some additional mild conditions.

In the case when we have an additional constraint, instead of the noncooperative game we can consider the cooperative game. It means that the modifications of choosen strategies are necessary. This type of game is called a hybrid game.

In the evolution case as it is L mode to H mode transitions, we can consider evolutionary stable strategies. First „player" chooses the magnetic field and the second „player" chooses the adequate pressure p including as the possible actuators (players) also the neutral beam injection(NBI), electron cyclotron current drive (ECCD), impurities of helium atoms, ICRF, internal coils and emission of neutrons. The Lawson criteria also should be achieved as one kind of constraint on equations.

The play is based on the principle of the action-reaction mechanism. The action (i.e. controllers) wants to obtain the maximum output energy with minimum of differences forces and the reaction (pressure) wants to obtain minimum output energy with maximum of differences forces. We choose as strategy for players : minimax(actuators)= maximin(observers,pressure) differences of energies. For such purpose the methods of fuzzy scaling have been applied. Presented model is not of general nature but it gives a new light on the problem of confinement degradation.

REFERENCES

[1] Rastovic,D., Applications of artificial intelligence and multi-variable control of chaos on tokamak equilibriums,Chapter in book:"*Glow discharges and tokamaks*",Nova Sci.Publishers,(in print),2010.

[2] Grad,H, and Rubin,H., *Hydromagnetic equilibria and force-free fields*, Proceedings of the 2nd UN Conf. On the Peaceful Uses of Atomic Energy,Vol.31,Geneva; IAEA,p.190,1958.

[3] Shafranov,V.D., Plasma equilibrium in a magnetic field, *Reviews of Plasma Physics*,Vol.2,New York:Consultants Bureau,p.103.

[4] Struchrup,H.and Weiss, W., Maximum of the local entropy production becomes minimal in stationary proceses, *Phys. Rev. Lett.*,80,5048-5051,1998.

[5] Rastovic,D.,Feedback stabilization of some classes of nonlinear transport equations, *Rend.Circ.Mat.Palermo*,51,325-332,2002.

[6] Rastovic,D., Vlasov-Poisson-Fokker-Planck equation and stabilization system, Anal.Univ.Vest Timisoara,ser.mat-inf.,1,42,141-148,2004.

[7] Rastovic,D., Transport theory and systems theory, N*uclear Technology and Radiation Protection*,1,20,50-58,2005.

[8] Amato,F.,Mattei,M. and Pironti,A.,Solution of the state feedback singular H^{∞} control problem for linear time-varying systems, *Automatica*, 36,1469-1479,2000.

[9] Rastovic,D., The Riemann-Roch theorem and a theorem on the dimension of the minimal relization of linear dynamical systems, *Arch.-Math.* (Brno),1,32,9-12,1996.

[10] Shaing,K.C., Toroidal momentum confinement in neoclassical quasilinear theory in tokamaks,*Physics of Plasmas*,8,1,193-200,2001.

[11] Rastovic,D.,Analytical description of long-pulse tokamaks, *J. Fusion Energy*,4,27,285-291,2008.

[12] Ardanuy,R. and Alcala,A., Minimax principle for stochastic differential games, *Extracta Matematicae*,2,6,184-186,1991.

[13] Amato,F., Mattei,M. and Pironti,A., Guaranteeing cost strategies for linear quadratic differential games under uncertain dynamics,*Automatica*,38,507-515,2002.

[14] Amato,F. and Pironti,A., H∞ optimal terminal state control for linear systems with lumped and distributed time delays,Automatica,35,1619-1624,1999.

[15] Ariola,M. and Pironti,A., H∞ optimal terminal control for linear systems with delayed states and controls, *Automatica*, 44,2676-2679,2008.

[16] Sen,A.,K., Nagashima M. and Longman, R.W., Optimal control of tokamak resistive wall modes in the presence of noise,*Phys. Plasmas*,10,4350-4357,2003.

[17] Fombellida,R.J. and Rincon-Zapatero,J.P., Markov perfect Nash equilibrium in stochastic differential games of a generalized Euler equations system,Working Paper 08-67,*Economic Series*(31),November,Madrid,2008.

[18] Rastovic,D.,Fuzzy scaling and stability of tokamaks, *J. Fusion Energy*,1,28,101-106,2009.

[19] Rastovic,D.,Fractional variational problems and particle in cell gyrokinetic simulations with fuzzy logic approach for tokamaks, *Nuclear Technology and Radiation Protection*,2,24,138-144,2009.

[20] Petrosyan,L.A. and Yeung,D.W.K., Subgame-consistent cooperative solutions in randomly furcating stochastic differential games, *Mathematical and Computer Modelling*, 45,1294-1307,2007.

In: Brownian Motion: Theory, Modelling and Applications ISBN: 978-1-61209-537-0
Editors: R.C. Earnshaw and E.M. Riley

Chapter 10

THERMOPHORESIS AND BROWNIAN DIFFUSION OF NANOPARTICLES IN A NONISOTHERMAL GAS FLOW

S. P. Fisenko and J. A. Khodyko
A.V. Luikov Heat and Mass Transfer Institute, National Academy of Sciences of Berlaus, Minsk, Belarus

Keywords: laminar flow, deposition, the Galerkin method, simulation, radius nanoparticle, free molecular regime.

INTRODUCTION

Transport processes of nanoparticles in gas flows have been considered in many distinguished papers [1 and references therein]. To mention that practically important process of Brownian deposition of nanoparticles on the wall was investigated in [1 – 9 and references therein]. At present time considerable attention is attracted to the specific of Brownian diffusion of nanoparticles in non-isothermal gas flows in the flow reactors. This interest is due to the ability to control movement and deposition of nanoparticles by changing the temperature of the reactor wall [3, 6-9]. The objective of our research is to investigate by the methods of mathematical modeling of the interference of thermophoresis and Brownian diffusion on the movement of nanoparticles with a laminar gas flow in a flow cylindrical reactor when the inlet temperature of the flow and the wall one are very different. The structure of the paper is the following. The mathematical model is described in the chapter 1. The semi-quantitative analysis based on the Galerkin method is given in next chapter. Then we present results of numerical simulation. Finally we summarized our main results and conclusions.

1. Mathematical Model

The system of equations describing the Brownian motion and thermophoresis of nanoparticles in non-isothermal laminar gas flow in a cylindrical reactor includes several equations. Among them the simplest equation is the equation for the averaged velocity u (z), which we determine from the continuity equation in integral form:

$$u(z) = u_0 \frac{T(z)}{T_0} \quad (1)$$

where T(z) is average gas temperature at a distance z from the reactor inlet. We assume also that the equation of state has the form for ideal gas, and that the velocity profile in the reactor very accurately is described by the Poiseuille profile. For simulation of field of temperature $T(r,z)$ we have equation

$$u(r,z)\frac{\partial T(r,z)}{\partial z} = \frac{1}{r\rho c}\frac{\partial}{\partial r}\left[\lambda\left(T(r,z)\right)r\frac{\partial T(r,z)}{\partial r}\right] \quad (2)$$

where λ is thermal conductivity of the gas flow, which depends on local gas temperature , ρ and c is the mass density and specific heat of gas.

The equation for calculating the numerical density of nanoparticles n(r, z) taking into account Brownian diffusion and thermophoresis in nonisothermal gas medium is still the subject of scientific debate. In particular, for a point particle in [7, 8 and references therein] it was obtained by the methods of nonequilibrium statistical mechanics that the diffusion flux of nanoparticles has two terms. The first term describes Brownian diffusion, while the second term describes the contribution of the temperature gradient. Analysis of this equation is given in [9]. This derivation is lustful only for small temperature gradient.

In this paper we investigate two-dimensional equation, which is valid for small concentrations of Brownian particles and explicitly takes into account the thermophoresis for large temperature gradients. This equation is

$$\frac{\partial\left(u(r,z)n(r,z)\right)}{\partial z} + v_{tf}\frac{\partial n(r,z)}{\partial r} = \frac{1}{r}\frac{\partial}{\partial r}\left[D_b(T(r,z))r\frac{\partial n(r,z)}{\partial r}\right], \quad (3)$$

where n(r,z) is the number density of nanoparticles, D_b is the Brownian diffusion coefficient, v_{tf} is the thermophoretic velocity of spherical nanoparticle in the free molecular regime [10]

$$v_{tr} = -\frac{3}{4}\frac{\eta\nabla T}{\rho T} = -K\nabla T,$$

and η is temperature dependent dynamic viscosity of gas.

In the free molecular regime taking into account the Einstein relation the coefficient of Brownian diffusion D_b can be written as [11]

$$D_b = kT(r,z)b,$$

with the mobility b of the nanoparticle nonlinearly depends on the temperature and inversely proportional to the square of the radius of the nanoparticle R_n

$$b = \frac{3}{16\pi R_n^2 P}\left(\frac{2\pi kT}{m}\right)^{0.5},$$

where P is total pressure in the system, m is mass of gas molecule. Influence of gas properties on the isothermal Brownian diffusion has been considered in [2, 4, 5].

Also we assume that after the collision with the reactor wall nanoparticle remains on it, the gas temperature equals the wall temperature (the regime of continuous medium). Then the boundary conditions to the system of equations (1 - 3) are:

$$n(R,z) = 0, \tag{4}$$

$$T(R,z) = T_w. \tag{5}$$

The condition of the cylindrical symmetry of the problem is expressed by two expressions:

$$\frac{\partial T(0,z)}{\partial r} = \frac{\partial n(0,z)}{\partial r} = 0. \tag{6}$$

We only consider the following initial conditions: the distribution of nanoparticles over the cross section of flow entering the reactor is uniform

$$n(r,0) = const, \tag{7}$$

inlet gas temperature is uniform over the cross section of flow and differs from the temperature of the wall

$$T(r,0) = T_0.$$

For similar isothermal problem different initial conditions are considered in [2]. Before proceeding to the numerical calculations it is very useful to make qualitative estimations [12, 13]. These results are presented at the next chapter.

2. Semi-Quantitative Analysis

At first we describe qualitative physical picture of the deposition of nanoparticles in a reactor. Let us consider an example where the temperature of the reactor wall is above the temperature of the incoming flow. At the entrance to the reactor a temperature gradient arises in the gas flow. Then it gradually decays. The thermophoretic velocity v_{th} is directly proportional to the temperature gradient in the gas. As a result the thermophoreic force affects the location of nanoparticles only on relatively short distance, while the Brownian diffusion "works" all the time. Nevertheless, thermophoresis distorts the initial distribution of nanoparticles, which Brownian motion tends to convert in the thermodynamic equilibrium distribution. It follows from (3), (4), and (6) that the spatially uniform distribution of nanoparticles is the equilibrium one. By virtue of the boundary condition (4), the equilibrium distribution of nanoparticles has the form n = 0.

It follows from the analysis of (3) that the influence of Brownian motion is a very significant factor of the deposition of nanoparticles on the wall (see also Figure 2) if the inequality is valid

$$A = \frac{D_b L}{R^2 u_0} \geq 1, \quad (8)$$

where L is length and R is radius of the reactor. With the concept of characteristic length of Brownian deposition in a cylindrical reactor, l_b [2, 4,],

$$l_b \approx \frac{4R^2 u_0}{\pi^2 D_b}, \quad (9)$$

the condition (8) can be rewritten as

$$\frac{L}{l_b} \geq 1,$$

in other words, the reactor should be relatively long for the manifestation of Brownian diffusion. Qualitative analysis of convective heat conductivity equation (2) shows that the characteristic length of change of the temperature field l_t is given by [14]

$$l_t = \frac{0.26 u_0 R^2 \rho c}{\lambda(T_w)}, \quad (10)$$

where λ is the thermal conductivity of gas, in the approximation of an ideal gas it does not depend on the total pressure in the reactor. The value of the characteristic length l_t considerably affects on the thermophoretic motion of nanoparticles. It follows from the qualitative analysis of the left side of equation (3) that if the inequality is valid

$$B = \frac{v_{th} l_t}{R u_0} \geq 1, \qquad (11)$$

then thermophoresis is substantial factor. It is worthy to note based on (10), that criterion B is not dependent on the velocity of gas flow. Following [2, 4] let us introduce important parameter γ(L), which is the ratio of nanoparticles, deposited on the wall of the reactor on the path L to the initial flux of nanoparticles. The parameter γ is calculated by using the expression

$$\gamma(L) = \frac{\int_0^R [u(0,r)n(0,r) - u(L,r)n(L,r)] rdr}{\int_0^R u(0,r)[n(0,r))] rdr}. \qquad (12)$$

The expression (12) takes into account both the change in flow velocity due to its temperature change and a change in the nanoparticles distribution forced by thermophoresis and by Brownian diffusion. For steady flow of nanoparticles the transmittance of a reactor is determined by the expression

$$1 - \gamma(L).$$

For a more detailed analytical analysis of the influence of Brownian motion and thermophoresis on the motion of nanoparticles in a cylindrical flow reactor we apply the Galerkin method [15]. We seek an approximate solution for the temperature field in the form of two term of the series expansion

$$T(r,z) = T_w + A_t J_0(br / R)$$
(13)

where J_0 is the Bessel function of the zeroth order, b is the smallest positive root of the equation

$$J_0(b) = 0,$$

well known that b is approximately equal to 2.48. Expression (13) exactly satisfies the boundary conditions (6). It can be shown that the solution of ordinary differential equations for the function A_t (z) is given by the expression

$$A_t(z) \sim \exp\left[-\frac{k_1}{k_0} z \right],$$

where the coefficients k_1 and k_0 are calculated by means of integrals. After numerical integration we have following expressions for these coefficients:

$$k_0 = u_0 R^2 \int_0^1 t\left(1-t^2\right) J_0^2(bt)dt \approx 0.1 u_0 R^2 ,$$

$$k_1 = b^2 \int_0^1 \frac{\lambda(T(r))}{\rho(T(r)c} t J_1^2(bt)dt \approx 0.13 \frac{b^2 \lambda(T_w)}{\rho(T_w)c} ,$$

where J_1 is the Bessel function of the first order. To note that $k_0/k_1 = l_t$. Thus, the temperature field in gas flow in the reactor is approximately described by the following expression:

$$T(r,z) \approx T_w + \left(T_0 - T_w\right) J_0(br/R) \exp\left(-z/l_t\right). \tag{14}$$

Now, taking into account (14), the thermophoretic velocity of nanoparticle can be represented as

$$v_{th} = -K\nabla T \approx -K \exp\left(-z/l_t\right)\left(T_0 - T_w\right)\frac{b}{R} J_1(br/R), \tag{15}$$

To note that on the axis of the reactor the thermophoretic velocity of nanoparticle is equal to zero, and decreases exponentially with increasing distance from the reactor inlet. Now we can start to analyze the equation (3) by the Galerkin method. Let us seek the solution of equation (3) in the simplest form, which nevertheless exactly satisfy to boundary conditions (4) and (6):

$$n(r,z) \approx A_n(z) J_0(br/R).$$

The unknown function A_n (z) is governed by the ordinary differential equation

$$k_0 \frac{dA_n}{dz} = -A_n\left[k_2 + k_3\right], \tag{16}$$

where coefficients are expressed by formulas:

$$k_3 = D_b(T_w) \int_0^1 t J^2{}_1(bt)dt \approx 0.13 D_b(T_w), \tag{17}$$

$$k_2 = \frac{KR\exp\left(-z/l_t\right)\left(T_0 - T_w\right)}{b} \int_0^1 t J^2{}_1(bt) J_0(bt)dt \approx 0.046 \frac{KR\exp\left(-z/l_t\right)\left(T_0 - T_w\right)}{b} \tag{18}$$

After integration of the equation (16), we have the formula with double exponent, which takes into account the influence of thermophoresis on nanoparticles distribution:

$$A_{\mathrm{n}}(t) = A_{\mathrm{n}}(0)\exp\left[-\frac{k_3}{k_0}z\right]\exp\left[\frac{0.023KR\exp\left(-z/l_{\mathrm{t}}\right)\left(T_0-T_{\mathrm{w}}\right)\rho(T_{\mathrm{w}})c}{\lambda(T_{\mathrm{w}})}\right].$$

Then the field of the number density of nanoparticles in reactor can be written as

$$n(r,z) = A_0 J_0(br/R)\exp\left[-\frac{z}{l_{\mathrm{b}}}\right]\exp\left[\frac{0.023KR\exp\left(-z/l_{\mathrm{t}}\right)\left(T_0-T_{\mathrm{w}}\right)\rho(T_{\mathrm{w}})c}{\lambda(T_{\mathrm{w}})}\right]$$

(19)

Thus for nonisothermal case the characteristic length l_{b} of the Brownian deposition of nanoparticles can be represented as (see also [2-4])

$$l_{\mathrm{b}} = \frac{0.14u_0R^2}{0.25D_{\mathrm{b}}(T_{\mathrm{w}})}, \tag{20}$$

In particular for reactor with for $T_{\mathrm{w}} = 583$ K , $T_0 = 283$K and $R_{\mathrm{n}} = 5$ nm calculated value $l_{\mathrm{b}} = 1.95$ m and the characteristic length of thermal field $l_{\mathrm{t}} = 0.008$ m. Clear that thermophoresis can enhance or diminish the nanoparticles deposition (see also Figure 4.). The weakening of the deposition occurs in the case when the wall temperature is significantly above the temperature of the gas flow.

It is worthy to note that in the isothermal approximation the integral parameter γ approximately expressed as

$$\gamma(L) = 1 - \exp\left[-\frac{L}{l_{\mathrm{b}}}\right]. \tag{21}$$

For sufficiently long reactor compared with the value of l_{b} the expression (21) provides good accuracy of the description of experimental data [4, 6].

For a full quantitative analysis of the interaction of Brownian diffusion and thermophoresis in nonisothermal flow in the reactor is necessary to use methods of numerical simulation. For plane flow this interaction was studied in [3]. For a cylindrical reactor some our numerical results are presented in the next section.

3. Simulation Results

The system of equations (1 - 3) with the boundary conditions (4 - 6) and initial conditions (7) was solved by "semi- discrete straight lines" method [13]. We made calculations for reactor parameter used in experiments [6]: the average inlet flow velocity $u_0 = 0.5$ m / s, the channel diameter is 13 mm, and the radius of the nanoparticles is 5 nm.

Results of simulation of the motion of Brownian nanoparticles in isothermal cylindrical reactor are shown in Figure 1.a. Obviously, for the isothermal case the influence of thermophoresis is zero. The ratio γ (150R) = 0.24. As can be seen the number density of nanoparticles drops quite fast in the vicinity of the reactor wall. The exponential dependence of the number density of nanoparticles versus path length is obvious.

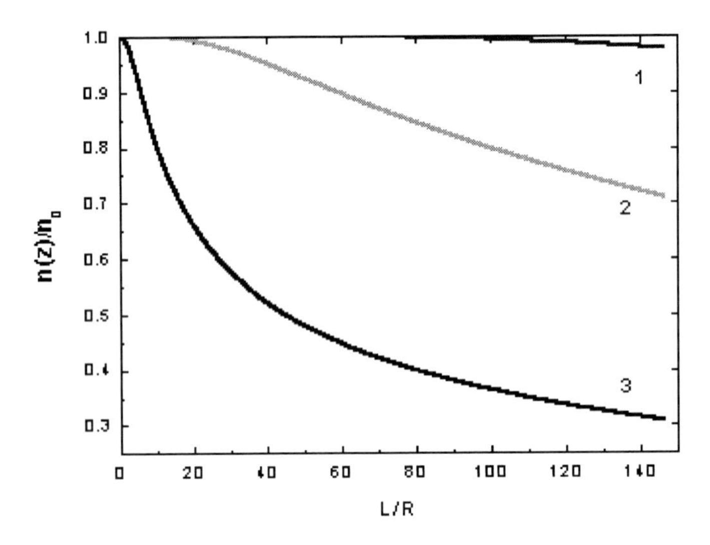

Figure 1.a. Dimensionless the number density of nanoparticles versus distance along axis of reactor 1 – 0.4 R, 2 – 0.7R, 3 – 0.9 R.

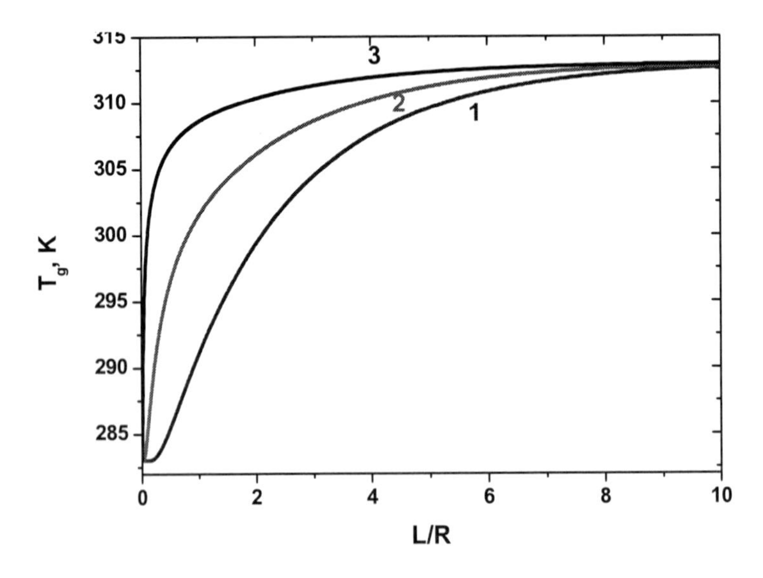

Figure 1.b. Gas temperature versus axial distance. 1 – 0.4 R, 2 – 0.7 R, 3 – 0.9 R.

The evolution of temperature of gas flow is displayed in Figure 1.b. Inlet temperature is equal to 283K. We see that on the way 10R the flow is practically isothermal one.

For isothermal reactor the change of γ versus the length of passage is shown in Figure 2. (curve 1).The use of criterion A is very effective. Curve 2 described the impact theBrownian diffusion and thermophoresis. The wall temperature is above the inlet flow temperature on 30K, therefore the temperature gradient is 4000 K /m. It is evident that thermophoresis with the direction of the temperature gradient decreases the deposition of nanoparticles, although the effect is quite small. Thus, the criterion A can accurately describe the isothermal Brownian deposition. According to (8) the criterion A depends on parameters of the problem :

$$A \sim \frac{T^{0.5}}{R^2 R_n^2 P u_0}.$$

When the wall temperature is above the temperature of the incoming thermophoresis reduces deposition of nanoparticles. It is important to note also that the negative effect of thermophoresis on the deposition strengthened by the circumstance that a hotter wall increases the average flow velocity. As our calculation shown this effect is not compensate by increasing the of Brownian diffusion coefficient. For short reactors the parameter γ is directly proportional to the coefficient of Brownian diffusion calculated at the wall temperature (see (8) and (17)). Thus γ is inversely proportional to the total pressure P in the system. It is evident that this qualitative conclusion is confirmed by the data shown in Figure 3. The calculation was performed for the isothermal case at temperature of 700K and for nanoparticles with a radius of 5 nm.

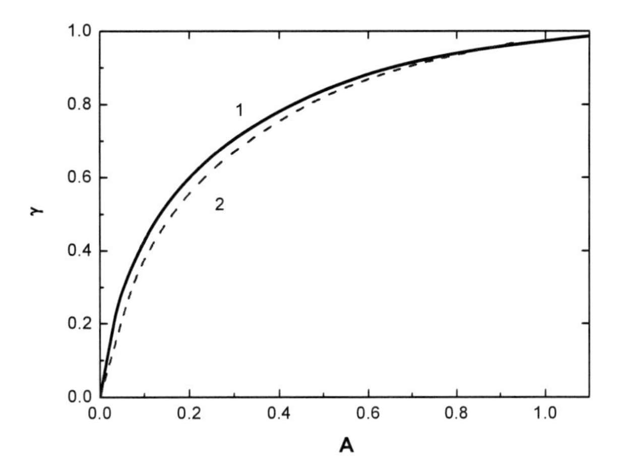

Figure 2. Parameter γ versus criterion A. 1 is isothermal problem, 2 wall temperature is higher on 30K of inlet flow temperature.

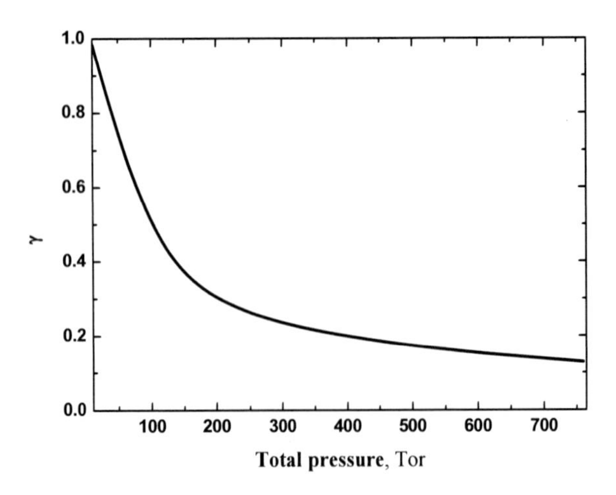

Figure 3. γ versus total pressure in reactor.

The influence of temperature differenced between the reactor wall and incoming flow on parameter γ is shown in Figure 4. T_w= 500 and total pressure 60 Tor. For isothermal case criterion A= 0.1 and γ=0.37. We see that for given reactor a change of temperature difference substantially affect on deposition of nanoparticles. The Criterion B is shown also in this picture. The distribution of the important ratio $nu\ (z,\ r)/n_0u_0$ is displayed in Figure 5. Calculations are made for typical experimental conditions: the wall temperature of the reactor is 400 K, the initial temperature of the gas flow is 300 K, pressure 60 Tor. Curve 2 shows the beginning of the deformation of the nanoparticles distribution under the influence of thermophoresis. We see that after equalizing temperature field the Brownian diffusion start to move nanoparticles to the wall.

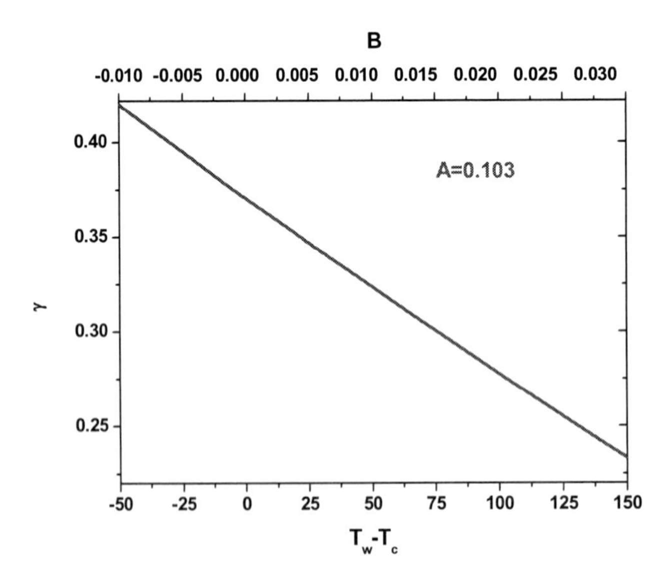

Figure 4. γ versus the temperature difference.

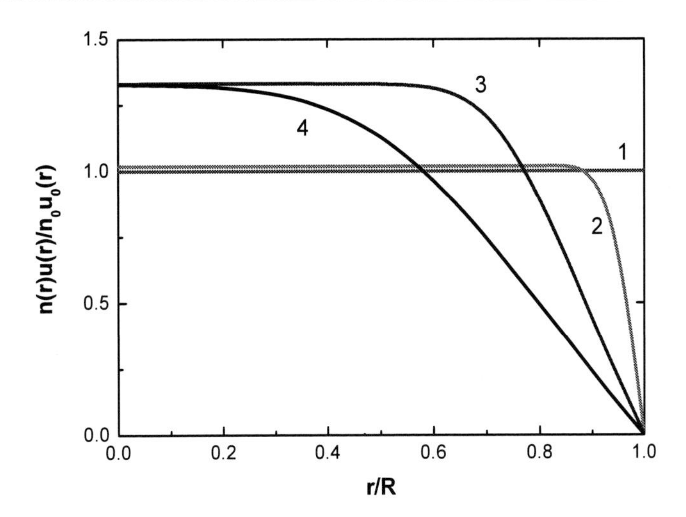

Figure 5. Normalized flux of nanoparticles verus radial position. 1 reactor inlet, 2 is on the axial distance 0.025 R, 3 is on the axial distance 15R, 4 is on the axial distance 146R. Criteria: A = 0.055, B = 0.033.

Unfortunately, we did not find experimental data which can be compared with our calculated one. We work in this direction.

4. Discussion of Results

We considered the Brownian diffusion and thermophoresis of spherical nanoparticles in a laminar flow incoming in flow reactor. When the temperature of the reactor wall is different from the inlet gas temperature a temperature gradient arises and, consequently, thermophoresis of nanoparticles.

Our mathematical model includes a two-dimensional equation of convective heat and two-dimensional equation of convective diffusion of nanoparticles (Smoluchowski equation) and the continuity equation for gas flow in integral form. We showed that there are two independent criteria to describe the interference of processes of Brownian diffusion and thermoforesis in this system: criterion A and criterion B (expressions (8) and (11)). Brownian diffusion and thermophoresis are considered in the approximation of the free molecular regime of kinetic theory of gases because the mean free pass of gas molecule is much larger than the nanoparticles radius.

It was shown that in the isothermal mode of reactor performance to describe the efficiency of the Brownian deposition we need only one criterion A. This conclusion is consistent with our previous results [2]. With increasing value of criterion A nanoparticles deposition increases also, see parameter γ in Figure 2. Numerical calculations and approximate analytical calculations showed that Brownian deposition of nanoparticles acts effectively, if the channel length L is greater than the characteristic length of Brownian deposition l_b (see equation (20)). Thus, when $L \gg l_b$, we have a complete deposition of particles on the channel walls (see equation (21)).

In the nonisothermal mode of reactor performance thermophoresis is faster process than Brownian diffusion, and it distorts the nanoparticles distribution function of radial coordinate.

If the criterion B <2 • 10^{-3}, the contribution of thermophoresis does not exceed a few percent of the value of γ. The simulation results shown that thermophoresis is significant factor only in the initial section of a flow reactor. Thus to suppress the deposition on the reactor walls is necessary to increase the rate of gas flow (criterion A) or increase the diameter of the channel (see criteria A and B and the corresponding Figure 2 and 4)

To note that there is a qualitative analogy between the problem under consideration and the problem of Brownian diffusion and drift of charged nanoparticles in a flow reactor with an external electric field [17]. Sure, there are close analogies between the movement of nanoparticles in the gas and liquid medium, but in a continuous medium thermophoresis is much weaker and deposition is governed by mainly Brownian diffusion. In particular, the radial profiles of nanoparticles in the channel, as shown in [18], can be qualitatively explained by the Brownian diffusion of nanoparticles on the wall.

In a gaseous environment usually there is a small deviation shape of nanoparticles from the spherical one. This circumstance induces a rotational motion of nanoparticles and rotational Brownian motion of nanoparticles. These phenomena can significantly change their deposition. The theoretical description of such phenomena requires a significant change in the mathematical description of the Brownian diffusion. For development of nanotechnology research activity at this direction is important and challenging.

REFERENCES

[1] Mädler L. and Friedlander S. K., Transport of Nanoparticles in Gases: Overview and Recent Advances //*Aerosol and Air Quality Research.* 2007, Vol. 7, No. 3. Pp. 304 – 342.

[2] Brin A. A., Fisenko S.P., Shnip A.I. Brownian Deposition of Nanoparticles from a Laminar Gas Flow Through a Channel // *Technical Physics*,. 2008. T. 53, № 9. P. 1141 – 1145.

[3] Fisenko S. P. and Shnip A. I. Deposition of nanoparticles on cold substrate from laminar gas flow. in "*Physics, Chemistry and Applications of Nanostructures*". Eds. V. E. Borisenko, S. V. Gaponenko, V. S. Gurin, Singapore, World Scientific: 2003. Pp. 291 – 293.

[4] Fisenko S.P. Effectiveness of Brownian deposition of nanoparticles from a gas flow in a tube//*J. Engineering Physics and Thermophysics*, 2010, т.83 №1 P.10- 14.

[5] Shimada M., Seto T., Okyuama K. Thermoporetic and evaporational losses of ultrafine particles in heated flow // *J. of American Institute of Chem. Engineers*. 1993. Vol. 39, №11. Pp. 1859–1869.

[6] Shimada M., Seto T., Okyuama K. Wall deposition of ultrafine aerosol particles by thermophoresis in nonisothermal laminar pipe flow of different carrier gas // *Japanese J. of Applied Physics*. 1994. Vol. 33. Pt.1, №2. Pp. 1174-1181.

[7] Zubarev D. N., Bashkirov A. G. Statistical theory of brownian motion in a moving fluid in the presence of a temperature gradient, *Physica A*, 1968, vol.39, pp.334-340.

[8] Bashkirov A.G. Nonequilibrium statistical mechanics of heterogeneous systems. II. Brownian motion of a large particle // *Theoretical and Mathematical Physics* 1981, т.44, p.623- 629.

[9] Rudyak V. Ya , Krasnolutskii S.L.. On the thermal diffusion of nanoparticles in gases // *Technical Physics*, 2010, V. 10. N 8.p.49-52.
[10] Talbot L., Cheng R. K, Schaefer R. W., and Willis D. R. Thermophoresis of particles in a heated boundary layer//*J. Fluid. Mech.* 1980. Vol. 101. Pp. 737–758.
[11] Kubo R. *Statistical Mechanics*, Amsterdam, North – Holland Publishing Co: 1965.
[12] Landau L.D. Liftshits E.M.. *Hydrodynamics*.Oxford, Pergamon Press, 1980.
[13] Krainov V. P., *Qualitative Methods in Physical Kinetics and Hydrodynamics*, American Institute of Physics, New York, 1992.
[14] Fisenko S. P., Brin A. A. Heat and mass transfer and condensation interference in a laminar flow diffusion chamber// *Intern. J. Heat and Mass Transfer*, 2006, v.49, issue 5/6, Pp. 1004-1014.
[15] Fletcher C. A. J. *Computational Galerkin Method*. New York, Springer: 1984.
[16] Verzhbitskii V. M., *Fundamentals of Numerical Methods* [in Russian], Moscow Vysshaya Shkola, 2002.
[17] Fisenko S.P. Brownian motion and the drift of charged nanoparticles in laminar gas flow in a plane channel //*J. Eng. Physics and Thermophysics*, 2009, v.82. p. 209-214.
[18] D. Wen, L. Zhang, Y. He, Flow and migration of nanoparticle in a single channel// *Heat Mass Transfer* (2009) vol. 45, pp.1061–1067.

In: Brownian Motion: Theory, Modelling and Applications ISBN: 978-1-61209-537-0
Editors: R.C. Earnshaw and E.M. Riley

Chapter 11

ON THE JOINT DISTRIBUTION OF QUADRATIC INTEGRALS OF BROWNIAN MOTION

Javier Villarroel and Mercedes Maldonado
University de Salamanca, Facultad de Ciencias,
Salamanca, Spain

1. Introduction

The paradigmatic model of a random process with continuous paths is the classical Brownian motion (BM) or Wiener process $\{W_t \equiv W(t), t \geq 0\}$. By this we understand a continuous-time Gaussian Markov process with independent increments and continuous sample paths, i.e., satisfying the following properties:

(1) the path $t \mapsto W_t$ is continuous with probability 1,

(2) $W_0 = 0$ with probability 1 (the process starts at the origin),

(3) For $s \leq t$ the increments of the process $W_t - W_s$ are homogeneous in time and satisfy the equality in distribution $W_t - W_s \overset{d}{\sim} W_{t-s} \sim \mathcal{N}(0, t-s)$ and hence are normally distributed with zero mean and variance the time-difference.

(4) $W_{t_2} - W_{t_1}$ is independent of $W_{t_4} - W_{t_3}$ whenever $t_1 \leq t_2 \leq t_3 \leq t_4$.

It follows from the above that

$$\Pr\left(W_t \in [x, x+dx]\right) = f(t,x)dx \text{ where } f(t,x) = (2\pi t)^{-\frac{1}{2}} e^{-x^2/(2t)}$$

$$\Pr\left(W_t \in [x, x+dx] | W_s = x_0\right) = \Pr\left(W_{t-s} \in [x, x+dx] | W_0 = x_0\right) = f(t-s, x-x_0)dx.$$

Here $f(t, x|s, x_0) \equiv f(t-s, x-x_0)$ is the density of the process W_t conditional on the past value $W_s = x_0$, $s \leq t$. Note that it is also the Greens function for the heat equation.

The history of Brownian motion as a mathematical model started in 1900 when L. Bachelier [1] proposed modelling the fluctuations in stock prices by Brownian motion. Subsequently A. Einstein in 1905, [3], reintroduced the model to explain the random motion of pollen particles in liquids (first observed empirically by R. Brown) and derived the partial differential equation that the relevant probability distributions satisfy. A rigorous mathematical proof of the existence of such a process was first given by N. Wiener in 1923, [16], by introducing convenient measures on an appropriate infinite-dimensional functional space. His proof was simplified and clarified by P. Levy [14]. The connection with the theory of partial differential equations (PDE) was clarified in [4, 5].

Related to BM one can define naturally interesting functionals, including $T_t \equiv \int_0^t \mathbf{1}_{\{W_s \geq 0\}} ds = \text{length}([0,t] \cap A)$, where $A = \{s \in \mathbb{R} : W_s \geq 0\}$, the total time that BM spends on the positive real-axis prior to t, studied by Levy himself who derived the celebrated Levy arc-sine law. Further interesting functionals are the first passage time to a point b: $\tau_b \equiv \inf\{t\colon W_t = b\}$ and the maximum to date $M_t \equiv \max_{0 \leq s \leq t} W_s$ (see Karatzas and Shreve [13] for the joint density of the processes (W_t, M_t)). However, the determination of the distribution of many functionals of BM is a non-trivial question whose answer remains unknown at this time.

In this paper we consider the functionals $Y_t \equiv \int_0^t W_s\, ds$ and $Z_t \equiv \int_0^t W_s^2\, ds$ and determine the joint density of (W_t, Y_t, Z_t). It turns out that this problem involves considering a more general one, namely that of finding the joint conditional density

$$\Pr(W_t \in dx, Y_t \in dy, Z_t \in dz | W_{t_0} = x_0, Y_{t_0} = y_0) \equiv \pi(t, x, y, z | t_0, x_0, y_0)\, dx\, dydz.$$

In section 2 appealing to the Kac-Feynmann formula we show that $\pi(t, x, y, z | t_0, x_0, y_0)$ is the Green's function—or fundamental solution, [7, 8]— for a certain partial differential operator. Before addressing the problem of determining such fundamental solution with all generality, in section 3 we consider the simpler bi-dimensional case (W_t, Z_t). The relevant PDE is solved and the result given in terms of a certain Laplace transform (results in that direction have also been given by Kac, [11, 12] or [9]). We further reduce down the Laplace inversion to a certain integral on the real axis, amenable to a numerical evaluation. The full three-dimensional problem is taken up in section 4 and the corresponding density is given, again in terms of a Laplace transform. We collect several relevant results from partial differential equations theory in an appendix.

2. Joint Density and Fundamental Solutions

Here we determine the joint density of the processes W_t, Y_t, Z_t where $Y_t \equiv \int_0^t W(s)\, ds$ and $Z_t \equiv \int_0^t W^2(s)\, ds$. Recall that

$$\Pr(W_t \in dx, Y_t \in dy, Z_t \in dz | W_{t_0} = x_0, Y_{t_0} = y_0) \equiv \pi(t, x, y, z | t_0, x_0, y_0)\, dx\, dy\, dz.$$

Note that the joint density of the processes W_t, Y_t, Z_t is simply $\pi(t, x, y, z | 0, 0, 0)$. The function $\pi(t, x, y, z | t_0, x_0, y_0)$ can be determined as follows. Let $s \geq 0$ and $\Theta : \mathbb{R}^2 \to \mathbb{R}$ be an arbitrary Borel-measurable function. We consider the expected value $\zeta(t_0, x_0, y_0; s)$ of $\Theta(W_t, Y_t) e^{-s(Z_t - Z_{t_0})}$ given the "starting values" at $t_0 : W_{t_0} = x_0, Y_{t_0} = y_0$

$$\zeta(t_0, x_0, y_0; s) \equiv \mathbb{E}\Big(\Theta(W_t, Y_t) e^{-s \int_{t_0}^t W_l^2\, dl} \Big| W_{t_0} = x_0, Y_{t_0} = y_0\Big).$$

It follows from direct definition that

$$\zeta(t_0, x_0, y_0; s) = \int dxdy \Theta(x, y) \Big(\int_0^\infty dz\, e^{-sz} \pi(t, x, y, z | t_0, x_0, y_0) \Big). \tag{2.1}$$

We next recall that according to the Kac-Feynmann theorem the function $\zeta(t_0, x_0, y_0; s)$ solves the terminal value problem in the backward variables t_0, x_0, y_0:

$$\Big(\frac{\partial}{\partial t_0} + \mathbf{D}_{x_0,y_0}\Big)\zeta(t_0, x_0, y_0; s) = 0; \ \lim_{t_0 \uparrow t} \zeta(t_0, x_0, y_0; s) = \Theta(x_0, y_0) \tag{2.2}$$

where we introduce the operator

$$\mathbf{D}_{x_0,y_0} \equiv \mathbf{A}_{x_0,y_0} - s x_0^2 \equiv \frac{1}{2}\frac{\partial^2}{\partial x_0^2} + x_0 \frac{\partial}{\partial y_0} - s x_0^2$$

and $\mathbf{A}_{x_0,y_0}$ is the infinitesimal generator of the bi-dimensional diffusion (W_t, Y_t). The latter is determined as follows (for this and other aspects of the connection between stochastic processes and the theory of partial differential equations see [2, 13]). Note that (X_t, Y_t) solves the stochastic differential equations

$$dX_t = 0dt + dW_t, \qquad dY_t = X_t dt + 0dW_t$$

with $(X_1, X_2) \equiv (X_t, Y_t)$, $(W_1, W_2) \equiv (W_t, 0)$ the former system reads in component form $dX_j(t) = \mu_j(X_k)\,dt + \sum_k \sigma_{jk}\, dW_k$ where the infinitesimal drift and infinitesimal variance matrix are given, respectively, by $\mu(x, y) = (0, x)$, $\sigma = \mathrm{Diag}(1, 0)$. The theory of stochastic differential equations yields then that $\mathbf{A}_{x_0,y_0} = \frac{1}{2}\frac{\partial^2}{\partial x_0^2} + x_0 \frac{\partial}{\partial y_0}$.

The fundamental solution or Green's function (see [7, 8, 13]) corresponding to Eq. (2.2) is a function $G(t, x, y; s|t_0, x_0, y_0)$ such that for fixed t, x, y solves the backward equations

$$\Big(\frac{\partial}{\partial t_0} + \mathbf{D}_{x_0,y_0}\Big)G(t, x, y; s|t_0, x_0, y_0) = 0 \tag{2.3}$$

while for fixed t_0, x_0, y_0 it solves the forward equations

$$\Big(\mathbf{D}^{\dagger}_{x,y} - \frac{\partial}{\partial t}\Big)G \equiv \Big(\frac{1}{2}\frac{\partial^2}{\partial x^2} - x\frac{\partial}{\partial y} - s\,x^2 - \frac{\partial}{\partial t}\Big)\, G(t, x, y; s|t_0, x_0, y_0) = 0. \tag{2.4}$$

Furthermore, it satisfies the boundary condition $\lim_{t_0 \to t} G(t, x, y; s|t_0, x_0, y_0) = \delta(x - x_0, y - y_0)$.

Formal differentiation under the integral sign shows that the solution to the *final* value problem Eq. (2.2) can be represented as

$$\int dx\, dy\, G(t, x, y; s|t_0, x_0, y_0)\Theta(x, y) = \zeta(t_0, x_0, y_0; s). \tag{2.5}$$

Then, for given function $\Theta : \mathbb{R} \to \mathbb{R}$ we have two representations for the solution of the *final* value problem, namely formulae (2.1) and (2.5). Thus,

$$\begin{aligned}
\int dx\, dy\, G(t, x, y; s|t_0, x_0, y_0)\Theta(x, y) &= \zeta(t_0, x_0, y_0; s) \\
&= \mathbb{E}\Big(\Theta(X_t, Y_t)e^{-s\int_{t_0}^t W_l^2\, dl}\Big|W_{t_0} = x_0, Y_{t_0} = y_0\Big) \\
&= \int_0^\infty dz e^{-sz}\Big(\int_{\mathbb{R}} dx dy \Theta(x, y)\pi(t, x, y, z|t_0, x_0, y_0)\Big) \\
&\equiv \int_0^\infty dz\, e^{-sz} q(t, z|t_0, x_0, y_0).
\end{aligned}$$

Hence, inverting a Laplace transform we find with $c > 0$

$$\int_{\mathbb{R}} dx\, dy\, \Theta(x, y)\pi(t, x, y, z|t_0, x_0, y_0) \equiv q(t, z|t_0, x_0, y_0)$$

$$= \frac{1}{2\pi i}\int_{c-i\infty}^{c+i\infty} e^{sz} ds \int dx\, dy\, G(t, x, y; s|t_0, x_0, y_0)\Theta(x, y)$$

$$= \int dx\, dy\, \Theta(x, y)\Big(\frac{1}{2\pi i}\int_{c-i\infty}^{c+i\infty} ds\, e^{sz} G(t, x, y; s|t_0, x_0, y_0)\Big)$$

which implies—since the function Θ is arbitrary— that

$$\pi(t, x, y, z|t_0, x_0, y_0) = \frac{1}{2\pi i}\int_{c-i\infty}^{c+i\infty} ds\, e^{sz} G(t, x, y; s|t_0, x_0, y_0). \tag{2.6}$$

3. Marginal Densities

We have just seen that the joint density of (W_t, Y_t, Z_t) follows by Laplace inversion of the fundamental solution to Eq. (2.3) $G(t, x, y; s|t_0, x_0, y_0)$. Before taking up the determination of such a function we first consider the simpler problem of obtaining the marginal conditional densities of (W_t, Z_t) defined as

$$\Pr(W_t \in dx, Z_t \in dz|W_0 = x_0) \equiv \varpi(t, x, z|x_0)\, dx\, dz.$$

Note that elementary probability yields that

$$\varpi(t, x, z|t_0, x_0) = \int \pi(t, x, y, z|t_0, x_0)\, dy.$$

Considering the Laplace transform

$$u(t, x; s|t_0, x_0) = \int_0^\infty dz\, e^{-sz}\varpi(t, x, z|t_0, x_0)$$

then the well known *Hadamard's method of descent* from partial differential equations theory yields that u solves the terminal value problem [10]

$$\left(\frac{1}{2}\frac{\partial^2}{\partial x^2} - s\,x^2 - \frac{\partial}{\partial t}\right) u(t, x; s|t_0, x_0) = 0,\ \lim_{t\to t_0} u(t, x; s|t_0, x_0) = \delta(x - x_0) \tag{3.1}$$

where $\delta(x - x_0)$ is a unit Dirac mass at x_0.

We now show how to solve this equation. Note first that u is homogeneous in time: $u(t, x; s|t_0, x_0) = u(t - t_0, x; s|0, x_0)$ and hence we can drop all reference to t_0. We make a change of dependent variable to introduce a function $v(t, x; s|x_0)$ as follows: define first

$\lambda(t)$ to be the solution to of $\frac{d^2\lambda}{dt^2} - 2s\lambda = 0$ satisfying the initial conditions $\lambda(0) = 1$; $\dot{\lambda}(0) = 0$, i.e., with $\beta \equiv (2s)^{\frac{1}{2}}$ then $\lambda(t) = \cosh(\beta t)$. Define next $v(t, x; s|x_0)$ via

$$u(t, x; s|x_0) = \sqrt{\lambda} \exp\left(-\frac{\dot{\lambda}}{2\lambda} x^2\right) v(t, x; s|x_0)$$

whereupon (3.1) yields that $v(t, x; s|x_0)$ solves the Cauchy problem

$$\left(\frac{1}{2}\frac{\partial^2}{\partial x^2} + \gamma(t)\frac{\partial(xv)}{\partial x} - \frac{\partial}{\partial t}\right) v(t, x; s|x_0) = 0, \quad \lim_{t\to 0} v(t, x; s|x_0) = \delta(x - x_0)$$

Here $\gamma(t) \equiv -\beta \tanh \beta t$. In terms of the Fourier transform:

$$\tilde{v}(t, \omega; s|t_0, x_0) = \int_{-\infty}^{\infty} e^{i\omega x} v(t, x; s|t_0, x_0)\, dx$$

the equation reads

$$\frac{\partial \tilde{v}}{\partial t} + \frac{s^2}{2}\tilde{v} + \gamma\omega\frac{\partial \tilde{v}}{\partial \omega} 0 = 0; \quad \lim_{t\to 0} \tilde{v}(t, \omega; s|t_0, x_0) = e^{ix_0\omega}. \tag{3.2}$$

The corresponding characteristic system is

$$dt = \frac{d\omega}{\gamma\omega} = -\frac{2}{\omega^2}\frac{d\tilde{v}}{\tilde{v}}$$

and hence the first integrals are

$$\psi_1(t, \omega) = \lambda(t)\,\omega, \qquad \psi_2(t, \omega) = \exp\left[\frac{\omega^2}{2}\Sigma^2(t)\right].$$

Therefore, by proposition 2, the solution of (3.2) is

$$\tilde{v}(t, \omega; s|t_0, x_0) = \exp\left[ix_0\omega\lambda(t) - \frac{\omega^2}{2}\Sigma^2(t)\right] \text{ where } \Sigma^2(t) = \beta^{-1}\cosh(\beta t)\sinh(\beta t,)$$

$$v(t, x; s|t_0, x_0) = (2\pi\Sigma^2)^{-\frac{1}{2}} \exp\left[-\frac{(x - \lambda(t)x_0)^2}{2\Sigma^2}\right].$$

Hence, if we have that $u(t, x; s|x_0) = u(t - t_0, x; s|0, x_0)$

$$u(t, x; s|0, x_0) = \frac{\beta}{\sqrt{\pi \sinh \beta t}} \exp\left[-\frac{\beta}{2}\coth \beta t \left(x - \frac{x_0}{\cosh \beta t}\right)^2 - \frac{\beta}{2}x_0^2 \tanh \beta t\right]$$

The Laplace transform of the joint density of (W_t, Z_t) follows letting $t_0 = 0$, $x_0 = 0$ and is given by

$$\Pr\Big(W_t \in [x, x{+}dx], Z_t \in [z, z{+}dz]\Big) = \frac{dxdz}{2\pi i}\int_{c-i\infty}^{c+i\infty} ds e^{sz} \frac{\beta}{\sqrt{\pi \sinh \beta t}} \exp\left[-\frac{\beta x^2}{2}\coth \beta t\right]$$

$$\Pr\left(Z_t \in [z, z+dz]\Big|W_{t_0} = x_0\right) = \frac{dz}{2\pi i}\int_{c-i\infty}^{c+i\infty} \frac{e^{sz}\,ds}{\sqrt{\cosh\beta(t-t_0)}} \exp\left[-\frac{\beta}{2}x_0^2 \tanh\beta t\right].$$

The marginal density of Z_t corresponds to the election $t_0 = 0, W_0 = 0 = x_0$ and is given by (see [15])

$$\Pr\left(Z_t \in [z, z+dz]\right) = dz\,\pi_Z(t, z|0) = \frac{dz}{2\pi i}\int_{c-i\infty}^{c+i\infty} \frac{e^{sz}\,ds}{\sqrt{\cosh\beta t}}.$$

The integrals can be reduced to one on the real axis as follows. Note first that the function $\sqrt{s}\tanh\sqrt{2st}$ is continuous across the real axis; further, if $s_k = -\frac{\pi^2(2k+1)^2}{8t^2}$ are the zeroes of $\cosh(\sqrt{2st})$ then as s describes a circle around s_k in a anti-clockwise manner $g(s) \equiv \cosh^{\frac{1}{2}}\left(\sqrt{2st}\right)$ changes sign going across the segments $A_{2n} = (s_{2n+1}, s_{2n})$, $n = 0\ldots\infty$ but remains unchanged across $A_{2n+1} = (s_{2n+2}, s_{2n+1})$. Consider a two-sheeted Riemann surface whose sheets are sewn by branch cuts across the segments A_{2n}. On this surface g is holomorphic. The integral is evaluated by closing the contour with a large half circle on the left half-plane of one of the sheets which then continues along both banks of the negative real axis on the same sheet, with branch cuts as indicated. The integral vanishes along the former while contributions from the upper and lower bank of A_{2n+1} cancel each other. Since on this surface the integrand is holomorphic, Cauchy's residue theorem implies that (recall that $\beta \equiv (2s)^{1/2}$)

$$\Pr\left(Z_t \in [z, z+dz]\Big|W_{t_0} = x_0\right) = 2dz\sum_{n=0}^{\infty}\int_{A_{2n}} \frac{\exp\left[sz - \frac{\beta}{2}x_0^2\tanh\beta(t-t_0)\right]ds}{\sqrt{\cosh\beta(t-t_0)}}, \quad t > 0.$$

We consider what happens if $t \to 0$. Note first that

$$A = \bigcup_{n=0}^{\infty} A_{2n} \underset{t\to 0}{\to} \emptyset,\ \cosh\beta t \underset{t\to 0}{\to} 1$$

hence, if $z > 0$ the continuity theorem for measures can be called upon to find

$$\lim_{t\to 0}\pi_Z(t, z|x_0) = 2\int_{\emptyset} e^{sz}\,ds = 0.$$

The result is easy to understand since $Z_0 = 0$ and hence all probability is concentrated at the origin so it does not have a density.

4. The Joint Density of (W_t, Y_t, Z_t)

We now consider the general problem of determining the joint probability density of (W_t, Y_t, Z_t): $\pi(t, x, y, z|t_0, x_0, y_0)$ given that W_t started initially from an an initial value x_0. Recall that we have to find the solution to Eq. (2.4)

$$\left(\frac{1}{2}\frac{\partial^2}{\partial x^2} - x\frac{\partial}{\partial y} - s\,x^2 - \frac{\partial}{\partial t}\right) G(t, x, y; s|t_0, x_0, y_0) = 0$$

with the boundary condition $\lim_{t_0 \to t} G(t, x, y; s|t_0, x_0, y_0) = \delta(x - x_0, y - y_0)$.

Thus $G(t, x, y; s|t_0, x_0, y_0) = G(t - t_0, x, y; s|0, x_0, y_0)$ and

$$\pi(t, x, y, z|t_0, x_0, y_0) = \frac{1}{2\pi i} \int_\Gamma G(t - t_0, x, y; s|0, x_0, y_0)\, e^{sz}\, ds,$$

Again letting $\beta \equiv (2s)^{\frac{1}{2}}$, $\lambda(t) = \cosh(\beta t)$ and

$$G(t, x, y; s|x_0, y_0) = \sqrt{\lambda(t)} \exp\left[-\frac{\dot{\lambda}(t)}{2\lambda(t)} x^2\right] \nu(t, x, y; s|x_0, y_0).$$

we see that this function solves (here $\gamma(t) = -\beta \tanh \beta t$)

$$\begin{aligned} &\frac{1}{2}\frac{\partial^2 \nu}{\partial x^2} - \frac{\partial \nu}{\partial t} + \gamma(t)\frac{\partial (x\nu)}{\partial x} - x\frac{\partial \nu}{\partial y} = 0, \\ &\lim_{t \downarrow t_0} \nu(t, x, y; s|x_0, y_0) = \delta(x - x_0, y - y_0). \end{aligned} \tag{4.1}$$

Thus in Fourier space

$$\tilde{\nu}(t, \omega_1, \omega_2) = \int_{\mathbb{R}^2} e^{i(\omega_1 x + \omega_2 y)} \nu(t, x, y; s|x_0, y_0)\, dx\, dy$$

we have to solve

$$\begin{aligned} &\frac{\partial \tilde{\nu}}{\partial t} + (\gamma(t)\omega_1 + \omega_2)\frac{\partial \tilde{\nu}}{\partial \omega_1} = -\frac{\omega_1^2}{2}\tilde{\nu}. \\ &\lim_{t \downarrow t_0} \tilde{\nu}(t, \omega_1, \omega_2) = e^{i(x_0\omega_1 + y_0\omega_2)} = h(\omega_1, \omega_2). \end{aligned} \tag{4.2}$$

which is a first order, quasi-linear partial differential equation with characteristic system

$$dt = \frac{d\omega_1}{\gamma(t)\omega_1 + \omega_2} = \frac{d\omega_2}{0} = -\frac{2\, d\tilde{\nu}}{\omega_1^2 \tilde{\nu}}.$$

Following appendix A we obtain adequate first integrals:

$$\psi_1(t, \omega_1, \omega_2) = \lambda(t)\omega_1 - \frac{\dot{\lambda}(t)}{2s}\omega_2, \qquad \psi_2(t, \omega_1, \omega_2) = \omega_2$$

$$\begin{aligned} \psi_3(t, \omega_1, \omega_2) &= \exp\left[\dot{\lambda}(t)\lambda(t)\frac{\omega_1^2}{4s} + \left(\dot{\lambda}(t)\lambda(t) - \frac{\dot{\lambda}(t)}{2\lambda(t)} - \dot{\lambda}(t) + s\,t\right)\frac{\omega_2^2}{4s^2}\right. \\ &\quad \left. + \left(1 + \lambda(t) - 2\lambda^2(t)\right)\frac{\omega_1\omega_2}{2s}\right]\tilde{\nu} \\ &= g(t, \omega_1, \omega_2)\tilde{\nu}. \end{aligned}$$

Note that despite the apparent singularity, g, is regular as $s \to 0$. Note also that $g(0, \omega_1, \omega_2) = 1$. By proposition 2 the solution of (4.2) is

$$\tilde{\nu}(t, \omega_1, \omega_2) = \exp\left[-\frac{1}{2}\left(\alpha_{1,1}\omega_1^2 + 2\alpha_{1,2}\omega_1\omega_2 + \alpha_{2,2}\omega_2^2\right)\right]$$
$$\exp\left[i\left(\lambda(t)x_0\omega_1 + \left(\frac{y_0 - \dot{\lambda}(t)}{2s}x_0\right)\omega_2\right)\right],$$

where

$$\alpha_{1,1} = \frac{\lambda(t)\dot{\lambda}(t)}{2s} \qquad \alpha_{1,2} = -\frac{1}{2s}\left(2\lambda^2(t) - \lambda(t) - 1\right)$$
$$\alpha_{2,2} = \frac{\dot{\lambda}(t)}{4s^2\lambda(t)}\left(2\lambda^2(t) - 2\lambda(t) - 1\right) + \frac{t}{2s} \tag{4.3}$$

Inverting the Fourier transform

$$\nu(t, x, y; s|x_0, y_0) = \frac{1}{2\pi\sqrt{|V|}} \exp\left[-\frac{1}{2|V|}\left(\alpha_{2,2}(x - \lambda(t)x_0)^2 + \alpha_{1,1}\left(y - y_0 + \frac{\dot{\lambda}(t)}{2s}x_0\right)^2\right.\right.$$
$$\left.\left. + \ 2\alpha_{1,2}(x - \lambda(t)x_0)\left(y - y_0 + \frac{\dot{\lambda}(t)}{2s}x_0\right)\right)\right], \tag{4.4}$$

with $|V| = \alpha_{1,1}\alpha_{2,2} - 2\alpha_{1,2}$.

Therefore

$$G(t, x, y; s|x_0, y_0) = \sqrt{\lambda(t)} \exp\left[-\frac{\dot{\lambda}(t)}{2\lambda(t)}x^2\right] \nu(t, x, y; s|x_0, y_0).$$

In particular, with $x_0 = y_0 = 0$ we have

$$G(t, x, y; s|0, 0) = \frac{\sqrt{\lambda(t)}}{2\pi\sqrt{|V|}} \quad \exp\left[-\frac{1}{2|V|}\left(\alpha_{2,2} - \frac{\dot{\lambda}(t)}{2\lambda(t)}\right)x^2 + \alpha_{1,1}y^2 + 2\alpha_{1,2}xy)\right].$$

and the joint density of $W_t, Y_t, Z_t)$ is finally found by Laplace inversion of this last expression:

$$\pi(t, x, y, z|0, 0, 0) = \frac{1}{2\pi i}\int_\Gamma G(t, x, y; s|0, 0, 0)\, e^{sz}\, ds,$$

5. Conclusion

In this work we have studied the distribution of the vector of stochastic processes (W_t, Y_t, Z_t) where W_t is a Brownian motion, $Y_t \equiv \int_0^t W_s\, ds$ and $Z_t \equiv \int_0^t W_s^2\, ds$ are, respectively, the integral and quadratic integral.

By appealing to the celebrated Kac-Feynmann formula we reduced the problem of finding the joint density of the above vector process to that of obtaining the fundamental solution for a certain partial differential operator with non-constant coefficients. We solve the latter problem by first using a certain point transformation which permits to get rid of the x-dependent terms; the procedure yields a constant coefficients, partial differential equation that we solve by joint Fourier-Laplace transformation.

Acknowledgement

The authors acknowledge support from MICINN under contracts No. MTM2009-09676 and from Junta de Castilla y León, SA034A08.

A. Partial Differential Equations of First Order

In this appendix we collect several useful results from the theory of partial differential equations.

Proposition 1 *Let* $x_1, x_2, \ldots, x_n, \phi$ *be a coordinate system in* $\mathbb{R}^{n+1}$ *and let* $R_j(x_1, \ldots, x_n, \phi)$, $j = 1, 2, \ldots, n$, $Q(x_1, \ldots, x_n, \phi)$ *be given functions. The solution of the* n- *dimensional, quasi-linear, non-homogeneous partial differential equation*

$$\sum_{j=1}^{n} R_j(x_i, \phi) \frac{\partial \phi}{\partial x_j} = Q(x_i, \phi) \tag{A.1}$$

is given by

$$\Phi\Big(\psi_1(x_i, \phi), \ldots, \psi_n(x_i, \phi)\Big) = 0,$$

where Φ *is an arbitrary function of class* C^1 *and the* ψ_j*'s are* n *independent first integrals of the characteristic system:*

$$\frac{dx_1}{R_1(x_i, \phi)} = \cdots = \frac{dx_n}{R_n(x_i, \phi)} = \frac{d\phi}{Q(x_i, \phi)}.$$

See [10] for the proof.

We also need the following interesting result.

Proposition 2 *Suppose that there exist a function* $g \colon \mathbb{R}^n \longrightarrow \mathbb{R}$ *and* n *independent first integrals* ψ_j, $j = 1, \ldots, n$ *of the characteristic system that satisfy*

$$\begin{aligned} &\psi_j(x_i, t, \phi) = \psi_j(x_i, t), \qquad \psi_j(x_i, t = 0) = x_j, \quad j = 1, \ldots, n-1, \; x_n \equiv t \\ &\psi_n(x_i, t, \phi) = g(x_i, t)\phi. \end{aligned}$$

Then the solution to the Cauchy problem for the non-homogeneous quasi-linear n *dimensional PDE* (A.1) *with initial condition*

$$\phi(x_1, \ldots, x_{n-1}, t = 0) = h(x_1, \ldots, x_{n-1});$$

is given by

$$\phi(x_1, \ldots, x_{n-1}, t) = \frac{g\Big(\psi_1(x_i, t), \ldots, \psi_{n-1}(x_i, t), 0\Big)}{g\Big(x_1, \ldots, x_{n-1}, t\Big)} h\Big(\psi_1(x_i, t), \ldots, \psi_{n-1}(x_i, t)\Big)$$

Proof. Recall that a general solution to Eq. (A.1) is given implicitly as $\Phi\{\psi_1, \ldots, \psi_n\} = 0$ or alternatively, by solving for one of the unknowns, as

$$\psi_n(x_i, t, \phi) = F(\overrightarrow{\psi}(x_i, t)).$$

Here $\overrightarrow{\psi} = (\psi_1, \ldots, \psi_{n-1})$ and $F\colon \mathbb{R}^{n-1} \longrightarrow \mathbb{R}$ is an arbitrary C^1 function. Hence by choosing the first integrals as indicated we have that

$$g(x_i, t)\phi(x_1, \ldots, x_{n-1}, t) = F(\psi_1(x_i, t), \ldots, \psi_{n-1}(x_i, t)).$$

This relationship implies initially ($t = 0$) that

$$g(x_1, \ldots, x_{n-1}, 0)h(x_1, \ldots, x_{n-1}) = F(\overrightarrow{\psi}(x_i, 0)) = F(x_1, \ldots, x_{n-1}).$$

Therefore

$$\phi(x_1, \ldots, x_{n-1}, t) = \frac{F(\overrightarrow{\psi}(x_i, t))}{g(x_i, t)} = \frac{g(\overrightarrow{\psi}(x_i, 0))}{g(x_i, t)} h(\overrightarrow{\psi}(x_i, t)).$$

References

[1] Bachelier (1900), *La Theorie de la Speculation*. Ann. Sci. Ecole Norm. Super., 3, 17 (Translated in: P. H. Cootner (ed.), The Random Character of Stock Market Prices, MIT Press, Cambridge, MA, 1988).

[2] R. Bhattacharya and E. Waymire (1990), *Stochastic processes with applications*. Wiley, NY.

[3] A. Einstein, *Ann. Physik* , **17**, 549 (1905).

[4] W. Feller, *Ann. Math.*, **55**, 468-519 (1952).

[5] W. Feller, *Ann. Math.*, **60**, 417-436 (1954).

[6] W. Feller, *Trans. Amer. Math. Soc.*, **77**, 1-31 (1954).

[7] A. Friedman (1964), *Partial differential equations of parabolic type*. Prentice Hall, N. J.

[8] A. Friedman (1964), *Stochastic differential equations, Vol. I and II*. Academic Press, N. Y.

[9] I. M. Gelfand, A. M. Yaglom, *Journal Mathematical Physics*, **1**, 48-69 (1960).

[10] F. John (1991), *Partial differential equations*. Springer Verlag, Berlin.

[11] M. Kac, *Proc. 2nd. Berkeley Symp. on Math. Stat. and Probability*, 189-215, University of California Press (1951).

[12] M. Kac, *Trans. Amer. Math. Soc.*, **65**, 1-13 (1949).

[13] I. Karatzas, S. E. Shreve (1991), *Brownian motion and stochastic calculus*. Springer.

[14] P. Lévy (1948), *Processus Stochastiques et Mouvement Brownien*. Gautier-Villars, Paris.

[15] J. Villarroel, *Stochastic Anal. Appl.*, **21 (6)**, 1391-1418 (2003).

[16] N. Wiener, *J. Math. Phys.*, **2**, 131-174 (1923).

In: Brownian Motion: Theory, Modelling and Applications ISBN: 978-1-61209-537-0
Editors: R.C. Earnshaw and E.M. Riley

Chapter 12

TWO EXTENSIONS OF FRACTIONAL BROWNIAN MOTION: APPLICATION TO TRABECULAR BONE RADIOGRAPHS

Rachid Jennane [1,*], ***Rachid Harba*** [1], ***Aline Bonami*** [2] ***and Eric Lespessailles*** [3]
[1]Institut PRISME-Université d'Orléans, France
[2]Laboratoire MAPMO-Université d'Orléans, France
[3]Equipe Inserm CHR d'Orléans, France

Abstract

In this chapter, starting from fractional Brownian motion of unique parameter H (0 < H < 1), we propose two variants which extend the classical fBm. First, the piece-wise fractional Brownian motion model of parameters H_o, H_i and γ presents two spectral regimes: it behaves like fBm of parameter H_o for low frequencies $|f| < \gamma$ and like fBm of parameter H_i for high frequencies $|f| > \gamma$. Second, the nth-order fractional Brownian motion enables the H parameter to be within the range $]n-1, n[$, where n is any strictly positive integer. The properties of these two processes are investigated. Special interest is given to their increments, which extend fractional Gaussian noises. Synthesis of realizations of such processes is discussed. Finally, an application of such processes for the characterization of trabecular bone radiographs is presented. The overall objective of this application is to provide methods helping in the early diagnosis of osteoporosis for elderly people.

Keywords: Fractals, fractional Brownian motion, Fractional Gaussian noises, bone, osteoporosis

1. Introduction

The main attraction of fractal geometry stems from its ability to describe the irregular or fragmented shape of natural features as well as other complex objects that traditional Euclidean geometry fails to analyze. This phenomenon is often expressed by spatial or time-domain statistical scaling laws. This concept enables a simple geometrical interpretation

*E-mail address: Rachid.Jennane@univ-orleans.fr

and is frequently encountered in a variety of fields, such as geophysics, biology or fluid mechanics. To this end, Mandelbrot introduced the notion of fractal sets [24], which enables to take into account the degree of regularity of the organizational structure related to the physical system's behavior.

Texture is known to provide information about depth and surface orientation and, as such, describes the content of both natural and artificial images. One particular textural model which has received recent interest, especially for its ability to model "natural" shapes and forms is that of fractional Brownian motions (fBm) [27]. This class of mathematical functions have been used to characterize the geometrical properties of sets. They have been used to generate images [13, 25, 24], classify texture [31, 30, 23], to describe the surface area of chemical reactants [35], and the length of coastlines [26]. FBm is governed by the single Hurst parameter, $H(0 < H < 1)$. When applied in image analysis, fractal geometry is often brought to the evaluation of the fractal dimension, D, from a given dataset. D is linked to the H parameter by: $H = E + 1 - D$, where E is the Euclidean dimension. It has been shown that H and the intuitive notion of roughness are closely linked: the lower H is, the higher is the roughness [31]. The interest of fBm is that it simply represents, with the H parameter, a wide class of $1/f^{\alpha}$ nonstationary signals with α in the range $]1,3[$. Another property of fBm is that it is statistically self-similar meaning that as scale is changing, the process looks like identical. For this reason, it has been applied in image processing for modeling texture having multiscale patterns such as natural scenes [31, 17]. Another interest of these models, is that they exhibit long-term memory. This notion has been widely encountered in experimental data sets (e.g., in hydrology [16] and in communications [6, 1]).

These three properties ($1/f$ spectral behavior, self-similarity, and long-term memory) are strictly equivalent for fBm and make it convenient for many applications. However, in some experimental cases, the classical definition of fBm is not sufficient to properly describe real data.

As a first example, short- and long-term correlations could be different. A unique frequency regime or self-similar behavior is not efficient to entirely describe the observed data. For example, in [37], aggregates of silica particles showed an $1/f$ spectral behavior with two slopes: a low- and a high-frequency regime separated by a cut-off frequency corresponding to the inverse mean size of particles. This kind of dual self-similarity, i.e., different long- and short-term correlations [12], was also shown on bone radiographs [15] and on smectite clay images [14]. A possible solution has been proposed to overcome this difficulty [20].

FBm-based models that are able to capture these long- and short term correlations were proposed in [11, 21]. They are either based on a filtered version of fBm or on a generalization of fBm from its structure function [39]. However, these models have no simple analytic definition, and some difficulties arise in the choice of the filter or in the range of variation of the process parameters. Other studies of fractal models were performed by the authors in [8] and [7] and, by others [4]. The change in the slope may be interpreted as a change in fractality according to the scale or the frequency, leading to the notion of a piecewise fractal process. Thus, it would be necessary to define a new continuous model based on fBm having this property. In [2], the general multifractional Brownian motion (gmBm) considered the parameter H as a continuous function of time. This extension leads to a more general model than that needed for the previous application. But this model does not lead to stationary increments nor to self-similar processes.

A second limitation of the model could be given. Values above 1 have been observed as, for example, in Nile River data ([5], p. 84) or in bone radiographs [22]. These two applications suggest the elaboration of a framework to take into account the fact that H could be greater than 1 for true data. The range for H has already been enlarged numerically [28] by incrementing/decrementing the modelized signal. On the other hand, a mathematical theory of generalized fBm has been developed in [3].

To give a new approach between an only numerical one and a purely theoretical one, this work presents two extensions of the classical fBm. First, the piece-wise fractional Brownian motion model (pfBm) of parameters H_o, H_i and γ and which presents two spectral regimes. Second, the nth-order fractional Brownian motion (nfBm) which enables the H parameter to be within the range $]n-1,n[$, where n is any strictly positive integer. The properties of these processes and their increments are presented. Finally, using radiograph bone images, an application of such processes for the biomedical diagnosis of osteoporotic patients is investigated.

2. Fractional Brownian Motion

The fractional Brownian motion (fBm) model proposed by Mandelbrot and Van Ness [27] is a stochastic fractal process governed by a unique parameter $0 < H < 1$, and is defined as:

$$B^H(t) - B^H(0) = \frac{1}{\Gamma\left(H+\frac{1}{2}\right)}\left[\int_{-\infty}^{0}\left((t-s)^{H-1/2} - (-s)^{H-1/2}\right) dB(s) + \int_0^t (t-s)^{H-1/2} dB(s)\right], \tag{1}$$

where t represents time, Γ represents the Gamma function and $B(s)$ is the standard Brownian motion.

A spectral definition of the fBm has been proposed by Reed *et al.* in [36], for every $-\infty < t < +\infty$:

$$B^H(t) = \frac{1}{2\pi}\int_{-\infty}^{+\infty}\left(\frac{1}{i\omega}\right)^{H+\frac{1}{2}}\left(e^{it\omega}-1\right) d\beta(\omega). \tag{2}$$

where $\beta(\omega)$ is the complex Brownian motion in frequency.

The convergence of (2) in the L^2 sense is ensured when $\omega \to 0$ for $H < 1$ and when $\omega \to \infty$ for $H > 0$. This process is real, Gaussian, and continuous.

The covariance function of fBm is:

$$E\left[B^H(t)\overline{B^H(s)}\right] = \frac{C_H}{2}\left[|t|^{2H} + |s|^{2H} - |t-s|^{2H}\right], \tag{3}$$

where $C_H = \Gamma(1-2H)(cos(\pi H))\pi H$ is the constant calculated in [36]. Note from this expression that fBm is not stationary. This is due to the special role played by the kernel

$e^{it\omega} - 1$, while a kernel based on $e^{it\omega}$ would have rendered a stationary process.

FBm being nonstationary, it is more convenient to study its increments, called fractional Gaussian noises (fGn), which turn out to be stationary. FGn's are defined as:

$$G_u^H(t) = B_H(t+u) - B_H(t), t \in \Re. \tag{4}$$

Their covariance function might be expressed both in time and in frequency as:

$$E\left[G_u^H(t)G_u^H(s)\right] = \frac{C_H}{2}\left[|\tau - u|^{2H} + |\tau + u|^{2H} - 2|\tau|^{2H}\right]. \tag{5}$$

It only depends on the lag, $\tau = |t-s|$ and therefore, shows that fGn is a stationary process.

It is also direct to see that its power spectral density (PSD) is:

$$F_u^H(\omega) = \frac{4sin^2(\frac{u\omega}{2})}{|\omega|^{2H+1}}. \tag{6}$$

The self-similarity [27] of fBm is expressed for $a \succ 0$ as:

$$\left\{B^H(t_0 + at) - B^H(t_0)\right\} \overset{law}{=} a^H \left\{B^H(t_0 + t) - B^H(t_0)\right\} \tag{7}$$

For time scales dilated by a positive factor a, the process looks identical to itself, with an amplitude factor a^H. All this means that whatever the observation scale and the time t_0 we start with, the process seems identical, i.e., presents the same roughness.

The next section defines a more general model that takes into account the limitations of the fBm model for true data, especially to capture long- and short-trem correlations that could be different.

3. Piece-Wise Fractional Brownian Motion

Dual self-similarity or different long- and short-term correlation can be defined in the same model as follows:

For $H_o < 1, 0 < H_i < 1, \gamma > 0$, and for every $-\infty < t < +\infty$, let $B^{H_o,H_i,\gamma}(t)$ be defined by the stochastical integral:

$$B^{H_o,H_i,\gamma}(t) = \frac{1}{2\pi}\int_{-\infty}^{+\infty} F^{H_o,H_i,\gamma}(\omega)\left(e^{it\omega} - 1\right) d\beta(\omega). \tag{8}$$

with

$$F^{H_o,H_i,\gamma}(\omega) = \frac{1_{[0,\gamma)}(|\omega|)}{(i\omega)^{H_o+(1/2)}} + \gamma_{H_i - H_o}\frac{1_{[\gamma,\infty)}(|\omega|)}{(i\omega)^{H_i+(1/2)}}, \tag{9}$$

where $\beta(\omega)$ is the complex brownian motion in frequency, and $1_I(w)$ is the indicator function over interval I (i.e., $1_I(\omega) = 1$ if $\omega \in I$, and $1_I(\omega) = 0$, otherwise). Then $B^{H_o,H_i,\gamma}(t)$ is called the piece-wise fractional Brownian motion (pfBm) model of parameters H_o, H_i, and γ.

This integral is convergent in the L^2 sense when $\omega \to 0$ if $H_o < 1$ and when $\omega \to \infty$ if $H_i > 0$. The continuity of $F^{H_o,H_i,\gamma}(\omega)$ is ensured at the cut-off frequency γ by the factor $\gamma^{H_i - H_o}$.

From the definition of pfBm, the variance of the process increments, which are also called the structure function $f(s)$ [39], easily results in:

$$\begin{aligned} f(s) &= Var\left[B^{H_o,H_i,\gamma}(t+s) - B^{H_o,H_i,\gamma}(t)\right] \\ &= \frac{1}{\pi^2}\int_{-\infty}^{+\infty} sin^2\left(\frac{\omega s}{2}\right)\left|F^{H_o,H_i,\gamma}(\omega)\right|^2 d\omega. \end{aligned} \quad (10)$$

From the definition of $F^{H_o,H_i,\gamma}(\omega)$, and by making the change of variable $\omega \to \omega s$, it follows that $f(s)$ behaves as $|s|^{2H_i}$ for fine scales, i.e., as $s \to 0$, and as $|s|^{2H_o}$ for large scales, i.e., as $s \to +\infty$. The covariance of pfBm directly results and clearly indicates that this process is nonstationary.

Thus, pfBm is real, Gaussian, nonstationary, and can be seen to be continuous. Note that pfBm is the classical fBm of parameter H when $H = H_o = H_i \in]0;1[$.

The nonstationarity of the pfBm makes it difficult to study such processes, for this reason it is more convenient to study piece-wise fractional Gaussian noises.

3.1. Piecewise Fractional Gaussian Noises

This section defines the increments of the pfBm process.

Let $B^{H_o,H_i,\gamma}(t)$ be a pfBm, and let its increments be defined by

$$G_u^{H_o,H_i,\gamma}(t) = B_{H_o,H_i,\gamma}(t+u) - B_{H_o,H_i,\gamma}(t), t \in \Re \quad (11)$$

where the lag $u \in \Re$ is fixed. The processes $\left\{G_u^{H_o,H_i,\gamma}(t)\right\}$ are called piecewise fractional Gaussian noises (pfGn) of parameters H_o, H_i, and γ.

Their cross-correlation function is given by:

$$G_{u,v}^{H_o,H_i,\gamma}(t,s) = E\left[G^{H_o,H_i,\gamma}(t)\overline{G^{H_o,H_i,\gamma}(s)}\right]. \quad (12)$$

pfGn are stationarity processes. Their PSD is given by:

$$4sin^2\left(\frac{u\omega}{2}\right)\left|F^{H_o,H_i,\gamma}(\omega)\right|^2. \quad (13)$$

At this point, we know that pfBm is a continuous, Gaussian, and nonstationary process, having stationary increments termed pfGn. To complete the description of the pfBm process, its self-similarity will be assessed, but unlike fBm, the two frequency regimes cannot lead to a unique self-similarity property but to two asymptotic self-similarity regimes.

3.2. Asymptotic Self-Similarity of pfBm

The process $B^{H_o,H_i,\gamma}(t)$ behaves like a fBm of parameter H_o for low frequencies and like a fBm of parameter H_i for high frequencies. More precisely, for any $t_0 \in \Re$, and $H_o, H_i \in]0,1[$, we have the following:

$$lim_{a\to 0}\left\{\frac{B^{H_o,H_i,\gamma}(t_0+at)-B^{H_o,H_i,\gamma}(t_0)}{a^{H_i}\gamma^{H_i-H_o}}\right\}_t \overset{law}{=} \left\{B^{H_i}(t_0+t)-B^{H_i}(t_0)\right\}_t, \tag{14}$$

and

$$lim_{a\to\infty}\left\{\frac{B^{H_o,H_i,\gamma}(t_0+at)-B^{H_o,H_i,\gamma}(t_0)}{a^{H_o}\gamma^{H_i-H_o}}\right\}_t \overset{law}{=} \left\{B^{H_o}(t_0+t)-B^{H_o}(t_0)\right\}_t. \tag{15}$$

A complete proof of this results can be found in [33]. To conclude with self similarity, it can be pointed out that classical fBm/fGn are self-similar, whereas pfBm/pfGn presents an asymptotical self-similarity regime.

3.3. Synthesis of Piece-Wise Fractional Brownian Motion

We know that for a large scale of observation $a \to \infty$, the process behaves like a self-similar process with parameter H_o. For a closer scale of observation when $a \to 0$, the process is governed by the self-similar parameter H_i. This finding can be noticed on the realizations of pfBm on figures 1 and 2. These realizations were obtained using the Fourier synthesis method [24]. For an illustration purpose, this classical method is suitable. The applicability of recent development of fast and exact synthesis method to the pfBm [34] is still an open problem. Two realizations of the process are presented with $H_o = 0.2$ and $H_i = 0.8$ for the first one and with $H_o = 0.8$ and $H_i = 0.2$ for the second one. For both signals, $\omega = 0.05$. Top drawings show the processes with a large scale, and it is clear that it behaves like H_o: rough for $H_o = 0.2$ and smooth for $H_o = 0.8$. On the opposite side, the bottom drawings show the behavior following H_i for a closer scale: smooth for $H_i = 0.8$ and rougher for $H_i = 0.2$.

The aim of the following section is the elaboration of a framework to take into account the fact that the H parameter can be greater than 1 for true data. More precisely, the purpose is to propose a generalization of fBm with higher range of variation for H parameter.

4. N-th Order Fractional Brownian Motion

The nth-order fractional Brownian motion (n-fBm) [32], B_H^n is defined as:

$$\begin{aligned} B_H^n(t) &= \frac{1}{\Gamma\left(H+\frac{1}{2}\right)}\left\{\int_{-\infty}^{0}\left[(t-u)^{H-1/2}-(-u)^{H-1/2}-\ldots\right.\right. \\ &\quad \left. -\left(H-\frac{1}{2}\right)\ldots\left(H-\frac{2n-3}{2}\right).(-u)^{H-n+1/2}\frac{t^{n-1}}{(n-1)!}\right]dB(u) \\ &\quad \left. +\int_{0}^{t}(t-u)^{H-1/2}dB(u)\right\}. \end{aligned} \tag{16}$$

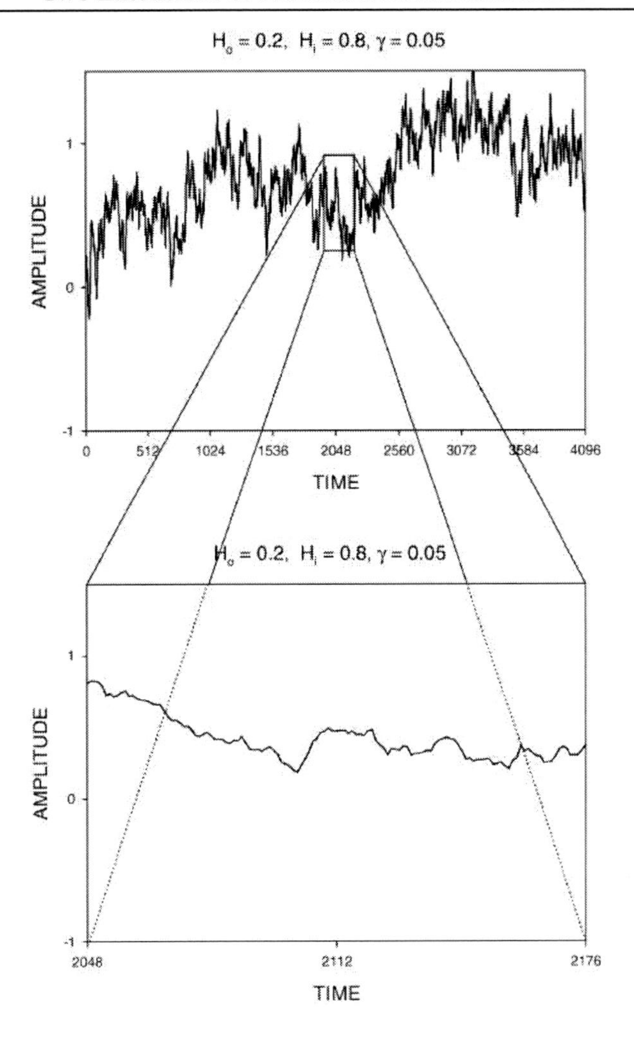

Figure 1. A realization of pfBm with $H_o = 0.2, H_i = 0.8$, $\gamma = 0.05$ presented across two scales. Large scale (Top). Small scale (Bottom).

where n is any strictly positive integer, $B(t)$ is the standard Brownian motion. This definition allows H to be in the range $]n-1,n[$.

When $n = 1$, this definition corresponds to fBm taking value 0 at time 0.

Remark 1: Note that from (16), $B_H^n(0) = 0$ and that the derivative of $B_H^n(t)$ is $B_{H-1}^{n-1}(t)$. In addition, the nth-order fractional Brownian motion can be seen as the antiderivative of order n of B_{H-n+1}^1 having derivatives up to order $n-1$ vanishing at 0. In particular, B_H^n has continuous derivative up to order $n-1$, and the last one has the same smoothness as the fBm of order $H-n+1$. When $n = 1$, equation (16) corresponds to fBm taking value 0 at time 0.

The n-fBm can also be defined using the spectral representation as:

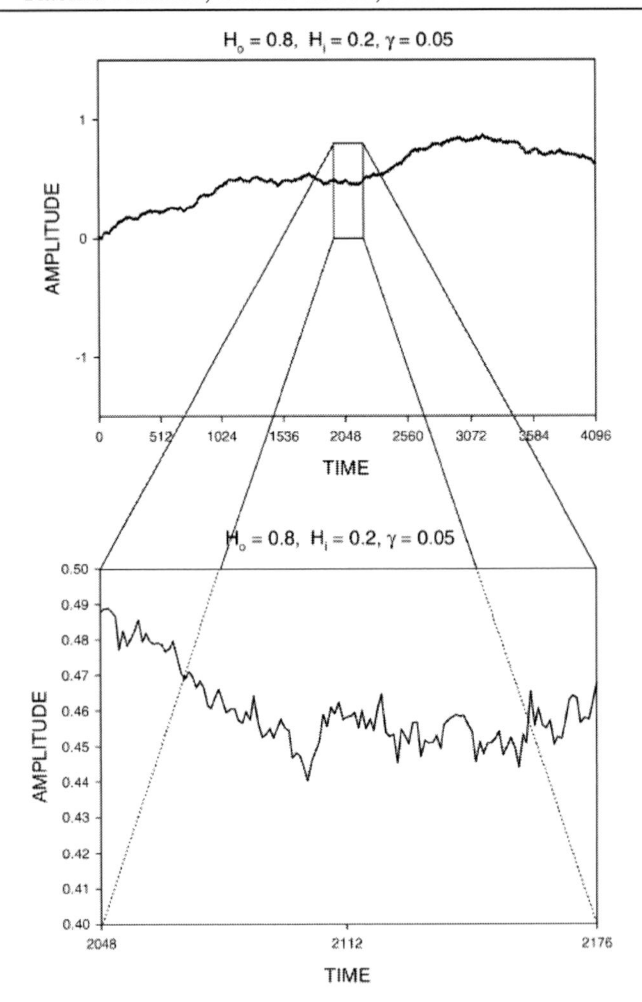

Figure 2. A realization of pfBm with $H_o = 0.8, H_i = 0.2$, $\gamma = 0.05$ presented across two scales. Large scale (Top). Small scale (Bottom).

$$B_H^n(t) = \frac{1}{2\pi}\int_{-\infty}^{+\infty} \frac{1}{(i\omega)^{H+1/2}} \left[e^{it\omega} - \sum_{k=0}^{n-1} \frac{(it\omega)^k}{k!} \right] d\beta(\omega). \tag{17}$$

where β is the complex Brownian process in frequency.

As the classical fBm, the n-fBm is a zero mean, self-similar, and nonstationary process.

Using Remark 1, we know that:

$$B_H^n(t) = \int_0^t B_{H-1}^{n-1}(u)du. \tag{18}$$

From bilinearity, we get the covariance function, $G_{H,n}(t,s)$ of n-fBm:

$$G_{H,n}(t,s) = E\left[B_H^n(t)B_H^n(s)\right] = \int_0^t \int_0^s G_{H-1,n-1}(u,v)dudv. \tag{19}$$

Using the expression of $G_{H-n+1,1}$, we get:

$$G_{H,n}(t,s) = (-1)^n \frac{C_H^n}{2}\left\{|t-s|^{2H} - \sum_{j=0}^{n-1}(-1)^j \binom{2H}{j}\left[\left(\frac{t}{s}\right)^j |s|^{2H} + \left(\frac{s}{t}\right)^j |t|^{2H}\right]\right\}, \tag{20}$$

where

$$\binom{\alpha}{j} = \frac{\alpha(\alpha-1)...(\alpha-(j-1))}{j!} \tag{21}$$

and for $n \geq 2$

$$C_H^n = \frac{C_{H-n+1}^1}{(2H)(2H-1)...(2H-(2n-3))} = \frac{1}{\Gamma(2H+1)\,|sin(\pi H)|}. \tag{22}$$

Equation (20) clearly proves that n-fBm is nonstationary.
Using the fact that

$$\sum_{j=0}^{n-1}\binom{2H}{j}(-1)^j = (-1)^{n-1}\binom{2H-1}{n-1} \tag{23}$$

for $t = s$, we get the variance, of B_H^n

$$V\left[B_H^n(t)\right] = C_H^n \binom{2H-1}{n-1} |t|^{2H}. \tag{24}$$

As for fBm, the variance of this nonstationary process follows a t^{2H} law.
From (20), with $t = at$, the self-similarity of n-fBm follows:

$$B_H^n(t) \stackrel{law}{=} a^{-H} B_H^n(at) \tag{25}$$

In the next section, we will be interested in nth-order fractional Gaussian noises.

4.1. N-th Order Fractional Gaussian Noises

In this section, we are interested in kth-order increments of the n-fBm process. We will see for which k the increments become stationary, leading to the definition of nth-order fractional Gaussian noises (n-fGn).

4.1.1. k-th Order Increments of n-fBm

For a function g, we denote its increments by:

$$\Delta_l g(x) = g(x+l) - g(x) \tag{26}$$

where l is the lag.

Recursively, the kth-order increments are defined as follows:

$$\Delta_l^{(k)} g(x) = \Delta_l^{(k-1)} g(x+l) - \Delta_l^{(k-1)} g(x). \tag{27}$$

The kth-order increments can be obtained as:

$$\Delta_l^{(k)} g(x) = \sum_{j=0}^{k} (-1)^{k-j} \binom{k}{j} g(x+jl). \tag{28}$$

The covariance function of the kth-order increments of n-fBm is obtained by:

$$G_{H,n}^{(k)}(t,s) = E\left[\Delta_{l,t}^{(k)} B_H^n(t) \Delta_{l,s}^{(k)} B_H^n(s)\right]. \tag{29}$$

We start from (20), which corresponds to $k=0$. We use linearity, which leads to

$$\begin{aligned} G_{H,n}^{(k)}(t,s) &= \frac{C_H^n}{2}(-1)^n \left\{ \Delta_{l,t}^{(k)} \Delta_{-l,t}^{(k)} \left(|t-s|^{2H}\right) \right. \\ &\quad - \sum_{j=0}^{n-1} \left[(-1)^j \binom{2H}{j} \Delta_l^{(k)}(t^j) \Delta_l^{(k)} \left(s^{-j}|s|^{2H}\right) \right. \\ &\quad \left.\left. + \Delta_l^{(k)}(s^j) \Delta_l^{(k)} \left(t^{-j}|t|^{2H}\right) \right] \right\}. \end{aligned} \tag{30}$$

4.1.2. n-th Order Stationary Increments of n-fBm

Lest us now define the n-th stationary increments of n-fBm, using the following remark:

$$\Delta_l^{(k)}(t^j) = 0, k \geq j+1. \tag{31}$$

We observe that for $k \geq n$, the expression (30) simplifies, and increments appear to be stationary. The case $k=n$ corresponds to the first k value rendering stationarity of the increments, namely, n-fGn.

We get the covariance of n-fGn for $\tau = (t-s)$

$$\begin{aligned} G_{H,n}^{(k)}(\tau) &= E\left[\Delta_l^{(n)} B_H^n(t-\tau) \Delta_l^{(n)} B_H^n(t)\right] \\ &= (-1)^n \frac{C_H^n}{2} \Delta_l^{(n)} \Delta_{-l}^{(n)} \left(|\tau|^{2H}\right). \end{aligned} \tag{32}$$

A more explicit expression of the covariance function of n-fGn can be written as follows:

$$G_{H,n}^{(k)}(\tau) = \frac{C_H^n}{2}(-1)^n \sum_{j=-n}^{n} (-1)^j \binom{2n}{n+j} |\tau+jl|^{2H}. \tag{33}$$

Let us discuss the physical meaning of the expression, the long-term dependence properties of fractional Gaussian noises (increments of fBm) are determined by the shape of their covariance functions at infinity. Thus, the behavior of $G_{H,n}^{(n)}(\tau)$ for $\tau \to +\infty$ is of interest. For $H \neq n-1/2$, it is equivalent to

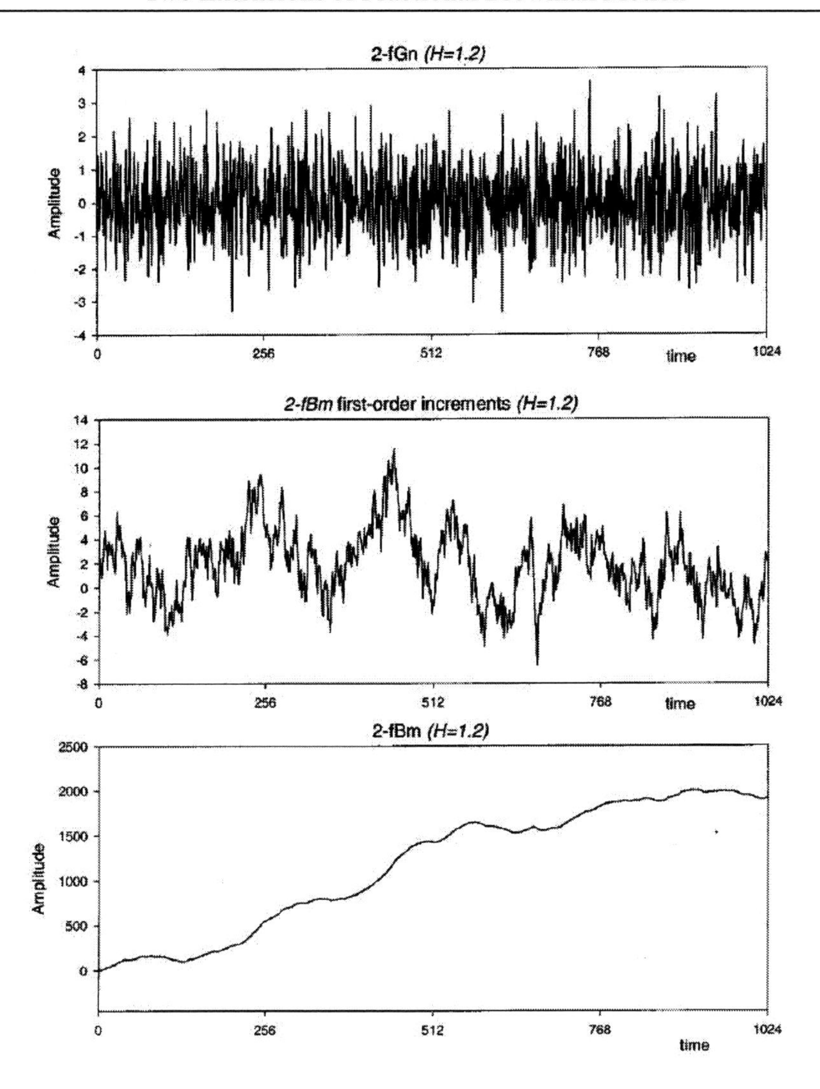

Figure 3. A realization of 1024 samples for $H = 1.2$. (From top to bottom) 2-fGn, 2-fBm first-order increments, and 2-fBm.

$$\frac{C_H^n}{2}(2H)(2H-1)...(2H-2n+2)(2H-2n-1+1)\,|\tau|^{2H-2n}\,l^{2n}. \tag{34}$$

The sign of $G_{H,n}^{(n)}(\tau)$ is the one of $2H-2n+1$. For $n-1 < H < n-1/2$, the process is negatively correlated. For $n-1/2 < H < n$, the process is positively correlated, and in addition, since it decays as a power function, it is said to be a long-term memory process [5]. The case when $H = n-1/2$ is a half integer is of interest. Denoting $G_n(\tau) = G_{n-1/2,n}^{(n)}(\tau)$, it is seen that $G_n(\tau) = 0$ for $|\tau| > n\,|l|$.

Finally, note that for $n = 1$, n-fGn is the classical fGn.

4.2. Power Spectral Density of *n*-fGn

The power spectral density of *n*-fGn, noted $I_{H,n}(\omega)$ is defined as:

$$\begin{aligned} I_{H,n}(\omega) &= \int_{-\infty}^{+\infty} G_{H,n}^{(n)}(\tau)e^{-i\omega\tau}d\tau \\ &= l^{2n}\left(\frac{sin\frac{\omega l}{2}}{\frac{\omega l}{2}}\right)^{2n} |\omega|^{2n-1-2h}, \end{aligned} \quad (35)$$

when $\omega \to 0$, $I(\omega) \to 1/\omega^{2H-2n+1} = 1/w^{\alpha}$ with $-1 < \alpha < 1$. At low frequencies, *n*-fGn behaves like fGn.

Although nonstationary, it would be tempting to evaluate the spectral properties of *n*-fBm. Roughly speaking, it can be said that the spectral shape of *n*-fBm is equal to $1/\omega^{2H+1}$. With $n-1 < H < n$, we get a $1/\omega^{\alpha}$ spectral shape with α in the range $]2n-1, 2n+1[$. These new models, with n being any strictly positive integer, allow a global range of variation for α in the range $]1, +\infty[$. It offers a larger framework than classical fBm for nonstationary $1/f^{\alpha}$ signals, where α can only vary in the range $]1, 3[$.

4.3. Synthesis of *n*-th Order Fractional Gaussian Noises

In this section, we propose a synthesis method for *n*-fGn. Without loss of generality, the case $n = 2$ is considered which means that H is considered in the range $]1, 2[$. This choice is motivated by two reasons. First, this case was the one that motivated the generalized model for values of H just above 1. Secondly, higher order *n*-fBm $(n > 2)$ will require much larger signal length than 2-fBm to illustrate their particular property of smoothness. Definition and properties of 2-fBm and 2-fGn are first briefly presented.

4.3.1. 2-fBm and Its Second-order Increments

Second-order fractional Brownian motion (2-fBm) is defined as:

$$\begin{aligned} B_H^2(t) &= \frac{1}{\Gamma(H+\frac{1}{2})}\left\{\int_{-\infty}^{0}\left[(t-u)^{H-1/2} - (-u)^{H-1/2} - \right.\right. \\ &\left.\left(H-\frac{1}{2}\right)(-u)^{H-3/2}\right] dB(u) \\ &\left. + \int_0^t (t-u)^{H-1/2} dB(u)\right\}. \end{aligned} \quad (36)$$

for $t \geq 0$, with $1 < H < 2$.

B_H^2 is Gaussian, continuous, zero-mean, and self-similar.
Second-order increments of 2-fBm (2-fGn), which are stationary, are defined as:

$$\Delta_l^{(2)} B_H^2(t) = B_H^2(t+2l) - 2B_H^2(t+l) + B_H^2(t). \quad (37)$$

Their autocorrelation function is given by:

$$G_{H,2}^{(2)} = E\left[\Delta_l^{(2)} B_H(t+\tau)\Delta_l^{(2)} B_H(t)\right]$$
$$= \frac{C_H}{2}\left[|\tau+2l|^{2H} - 4|\tau+l|^{2H} + 6|\tau|^{2H} - 4|\tau-l|^{2H} + |\tau-2l|^{2H}\right], \quad (38)$$

and the corresponding spectrum $I(\omega)$ is given by:

$$I(\omega) = l^4 \left(\frac{sin\frac{\omega l}{2}}{\frac{\omega l}{2}}\right)^4 |\omega|^{3-2H}. \quad (39)$$

4.3.2. Synthesis of 2-fGn Realizations

As 2-fBm is a nonstationary process, its synthesis is difficult. For this reason, we consider the synthesis of its second order increments which are stationary. The Cholesky decomposition method [23] can be used. The covariance matrix, R of 2-fGn is:

$$R[i,j] = (G_2[i-j])_{i,j=0,\ldots,N-1}. \quad (40)$$

It is a definite, positive, and Toeplitz matrix. R could be decomposed using the Cholesky decomposition method as:

$$R = LL^T. \quad (41)$$

where L is a lower triangular matrix, and T stands for the matrix transposition operator. Consider the transform $V = LY$, with Y being a white Gaussian noise vector. Then, $E[VV^T] = R$, proving that V has the same second-order statistics as 2-fGn.

For a Gaussian process, a synthesis method is exact if the second-order statistics of the generated data are equal to those of the process. This is clearly the case for the Cholesky method described above since second-order statistics of V are the same as those of the 2-fGn process.

For the reconstruction of the 2-fBm signal, B_H^2 depends not only on the 2-fGn process but on the first order increment, $\Delta^{(1)}B_H^2$ as well. In order to take advantage of the Toeplitz matrix, we prefer to replace in simulations the 2-fBm by its approximation, which has same second-order increments such that $\Delta^{(1)}B_H^2(0) = 0$. Moreover, the 2-fGn process is stationary and, thus, is easier to use in practice than its related 2-fBm nonstationary model. The approximated $B_H^2(m)$ process is recovered from $\Delta^{(l)}B_H^2(m)$ in the following way. Start with $\Delta^{(2)}B_H^2(m)$, which are the first increments of $B_H^2(m)$ deduced from (27):

$$\Delta^{(1)}B_H^2(m) = \Delta^{(2)}B_H^2(m-1) + \Delta_1^{(1)}B_H^2(m-1), \quad (42)$$

with the initial condition $\Delta^{(1)}B_H^2(0) = 0$. This leads to

$$B_H^2(m) = \Delta^{(1)}B_H^2(m-1) + B_H^2(m-1). \quad (43)$$

with $B_H^2(0) = 0$.

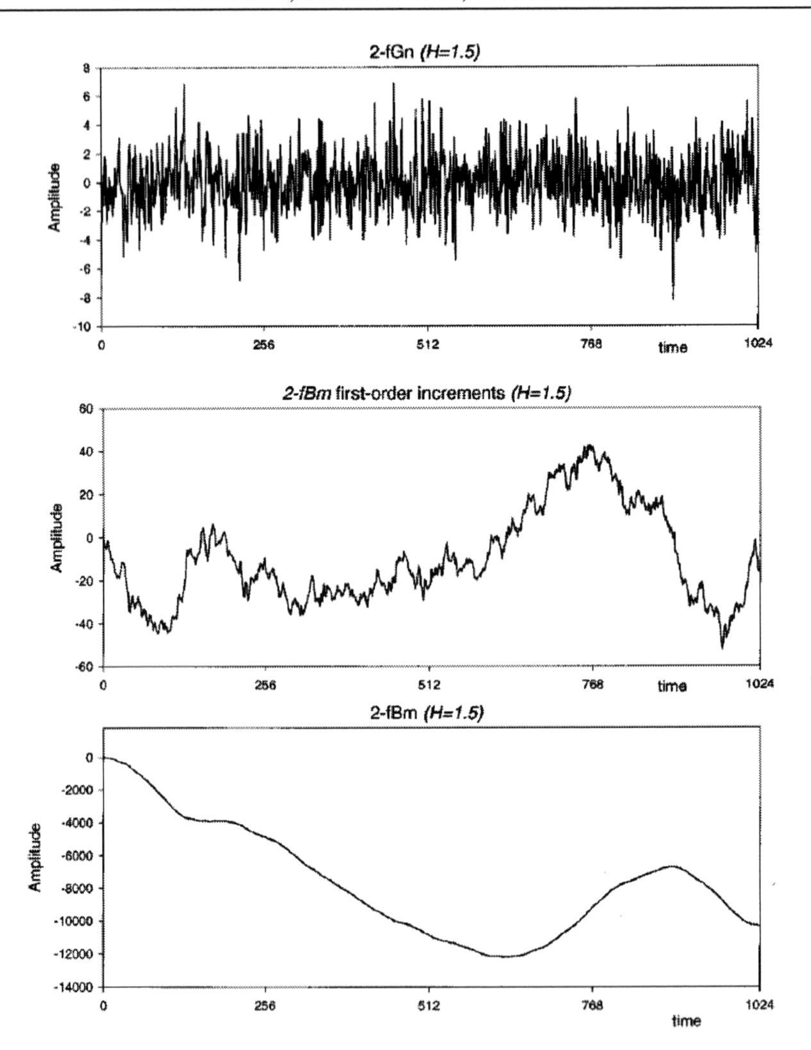

Figure 4. A realization of 1024 samples for $H = 1.5$. (From top to bottom) 2-fGn, 2-fBm first-order increments, and 2-fBm.

For $N = 1024$, realizations of 2-fGn, the increments of 2-fBm, and 2-fBm are presented for $H = 1.2$, $H = 1.5$, and $H = 1.8$ in figures 3, 4, and 5 respectively. The general appearance of 2-fGn (top drawings) looks like the classical fGn, whereas increments of 2-fBm (middle drawings) are very much like fBm. On the other hand, 2-fBm (bottom drawings) exhibits smoother behaviors than classical fBm. Namely, they are able to model nonstationary signals having very soft shapes.

4.3.3. Analysis of 2-fBm Realizations

Among estimators for the H parameter of fBm, the maximum likelihood estimator (MLE) is often considered to be the best one. It has been shown in [9] that it is asymptotically efficient for fBm (ie. unbiased and reaches the Cramer-Rao lower bound (CRLB)) and that estimates are asymptotically Gaussian distributed, with a rate of convergence $\sqrt{N}$. The

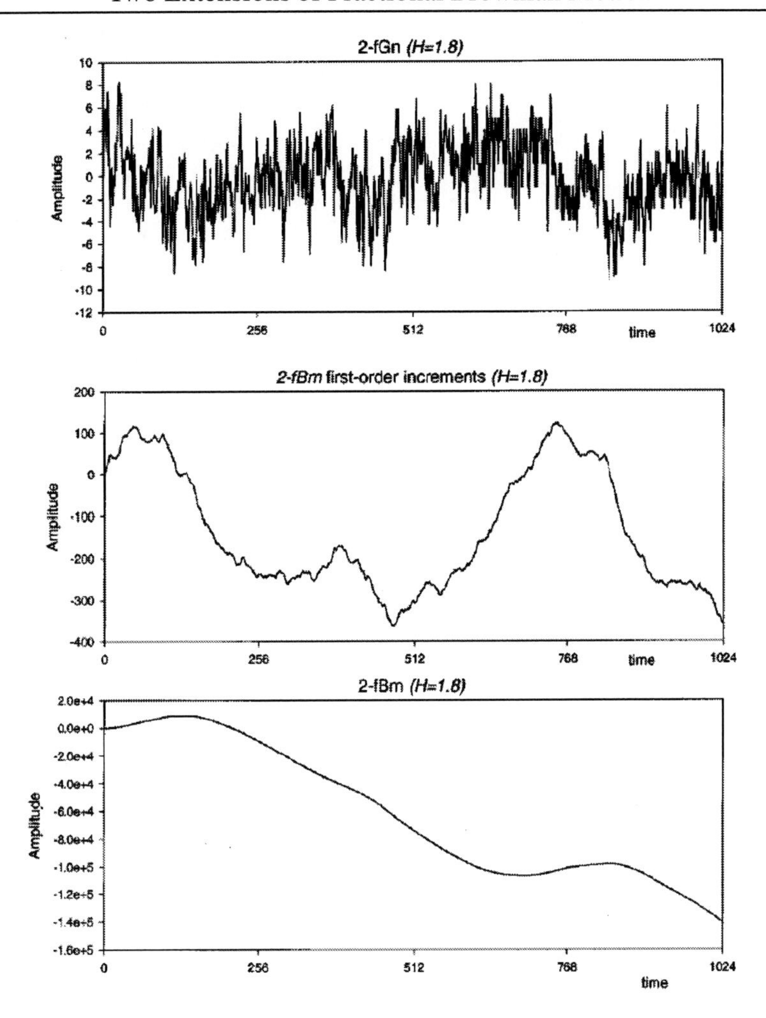

Figure 5. A realization of 1024 samples for $H = 1.8$. (From top to bottom) 2-fGn, 2-fBm first-order increments, and 2-fBm.

Implementation of MLE for 2-fGn is completely described in [32].

5. Application

Osteoporosis is considered as a public health issue [19]. Prevention of fracture results in finding which populations are at risk for fracture. Osteoporosis is defined as a disease characterized by low bone mass and microarchitectural alterations of bone tissue, leading to enhance bone fragility and consequent increase in fracture risk [29]. The bone mass density (BMD) is evaluated in the clinical routine while the microarchitectural modifications are not. The development of a useful microarchitecture indicator providing an appropriate risk factor of osteoporotic fractures would lead to a better diagnosis of osteoporosis [10]. This indicator should be independent from BMD and thus yield complementary information versus BMD to osteoporotic bone changes.

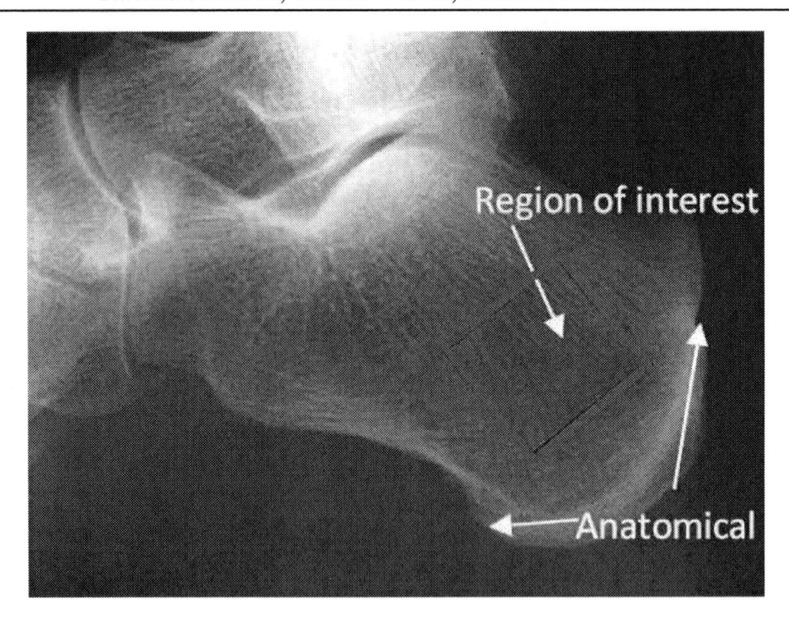

Figure 6. Calcaneus radiograph and its region of interest (red square).

Texture analysis applied to trabecular bone images offers the ability of exploiting the information present on conventional radiographs.

Calcaneus radiographs were performed after a standardized protocol. We used X-ray clinical apparatus with a tungsten tube and an aluminum filter of 1-mm thickness. The tube voltage was fixed to 36 kV and the exposure condition was 18 mA, with an exposure time of 0.08 s. The source-calcaneus distance was settled at 1 m, and the calcaneus was placed in contact with the sensor. An image of a region of interest (256 x 256 pixels) from a calcaneus radiography is extracted thanks to two anatomical marks figure 6. This region was scanned with 256 gray levels to obtain a trabecular bone texture image as presented in figure 7. Our technique was tested on a population composed of 77 radiographs provided by the medical stuff at the hospital of Orleans. Among these subjects, there were 38 patients with osteoporotic fractures (vertebral fracture) and 39 control cases. Because age has an influence on bone density and on trabecular bone texture analysis, the fracture cases were age-matched with the control cases.

The calcaneus (heel bone) as shown in figure 6 from a bone radiograph, presents a complex and textured image which is well suited to measure trabecular bone modifications. This bone is submitted to compression and tension forces produced by the walking and by the gravity. The evolution of the orientations of the trabeculae enables quantifying the degree of deterioration of the bone. For these reasons, first, we have analyzed these data with the classical fBm model. To take into account the trabecular bone modifications transposed on the radiographic images, we performed an oriented fractal analysis. For each direction, we analyzed several lines with the same orientation. The result in one direction was the average of the values for each line. This unidirectional evaluation was repeated in 36 directions. The final result for an image, is calculated as the mean value of the results of the 36 directions. The maximum likelihood estimator (MLE) [27, 23] was used to obtain the value of the H parameter of each line. The MLE was chosen because it is an asymptiotically unbiased

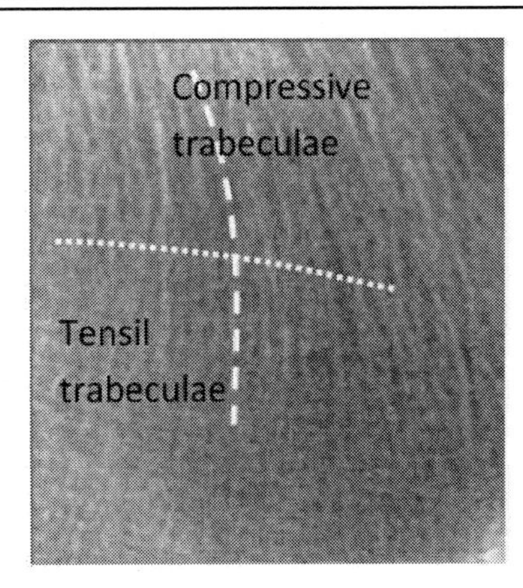

Figure 7. Region of interest measuring 256 x 256 pixels (2.7 x 2.7 cm^2) used for processing.

estimator which has the lowest theoretical variance. Furthermore, this method was tested on synthetic data and shows excellent results for short data length [18]. The mathematical background of this method can be found in [23].

However, A precise analysis of the mean periodogram performed on the lines of trabecular bone images presents two distinct frequency regimes as shown on figure 8. These two regimes are separated by a cut-off frequency γ. The low frequencies $|f| < \gamma$, or, equivalently, the large scales correspond to the intertrabecular area. While the high frequencies $|f| > \gamma$, or, equivalently, the small scales correspond to the trabecular area. This finding confirm that the pfBm model is more suited for such data than the standard fBm model. According to figure 8, γ was fixed to value of 26. This can be explained by the repartition of the bone trabeculae which mean thickness is about 100 to 200 μm separated by a distance of about 400 to 500 μm. According to the sensor's resolution used in this experiment, $\gamma = 26$ corresponds to the mean intertrabecular distance.

Then, the two extended models (pfBm and nfBm) were applied on these data to characterize the trabecular bone modifications, and check if it yields evaluating bone changes. The parameters H_o and H_i were measured by linear regression on the spectral density of the increments of the lines of the different images. Indeed, the non uniform changes due to osteoporosis induce variations of the degree of anisotropy.

A bilateral hypothesis test [38] was applied to evaluate the discrimination power of the H, H_o and H_i parameters. The nfBm model, with $n = 2$ was used when values over 1 for the H_o or H_i parameters were found. In this case, the 2fGn was used. Any obtained value is said to separate the 2 populations if the Student's |t| value exceeds 2.69 for a high level of significance of 0.01. For the H parameter of the classical fBm, t is equal to 4.09. For H_o and H_i, we found, 0.909 and 5.39 respectively.

This demonstrates clearly that the H_i parameter in high frequencies discriminates better the two populations. In conclusion, over a reasonable range of spatial frequencies, a fractal-based model has been validated and can be used to describe trabecular bone data in a concise scale invariant fashion.

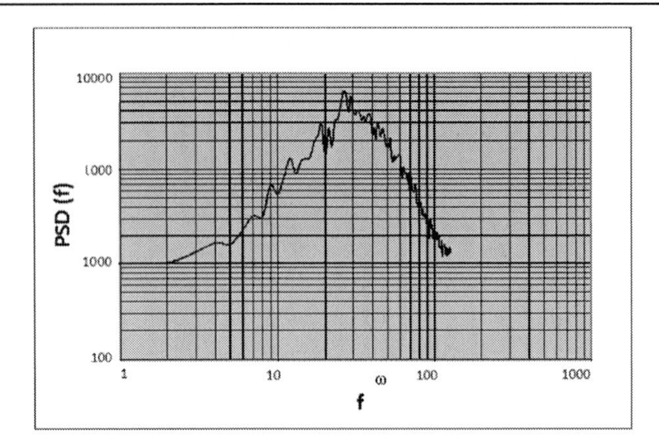

Figure 8. A representative mean periodogram (power spectral density or PSD) of the lines of an X-ray image.

6. Conclusion

In this work, the main properties of two extensions of the fractional Brownian motion model are presented and studied.

First, the piece-wise fractional Brownian model (pfBm) with parameters $0 < H_o, H_i < 1$, and $\gamma > 0$ is presented. This process presents two distinct frequency or self-similar regimes: at low frequencies $|\omega| < \gamma$ or, equivalently, at large scale, the process can be considered to be self-similar with parameter H_o, whereas for high frequencies $|\omega| > \gamma$ or small scale, H_i governs the process. For $H = H_o = H_i$ or limit cases $\gamma \to 0$ and $\gamma \to \infty$, pfBm reduces to classical fractional Brownian motion such that $0 < H < 1$. In addition, pfBm is real, continuous, Gaussian, and nonstationary but has continuous, Gaussian, and stationary increments.

Second, a generalization of fractional Brownian motion called nth-order fractional Brownian motion (n-fBm) is investigated. The H parameter governing n-fBm can vary in the range $]n-1, n[$, where n can be any strictly positive integer. Properties of this new non-stationary processes have been presented. Their nth-order increments are stationary and, thus, were called nth-order fractional Gaussian noises. n-fBm offers a larger framework than fBm to model $1/f^{\alpha}$ nonstationary signals with α within $]1, +\infty[$. These new processes allow the study of a wider range of behavior than classical fBm/fGn. Synthesis methods were discussed for such processes

An application of these processes have been applied to the field of biomedical imaging. Trabecular bone radiographs from two populations composed of osteoporotic patients on one side and healthy subjects on the other side have been characterized using the standard fBm and its extensions, pfBm and $n$fbm. Results demonstrates that the pfBm and n-fBm models, more suited for such data, better discriminate the two populations than the classical fBm. This fractal evaluation of trabecular bone architecture may have applications in a variety of tasks of clinical interest.

References

[1] P. Abry and D. Veitch. Wavelet analysis of long-range-dependent traffic. *IEEE Trans. Inform. Theory*, **44**:1–15, Feb. 1998.

[2] A. Ayache and J. Lévy-Véhel. The generalized multifractional brownian motion. *Statist. Inference Stochastic Process.*, **3**(1-2):7–18, 2000.

[3] L. Bel, G. Oppenheim, L. Robbiano, and M. C. Viano. Distribution processes with stationary fractional increments. In *Proc. ESAIM*, pages 43–54, http://www.emath.fr/Maths/Proc/Vol. 5 1998.

[4] A. Benassi and S. Deguy. *Multi-Scale Fractional Brownian Motion: Definition and Identification*. Clermont-Ferrand, France: Blaise Pascal Univ. Press, 1999.

[5] J. Beran. *Statistics for Log-Memory Processes*. London, U.K: Chapman and Hall, 1994.

[6] J. M. Berger and B. B. Mandelbrot. A new model for the clustering of errors on telephon circuits. *IBM J. Res. Develop.*, **7**:224–236, 1963.

[7] M. Cintract. *Définition d'un modèle fractal par zone: Synthèse, analyze et application*. Ph.D. dissertation, Univ. d'Orléans, Orleans, France, 1999.

[8] M. Cintract and R. Harba. Définition d'un modèle fractal par zone: Synthèse, analyze et application. In *in Proc. AGIS*, pages 289–294, 1997.

[9] R. Dahlaus. Efficient parameter estimation for self-similar processes. *Ann. Stat.*, **17**(4):1749–1766, 1989.

[10] D. W. Dempster. The contribution of trabecular architecture to cancellous bone quality. *J Bone Miner Res*, **15**(1):20–23, Jan 2000.

[11] M. Deriche and A. H. Tewfik. Signal modeling with filtered discrete fractional noise processes. *IEEE Trans. Signal Process.*, **41**(9):1239–2849, Sep. 1993.

[12] J. Feder. *Fractals*. New York: Plenum, 1988.

[13] A. Fournier, D. Fussell, and L. Carpenter. Computer rendering of stochastic models. *Commun. ACM*, **25**(6):371–384, 1982.

[14] R. Harba, M. Cintract, M. Zabat, and H. Hamme. Fractal estimation in a given frequency range: Application to smectite images. In *EUSIPCO Signal Process.*, pages 2485–2488, September 1998.

[15] R. Harba, G. Jacquet, R. Jennane, T. Loussot, C. L. Benhamou, E. Lespessailles, and D. Tourlière. Determination of fractal scales on trabecular bone x-ray images. *Fractals*, **2**:451–456, 1994.

[16] H. E. Hurst, R. P. Black, and Y. M. Simaka. *Long Term Storage: An Experimental Study*. London, U.K.: Constable, 1965.

[17] R. Jennane and R. Harba. Fractional brownian motion: A model for image texture. In *Proc. EUSIPCO Signal Processing*, pages 1389–1392, September 1994.

[18] R. Jennane, R. Harba, and G. Jacquet. Méthodes d'analyse du mouvement brownien fractionnaire : théorie et résultats comparatifs. *Traitement du Signal*, **18**(5-6):419–436, 2001.

[19] O. Johnell. The socioeconomic burden of fractures: today and in the 21st century. *Am J Med.*, **103**(2A)(2OS-25S):discussion 25S–26S, Aug 1997.

[20] L. M. Kaplan and C. C. Kuo. Extending self-similarity for fractional brownian motion. *IEEE Trans. Signal Processing*, **42**:3526–3530, Dec. 1994.

[21] R. L. Kashyap and P. M. Lapsa. Synthesis and estimation of random fields using long-correlation models. *IEEE Trans. Pattern Anal. Machine Intell.*, **PAMI-6**:800–809, 1984.

[22] T. Loussot, R. Harba, G. Jacquet, C. L. Benhamou, E. Lespesailles, and A. Julien. An oriented fractal analysis for the characterization of texture: Application to bone radiographs. In *EUSIPCO Signal Process.*, pages 371–374, September 1996.

[23] T. Lundahl, W. J. Ohley, S. M. Kay, and R. Siffert. Fractional brownian motion: A maximum likelihood estimator and its application to image texture. *IEEE-Trans. On Med. Imag.*, **MI-5**(3):518–523, Sept. 1986.

[24] B. B. Mandelbrot. *Fractals: Form, Chance and Dimension.* New York: Freeman, 1977.

[25] B. B. Mandelbrot. *The Fractal Geometry of Nature.* New York: Freeman, 1982.

[26] B. B. Mandelbrot. Stochastic models for the earth's relief, the shape and the fractal dimension of the coastlines and the number-area rule for islands. *Proc. Nat. Acad. Sci.*, **72**(10):3825–3828, Oct. 1975.

[27] B. B. Mandelbrot and J. W. V. Ness. Fractional brownian motion, fractional noises and applications. *SIAM Rev*, **10**:422–438, 1968.

[28] A. Manzanera, T. Bernard, F. Preteux, and B. Longuet. Medial faces from a concise 3d thinning algorithm. In *IEEE International Conference on Computer Vision (ICCV'99)*, pages 337–343, 1999.

[29] NIH. Consensus development conference: diagnosis, prophylaxis, and treatment of osteoporosis. *Am J Med*, **94**(6):646–650, Jun 1993.

[30] S. Peleg, J. Naor, R. Hartley, and D. Avnir. Multiple resolution texture analysis and classification. *IEEE-Trans. Pattern Anal. Machine Intell.*, **(4)**:518–523, July 1984.

[31] A. P. Pentland. Fractal-based description of natural scenes. *IEEE-Trans. Pattern Anal. Machine Intell.*, **(6)**:661–674, Nov. 1984.

[32] E. Perrin, R. Harba, C. Berzin-Joseph, I. Iribarren, and A. Bonami. nth-order fractional brownian motion and fractional gaussian noises. *IEEE Transactions on Signal Processing*, **49**(5):1049–1059, 2001.

[33] E. Perrin, R. Harba, R. Jennane, and I. Iribarren. Piecewise fractional brownian motion. *IEEE Transactions on Signal Processing*, **53**(3):1211–1215, mars 2005.

[34] E. Perrin, R. Harba, R. Jennane, and I. Iribarren. Fast and exact synthesis for 1-d fractional brownian motion and fractional gaussian noises. *IEEE Signal Process. Lett.*, **9**(11):382–384, Nov. 2002.

[35] P. Pfeeifer, D. Avinir, and A. Farin. Ideally irregular surfaces of dimension greater than two in theory and practice. *Surface Sci.*, **126**:569–572, 1983.

[36] I. S. Reed, P. C. Lee, and T. K. Truong. Spectral representation of fractional brownian motion in dimensions and its properties. *IEEE Trans. Inf. Theory*, **41**(n° 5):1439–1451, sept. 1995.

[37] D. Schaefer, J. E.Martin, P.Wiltzius, and D. S. Cannell. Fractal geometry of colloid aggregates. *Phys. Rev. Lett.*, **52**(5):2371–2374, 1984.

[38] M. R. Spiegel. *Theory and problems of probability and statistics.* Schaum, McGraw-Hill, New York, 1975.

[39] A. M. Yaglom. *Correlation Theory of Stationary and Related Random Functions.* New York: Springer-Verlag, 1987.

Reviewed by:

- Mohamed A. Deriche, King Fahd University of Petroleum and Minerals, Saudi Arabia, {*mderiche@kfupm.edu.sa*}
- William J. Ohley, University of Rhodes Island, USA, {*ohley@ele.uri.edu*}

In: Brownian Motion: Theory, Modelling and Applications ISBN: 978-1-61209-537-0
Editors: R.C. Earnshaw and E.M. Riley

Chapter 13

BROWNIAN MOTION IN AN ELECTROMAGNETIC FIELD

J. I. Jiménez-Aquino[a,*]***, R. M. Velasco***[a]***,***
F. J. Uribe[a] ***and M. Romero-Bastida***[b]

[a] Departamento de Física, Universidad Autónoma Metropolitana-Iztapalapa
México, Distrito Federal, México

[b] SIP-ESIME Culhuacán, Instituto Politécnico Nacional
Col. San Francisco Culhuacán
Delegación Coyoacán, Distrito Federal México

PACS 05.45-a, 05.70.Ln,02.50.-r, 52.2540.Gj

Keywords: Brownian motion, Langevin equation, Fokker-Planck equation, Fokker-Planck-Kramers equation.

1. Introduction

After Einstein's theoretical explanation [1] of Brownian motion in 1905, and a few years latter by Langevin [2], and then experimentally corroborated by J. Perrin [3], an amazing number of works on the same topic has been broadly extended to other branches of science. Nowadays, the theory of Brownian motion continues to be of great interest in physics, chemistry, biology, and actually finds interesting applications in astronomy, economy, physiology, medicine, and many other fields you could think about. On the other hand, it seems that the study of Brownian motion in an electromagnetic field, which is our main interest in this work, was initiated by Taylor [4] in 1961 in his work entitled "Diffusion of a plasma in a crossed magnetic field" and immediately continued in the next two years by Kurşunoğlu [5, 6], and four years latter by Sturrock [7]. After these pioneers studies, the problem of Brownian motion in an electromagnetic field was and continues to be an interesting subject of research in other physical contexts [8]- [18]. The main approaches to study Brownian motion in a general way go through two theoretical points of view: one

*E-mail address: ines@xanum.uam.m

is given through the Langevin equation and its possible generalizations and the other is the Fokker-Planck (FP) or Fokker-Planck-Kramers (FPK) equations. The FP equation is a partial differential equation for the velocity-space transition probability density, whereas the FPK is the one associated with the phase-space transition probability density. The solution and applications of both partial differential equations for ordinary Brownian motion have widely been studied in various references, see for instance [19]- [24]. To the best our knowledge, the study of solutions of both equations for the Brownian motion in an electromagnetic field was initiated by Ferrari [9] in 1990. In his work, Ferrari starts with the kinetic Boltzmann equation to establish the FP and FPK equations for a heavy ion in light gases playing the role of a thermal bath, in the presence of an electromagnetic field [9]. He succeeded in solving explicitly the FP equation for the heavy ion, using a transformation which allows this equation to be written into a similar way as a field-free equation. However, the proposal could not be extended to solve explicitly the FPK equation for the charged particle. In 2001 Czopnik and Garbaczewski [10] proposed a Gaussian distribution function for the correlation function of the relevant variables to solve the equation. Such a solution was the ansatz followed to integrate the equation and it is closely related to the one found by Chandrasekhar in his celebrated paper [19]. In 2005 a combination of both Czopnik and Garbaczewski's [10] rotated Stokes force and Ferrari's [9] gauge was used by Simões and Lagos [11], to propose another alternative method of solution to FPK equation for a heavy ion embedded in a fluid in the presence of external mechanic and electromagnetic fields. The method relies upon a transformation of FPK equation into a field-free equation, in a similar way as that advanced by Ferrari [9]. However, instead of an explicit solution for the FPK equation, a Gaussian distribution function was proposed. It was until 2006 that a new method introduced by Jiménez-Aquino and Romero-Bastida [13, 14] allowed to solve the problem by means of a transformation based on rotation matrices in the Langevin equation. This methodology causes the Langevin equation to look very similar to the one describing ordinary Brownian motion. Indeed, the method allows us to solve explicitly not only the FP equation, but also the FPK equation in a simple way, following closely the Chandrasekhar's methodology [19]. After the paper by Jiménez-Aquino and Romero-Bastida [13, 14], several problems and applications of Brownian motion in an electromagnetic field have been worked with the rotation matrix method. Some examples are: fluctuation theorems, Jarzynski equality [53, 64–66, 68], and detection of weak signals in an electromagnetic field [77].

Recently, the fluctuation Theorems (FTs) and related research have been of great interest in nonequilibrium statistical physics of small systems in which the fluctuations play a fundamental role [27]- [68]. These theorems involve a wide class of systems as well as several equilibrium and nonequilibrium quantities, including Helmholtz free energy [30, 31, 53, 65], work [40, 53, 65, 68], heat [41], and entropy production [33, 34, 64]. Basically, the FTs explain how macroscopic irreversibility appears naturally in systems that obey time-reversible microscopic dynamics and they have been relevant for the extrapolation of thermodynamic concepts to nanosized systems that are of interest to biologists, physicists, and engineers. They hold for the steady state [29, 40] and transient situations [27, 33, 40], and have been corroborated by both computer simulations [27, 28, 36, 38, 50, 52] and experiments [38, 42, 43, 50]. In fact, the first experiment performed to demonstrate the fluctuation theorem was reported by Wang *et al.* [38], for a small system over short times. The experiment confirms a theoretically predicted work relation associated to the integrated transient fluctuation theorem (ITFT), which gives an expression for the ratio between the probability to find positive and negative values of the fluctuations in the total work done on the system for a given time in a transient state. Also, in the context of the FTs, the somewhat unexpected Hall-type fluctuation relation for a two-dimensional system of non-interacting electrons under the influence of crossed electric and magnetic fields in a high friction limit with the surrounding medium, has been established for first time by Roy and Kumar [62]. The other quantity related with the work-fluctuation theorems is the Jarzynski equality (JE), which relates averages of non-equilibrium quantities with the equilibrium free energy differences between equilibrium states [30]. This equality has been verified experimentally in biological systems [39]. The generalizations to arbitrary transitions between non-equilibrium stationary states [37] has also been verified in the experiment [43]. Recently, the transient fluctuation theorem (TFT) has been proved for the Brownian motion of a classical harmonic oscillator under the action of a magnetic field [53,64,68], and the JE has been used in Ref. [53,65] to show its consistency with the Bohr-van Leeuwen (BvL) theorem on the absence of orbital diamagnetism in a classical system of charged particles in thermodynamic equilibrium [69]. As far as we known the fluctuation relations and the JE for a charged harmonic oscillator in an electromagnetic field have not been tested experimentally.

On the other hand, among the variety of physical phenomena in which the stochastic fluctuations play also a fundamental role are those which have been called to as dynamical relaxation of nonlinear systems. The stochastic resonance and the decay of unstable states are just two examples of such a variety of phenomena. In 1989, it was shown by Vemuri and Roy [71] that weak optical signals can be detected via the transient dynamics of a laser (the switch-on process of the laser) much in the same way as the superregenerative detection in radar receivers. The physical idea behind this proposal is that weak signals are greatly amplified when used to trigger the decay of an unstable state. This fact was immediately supported theoretically through the statistic of the first passage time distribution [72,74], and experimentally corroborated by Littler *et al.* [73]. The switch-on process of a laser is an example of the decay of unstable state which can be characterized by studying properties of the initiation time. This is the interval of time taken by the laser intensity to reach a prescribed reference value of a few percent of the steady state intensity. The unavoidable presence of internal noise causes the decay of the initial unstable state of almost zero intensity to the final state of finite intensity [72]. The presence of an external signal would also affect this initiation time accelerating the process. It was shown that from measurements of the mean initiation time and its variance one can get some information about the strengths of the internal noise and the injected signal. In particular, the transient regime of the laser can used as a superregenerative detector of very small injected signals [71]. The detection of weak optical signals has been demonstrated experimentally by measurements of the statistic of the initiation time in an argon laser under the influence of an attenuated He-Ne laser which produces the injected signal [73]. Inspired by the works done by Vemuri and Roy [71], Balle *et al.* [72], and Dellunde *et al.* [75], an alternative physical mechanism for the detection of weak and large amplitudes of an electric field has recently been proposed by Jiménez-Aquino and Romero-Bastida [77]. It consists in the decay process of a Brownian charged particle, from the unstable state in a two-dimensional bistable potential, and under the action of a constant electromagnetic field. In a similar way as in the laser, the weak signal can be detected when used to trigger the decay process of the charged Brownian particle. The decay process is also characterized through the statistics of the mean first passage time (MFPT) distribution, and the criterion of detection is established through the statistics of the MFPT [77]. We say that this process is an alternative one, because there exists another physical mechanism (phenomenon) known as stochastic resonance (SR), first introduced by Benzi *et al.* [78] in 1981, which is capable of detecting and transmitting efficiently weak signals information in nonlinear systems due to the presence of random noise. The main characteristic is the stochastic relaxation in modulated bistable systems. The phenomenon shows the role played by the noise as it contributes to enhance the response of a bistable system to weak signals. The detection of weak signals, in the decay process of unstable states is something similar (not equal) to SR process, because both are basically related to stochastic relaxation in nonlinear systems in which the cooperative effect between weak signals and surrounding noise plays a fundamental role. The SR has widely been studied in physics, chemistry, and recently with mayor interest in biology and medicine [78]- [95]. The decay process of an unstable state which is an alternative SR-like mechanism, opens new roads of investigation. It may serve to explore the detection process of weak signals in some other systems different of lasers, as for example the stochastic relaxation in a bistable magnetic system [80], or that in a single bistable neuron [95], ion channels [90], among others, in which the decay of the unstable state in the presence of weak electromagnetic field becomes of practical interest.

2. Langevin and Fokker-Planck Equations for Several Variables

It has been well established [21] that for N stochastic variables $\{\xi\} = \xi_1, \xi_2, \ldots, \xi_n$ the general Langevin equations have the form $(i = 1, 2, \ldots, N)$

$$\dot{\xi}_i(t) = h_i(\{\xi\}, t) + g_{ij}(\{\xi\}, t)\Gamma_j(t)\,, \tag{1}$$

where $h_i(\{\xi\}, t)$ and $g_{ij}(\{\xi\}, t)$ are in general nonlinear functions of $\{\xi\}$. In this section we use the Einstein's summation convention. The $\Gamma_j(t)$ are Gaussian random variables with zero mean value and correlation functions

$$\langle \Gamma_j(t) \rangle = 0, \qquad \langle \Gamma_i(t)\Gamma_j(t') \rangle = 2\delta_{ij}\delta(t - t')\,. \tag{2}$$

On the other hand, the Fokker-Planck or forward Kolmogorov equation for the transition probability density $P(\{x\}, t|\{y\}, t')$ associated with the Langevin Eq. (1), together with the property (2) reads as $(t \geq t')$

$$\frac{\partial P}{\partial t} = -\frac{\partial}{\partial x_i} D_i(\{x\}, t)\, P + \frac{\partial^2}{\partial x_i \partial x_j} D_{ij}(\{x\}, t)\, P\,, \tag{3}$$

where $D_i(\{x\}, t)$ is the drift coefficient and $D_{ij}(\{x\}, t)$ the diffusion coefficient defined respectively as

$$D_i(\{x\}, t) = h_i(\{x\}, t) + g_{kj}(\{x\}, t)\frac{\partial}{\partial x_k} g_{ij}(\{x\}, t)\,, \tag{4}$$

$$D_{ij}(\{x\}, t) = g_{ik}(\{x\}, t) g_{jk}(\{x\}, t)\,, \tag{5}$$

with the initial condition

$$P(\{x\}, t'|\{y\}, t') = \delta(\{x\} - \{y\}) . \tag{6}$$

Here $\xi_k(t) = x_k$ for $k = 1, 2, \ldots, N$ represents the sharp value of the variable $\xi(t)$ at time t. If we multiply Eq. (3) by an initial probability density $\mathcal{W}(\{y\}, t')$ and integrate over y we obtain the Fokker-Planck equation for the probability density $\mathcal{W}(\{x\}, t)$ for all times $t > t'$, that is

$$\frac{\partial \mathcal{W}}{\partial t} = -\frac{\partial}{\partial x_i} D_i(\{x\}, t)\, \mathcal{W} + \frac{\partial^2}{\partial x_i \partial x_j} D_{ij}(\{x\}, t)\, \mathcal{W} . \tag{7}$$

The complete information of a Markov process is contained in the joint probability distribution $\mathcal{W}_2(\{x\}, t; \{y\}, t')$ which can be expressed as

$$\mathcal{W}_2(\{x\}, t; \{y\}, t') = P(\{x\}, t|\{y\}, t')\mathcal{W}(\{y\}, t') . \tag{8}$$

If the drift and diffusion coefficients do not depend on time, a stationary solution may exist. In this case, the transition probability density P depends only on the time difference $t - t'$, and we may write for $t \geq t'$ the joint probability distribution in the stationary state as follows

$$\mathcal{W}_2(\{x\}, t; \{y\}, t') = P(\{x\}, t - t'|\{y\}, 0)\mathcal{W}_{st}(\{y\}) . \tag{9}$$

A particular case of Eq. (1) is the Ornstein-Uhlenbeck [20] Process, which is described by

$$\dot{\xi}_i(t) = -\gamma_{ij}\xi_j + \Gamma_i(t) , \tag{10}$$

with delta correlated Gaussian distributed force

$$\langle \Gamma_j(t) \rangle = 0, \qquad \langle \Gamma_i(t)\Gamma_j(t') \rangle = q_{ij}\delta(t - t') , \tag{11}$$

where the coefficients $q_{ij} = q_{ji}$ which describe the strength of the noise do not depend on the variables ξ_k, and γ_{ij} are the components of a constant matrix. For this type of process, the drift coefficient is linear in the variable x and the matrix containing the diffusion coefficients D_{ij} is constant. Besides it is a symmetric matrix, so that

$$D_i = -\gamma_{ij}x_j, \qquad D_{ij} = D_{ji} . \tag{12}$$

2.1. Transformation of Variables

When a transformation of variables is performed in the Langevin equations, Eqs. (1), the drift and diffusion coefficients must be calculated accordingly. If we introduce new variables ξ' in Eq. (1) such that

$$\xi' = \xi'(\{\xi\}, t) , \tag{13}$$

then the Langevin equation in the new variables transforms as

$$\dot{\xi}'_i(t) = h'_i(\{\xi'\}, t) + g'_{ij}(\{\xi'\}, t)\Gamma_j(t) , \tag{14}$$

where

$$h_i' = \frac{\partial \xi_i'}{\partial t} + \frac{\partial \xi_i'}{\partial \xi_k} h_k \, ; \qquad g_{ij}' = \frac{\partial \xi_i'}{\partial \xi_k} g_{kj} \, . \tag{15}$$

Hence, the transformed drift and diffusion coefficients (writing x' instead of ξ' in the argument) are given by

$$D_i' = \left(\frac{\partial x_i'}{\partial t} \right) + \frac{\partial x_i'}{\partial x_k} D_k + \frac{\partial^2 x_i'}{\partial x_r \partial x_k} D_{rk} \, , \tag{16}$$

$$D_{ij}' = \frac{\partial x_i'}{\partial x_r} \frac{\partial x_j'}{\partial x_k} D_{rk} \, , \tag{17}$$

where D_i and D_{ij} are given respectively by Eqs. (4) and (5). The Fokker-Planck Eq. (7) is also transformed in the new variables as

$$\left(\frac{\partial \mathcal{W}'}{\partial t} \right)_{x'} = - \frac{\partial}{\partial x_i'} D_i' \mathcal{W}' + \frac{\partial^2}{\partial x_i' \partial x_j'} D_{ij}' \mathcal{W}' \, , \tag{18}$$

where the transformed drift D_i' and diffusion D_{ij}' coefficients are taken from Eqs. (16) and (17). The probability density $\mathcal{W}$ is transformed as $\mathcal{W}' = J\mathcal{W} = \mathcal{W}/J'$, where J is the Jacobian of the transformation defined by

$$J \equiv d^N x / d^N x' = |\mathrm{Det}(\partial x_i / \partial x_j')| = 1/J' = 1/|\mathrm{Det}(\partial x_i' / \partial x_j)| \, , \tag{19}$$

and $d^N x$ and $d^N x'$ are the corresponding volume elements.

3. Brownian Motion for a Free Particle

The three-dimensional Brownian motion for the velocity $\mathbf{v}(t)$ in the absence of external forces can be described with the Langevin Eq. (10) and it is given by

$$m\dot{\mathbf{v}} = -\alpha \mathbf{v} + \boldsymbol{\zeta}(t) \, , \tag{20}$$

where m is the mass of the Brownian particle, $-\alpha\mathbf{v}$ is the systematic force representing the dynamical friction and α the friction coefficient. This equation can also be written as

$$\dot{\mathbf{v}} = -\gamma \mathbf{v} + \mathbf{f}(t) \, , \tag{21}$$

where $\gamma = \alpha/m$ and $\mathbf{f}(t) = \boldsymbol{\zeta}(t)/m$. The components of vector $\mathbf{f}(t)$ satisfies the same properties as that of Eq. (11)

$$\langle f_i(t) f_j(t') \rangle = 2\lambda \, \delta_{ij} \delta(t - t') \, , \tag{22}$$

λ being the noise intensity which satisfies the fluctuation-dissipation relation $\lambda = \gamma k_B T/m$, with k_B the Boltzmann constant and T the temperature of the bath. In this case $g_{ij} = \sqrt{2}\, \delta_{ij}$ and the drift and diffusion coefficients can be written as

$$D_i = -\gamma v_i, \qquad D_{ij} = \lambda \, \delta_{ij} \, . \tag{23}$$

3.1. The Fokker-Planck Equation

The Fokker-Planck equation for the transition probability density $P(\mathbf{v}, t|\mathbf{v}_0)$ of the velocity $\mathbf{v}$ at time t conditioned by the initial data $\mathbf{v}_0$ at time $t = 0$, as required by Eq. (3) is given as follows

$$\frac{\partial P}{\partial t} = \gamma \nabla_{\mathbf{v}} \cdot (\mathbf{v}P) + \lambda \nabla^2_{\mathbf{v}} P . \tag{24}$$

The transition probability density satisfies the initial condition

$$\lim_{t\to 0} P(\mathbf{v}, t|\mathbf{v}_0) = \delta(\mathbf{v} - \mathbf{v}_0) = \delta(v_x - v_{0x})\delta(v_y - v_{0y})\delta(v_z - v_{0z}) . \tag{25}$$

Following Chandrasekhar's methodology [19], the general solution of Eq. (24) is connected with its associated first-order equation

$$\frac{\partial P}{\partial t} - \gamma \nabla_{\mathbf{v}} \cdot (\mathbf{v}P) = 0 . \tag{26}$$

The general solution of this first-order equation involves the three first integrals of the Lagrangian subsidiary system

$$\dot{\mathbf{v}} = -\gamma \mathbf{v} , \tag{27}$$

which can written as

$$\mathbf{v} e^{\gamma t} = \mathbf{I}_1 , \tag{28}$$

and $\mathbf{I}_1 = \mathbf{v}_0$ is a constant. The solution of Eq. (24) comes after the three following steps. First, a new space is introduced, defined through the vector $\mathbf{p} = (p_1, p_2, p_3) = \mathbf{v}e^{\gamma t}$, then Eq. (24) becomes

$$\frac{\partial P}{\partial t} = 3\gamma P + \lambda e^{2\gamma t} \nabla^2_{\mathbf{p}} P . \tag{29}$$

As a second step, an auxiliary function is defined as $\chi = P e^{-3\gamma t}$ allowing Eq. (29) be transformed into

$$\frac{\partial P}{\partial t} = \lambda e^{2\gamma t} \nabla^2_{\mathbf{p}} P , \tag{30}$$

which is the third step and it is easily solved. So that, the solution of Eq. (24) is then established as

$$P(\mathbf{v}, t|\mathbf{v}_0) = \frac{1}{[2\pi\lambda(1 - e^{-2\gamma t})/\gamma]^{3/2}} \exp\left(- \frac{\gamma |\mathbf{v} - e^{-\gamma t}\mathbf{v}_0|^2}{2\lambda(1 - e^{-2\gamma t})} \right) . \tag{31}$$

This transition probability density becomes the usual Maxwellian distribution function as the time $t \to \infty$, so

$$P(\mathbf{v}) = \left(\frac{m}{2\pi k_B T} \right)^{3/2} \exp\left(- \frac{m|\mathbf{v}|^2}{2k_B T} \right) , \tag{32}$$

which corresponds to the stationary distribution and $\lambda = \gamma k_B T/m$ has been used. We must notice that the solution (31) can also be written as the product of three independent probability densities, that is, $P(\mathbf{v}, t|\mathbf{v}_0) = P(v_x, t|v_{0x})P(v_y, t|v_{0y})P(v_z, t|v_{0z})$, where

$$P(v_i, t|v_{0i}) = \frac{1}{\sqrt{2\pi\lambda(1 - e^{-2\gamma t})/\gamma}} \exp\left(- \frac{\gamma(v_i - e^{-\gamma t}v_{0i})^2}{2\lambda(1 - e^{-2\gamma t})} \right) , \tag{33}$$

and $i = x, y, z$.

3.2. Fokker-Planck-Kramers Equation for Free Brownian Motion

The phase-space Langevin equation for free Brownian motion takes into account the position vector $\mathbf{r} = (x, y, z)$ and the velocity $\mathbf{v}$, in such a way that the set of Langevin equations reads

$$\dot{\mathbf{r}} = \mathbf{v}, \tag{34}$$

$$\dot{\mathbf{v}} = -\gamma\mathbf{v} + \mathbf{f}(t), \tag{35}$$

where the noise term $\mathbf{f}(t)$ satisfies the same properties of Gaussian white noise as given before. Equations (34) and (35) constitutes a set of six stochastic differential equations similar to that given by Eq. (10), with the vector $\{x\} = (x_1, x_2, x_3, x_4, x_5, x_6) = (x, y, z, v_x, v_y, v_z)$. Taking into account Eq. (3), it is now easy to construct the phase-space Fokker-Planck equation, named also to as FPK equation. This equation associated with the phase-space transition probability density $P(\mathbf{r}, \mathbf{v}, t|\mathbf{r}_0, \mathbf{v}_0)$ governing the probability of the occurrence of the velocity $\mathbf{v}$ and the position $\mathbf{r}$ at time t given that $\mathbf{v} = \mathbf{v}_0$ and $\mathbf{r} = \mathbf{r}_0$ at $t = 0$, can be written as

$$\frac{\partial P}{\partial t} + \mathbf{v} \cdot \nabla_{\mathbf{r}} P = \gamma \nabla_{\mathbf{v}} \cdot (\mathbf{v}P) + \lambda \nabla_{\mathbf{v}}^2 P. \tag{36}$$

The initial condition in this case is given as

$$\lim_{t \to 0} P(\mathbf{r}, \mathbf{v}, t|\mathbf{r}_0, \mathbf{v}_0) = \delta(\mathbf{v} - \mathbf{v}_0)\delta(\mathbf{r} - \mathbf{r}_0), \tag{37}$$

and $\delta(\mathbf{r} - \mathbf{r}_0) \equiv \delta(x - x_0)\delta(y - y_0)\delta(z - z_0)$. In a similar way, as Chandrasekhar proceeded [19], the solution of Eq. (36) is connected with the solution of the first-order equation

$$\frac{\partial P}{\partial t} + \mathbf{v} \cdot \nabla_{\mathbf{r}} P - \gamma \nabla_{\mathbf{v}} \cdot (\mathbf{v}P) = 0. \tag{38}$$

The general solution of this equation can be expressed in terms of six independent integrals of the subsidiary system

$$\dot{\mathbf{v}} = -\gamma\mathbf{v}, \qquad \dot{\mathbf{r}} = \mathbf{v}, \tag{39}$$

namely

$$\mathbf{v}e^{\gamma t} = \mathbf{I}_1, \qquad \mathbf{r} + \gamma^{-1}\mathbf{v} = \mathbf{I}_2, \tag{40}$$

where $\mathbf{I}_1$ has been defined before and $\mathbf{I}_2 = \mathbf{r}_0 + \gamma^{-1}\mathbf{v}_0$. The explicit solution of Eq. (36), implies more elaborated algebra than the one needed to solve the FP equation in the previous section. The explicit and detailed calculation is given in Chandrasekhar paper [19]. Here we briefly establish that the solution of Eq. (36), by defining $P(\mathbf{R}, \mathbf{S}) \equiv P(\mathbf{r}, \mathbf{v}, t|\mathbf{r}_0, \mathbf{v}_0)$, is given by

$$P(\mathbf{R}, \mathbf{S}) = \frac{1}{8\pi^3(FG - H^2)^{3/2}} \exp\left(- \frac{(F|\mathbf{S}|^2 - 2H\mathbf{R} \cdot \mathbf{S} + G|\mathbf{R}|^2)}{2(FG - H^2)} \right), \tag{41}$$

where the variables

$$\mathbf{R} = \mathbf{r} - \mathbf{r}_0 - \gamma^{-1}\mathbf{v}(1 - e^{-\gamma t}), \tag{42}$$

$$\mathbf{S} = \mathbf{v} - \mathbf{v}_0 e^{-\gamma t}, \tag{43}$$

are given in terms of position and velocity. The functions $F,\ G,\ H$ are time dependent and given as follows

$$F = \frac{\lambda}{\gamma^3}(2\gamma t - 3 + 4e^{-\gamma t} - e^{-2\gamma t}), \quad (44)$$

$$G = \frac{\lambda}{\gamma}(1 - e^{-2\gamma t}), \quad (45)$$

$$H = \frac{\lambda}{\gamma^2}(1 - e^{-\gamma t})^2, \quad (46)$$

$$FG - H^2 = (ab - h^2)e^{-2\gamma t}, \quad (47)$$

besides the auxiliary quantities $a,\ b,\ h$ are also functions of time such that

$$a = \frac{2\lambda}{\gamma^2}t, \qquad b = \frac{\lambda}{\gamma}(e^{2\gamma t} - 1), \qquad h = \frac{2\lambda}{\gamma^2}(1 - e^{\gamma t}). \quad (48)$$

It is important to remark that the variables $\mathbf{S}$ and $\mathbf{R}$ represent a rewritting for the solutions given in Eq. (40). We also must notice that the solution (41) can be written as the product of three independent probability densities, namely $P(\mathbf{R},\mathbf{S}) = P(R_x, S_x)P(R_y, S_y)P(R_z, S_z)$, such that

$$P(R_i, S_i) = \frac{1}{2\pi\sqrt{(FG - H^2)}}\exp\left(- \frac{(F\,S_i^2 - 2H\,R_iS_i + G\,R_i^2)}{2(FG - H^2)} \right), \quad (49)$$

where $\mathbf{R} = (R_x, R_y, R_z)$ and $\mathbf{S} = (S_x, S_y, S_z)$ and $i = x, y, z$. Once the probability density (41) has been obtained, it is possible to calculate the marginal transition probability density $P(\mathbf{R}) \equiv P(\mathbf{r}, t|\mathbf{r}_0, \mathbf{v}_0)$ governing the probability of the position $\mathbf{r}$ at time t, given that the particle is at $\mathbf{r}_0$ with velocity $\mathbf{v}_0$ at $t = 0$. This can be done through the integration

$$P(\mathbf{R}) = \int P(\mathbf{R},\mathbf{S})\, d\mathbf{S}. \quad (50)$$

After a long algebra it reduces to

$$P(\mathbf{r}, t|\mathbf{r}_0, \mathbf{v}_0) = \left(\frac{1}{2\pi F}\right)^{3/2} \exp\left(- \frac{|\mathbf{r} - \mathbf{r}_0 - \gamma^{-1}\mathbf{v}_0(1 - e^{-\gamma t})|^2}{2F} \right), \quad (51)$$

which becomes the usual solution for $\gamma t \gg 1$

$$P(\mathbf{r}, t|\mathbf{r}_0) = \frac{1}{(4\pi Dt)^{3/2}} \exp\left(- \frac{|\mathbf{r} - \mathbf{r}_0|^2}{4Dt} \right), \quad (52)$$

corresponding to the stationary distribution and $D \equiv \lambda/\gamma^2 = k_B T/m\gamma$ is the Einstein's diffusion constant.

4. Brownian Motion in a Magnetic Field

This section concerns the study of a swarm of independent charged Brownian particles in the presence of a constant magnetic field. Due to the particles independence, the study focuses only on one particle of charge q and mass m. In this case, the corresponding Langevin equation reads [10, 13]

$$\dot{\mathbf{v}} = -\gamma \mathbf{v} + \frac{q}{mc}\mathbf{v} \times \mathbf{B} + \mathbf{f}(t)\,, \tag{53}$$

where again $\mathbf{f}(t)$ satisfies the same properties given in Eqs. (22). Here, and in what follows of this chapter we will assume that the magnetic field is constant and points along the z-axis of a Cartesian frame of reference, that is $\mathbf{B} = (0, 0, B)$, where B is the strength of the magnetic field. In this case, the stochastic process described by Eq. (53) can be split into two independent processes: one takes place on the x-y plane, perpendicular to the magnetic field, and the other one takes place along the z-axis parallel to this field. The stochastic process along the z-axis is not coupled with those in the plane x-y, and the corresponding Langevin equation is written as $\dot{v}_z = -\gamma v_z + f_z(t)$, which is the same as the z-component of Eq. (21). On the other hand, the planar stochastic process in the plane x-y becomes described by the Langevin equation

$$\dot{\mathbf{u}} = -\gamma \mathbf{u} + W\mathbf{u} + \overline{\mathbf{f}}(t) = -\Lambda \mathbf{u} + \overline{\mathbf{f}}(t)\,, \tag{54}$$

where the vectors $\mathbf{u} \equiv (v_x, v_y) = (u_x, u_y)$ and $\overline{\mathbf{f}}(t) \equiv (f_x, f_y)$, both lie on the x-y plane, W is an antisymmetric matrix, and $\Lambda = \gamma \mathbf{I} - W$, with $\mathbf{I}$ the unit matrix, such that

$$W = \begin{pmatrix} 0 & \Omega \\ -\Omega & 0 \end{pmatrix}, \qquad \Lambda = \begin{pmatrix} \gamma & -\Omega \\ \Omega & \gamma \end{pmatrix}, \tag{55}$$

and $\Omega = qB/mc$ is the cyclotron frequency. Notice should be made about the coupling between the components v_x, v_y in Eq. (54), which is due to the presence of the magnetic field. As a first step in the study for such process we wonder if the magnetic field modifies the fluctuation dissipation relation $\lambda = \gamma k_B T/m$. In fact, it is easily proven that this is not the case. The proof goes as follows: the solution of Eq. (54) reads

$$\mathbf{u}(t) = \mathrm{e}^{-\Lambda t}\mathbf{u}_0 + e^{-\Lambda t}\int_0^t e^{\Lambda s}\,\overline{\mathbf{f}}(s)\,ds\,, \tag{56}$$

where $e^{-\Lambda t} = e^{-\gamma t}e^{Wt}$. It can be shown that $e^{Wt} = \mathcal{R}(t)$ is an orthogonal rotation matrix [26], which satisfies the property $\mathcal{R}^{-1}(t) = \mathcal{R}^T(t)$, i.e., its inverse is the transpose matrix such that $e^{-Wt} = \mathcal{R}^{-1}(t)$ where

$$\mathcal{R}(t) = \begin{pmatrix} \cos \Omega t & \sin \Omega t \\ -\sin \Omega t & \cos \Omega t \end{pmatrix}. \tag{57}$$

The solution given in Eq. (56) can also be written as

$$\mathbf{u}(t) = \mathrm{e}^{-\gamma t}\mathcal{R}(t)\mathbf{u}_0 + e^{-\gamma t}\mathcal{R}(t)\int_0^t e^{\gamma t'}\mathcal{R}^{-1}(t')\,\overline{\mathbf{f}}(t')\,dt'\,, \tag{58}$$

where where $\mathbf{u}_0 = \mathbf{u}(0)$ is the initial condition in the x-y plane. This solution allows the calculation of the two-times velocity correlation function assuming that $\langle \mathbf{u}_0\,\overline{\mathbf{f}}(t)\rangle = 0$ for any $t \geq 0$. In terms of components we have

$$\langle u_i(t_1)u_j(t_2)\rangle = e^{-\gamma(t_1+t_2)}\mathcal{R}_{ik}(t_1)\mathcal{R}_{jl}(t_2)\langle u_{0k}u_{0l}\rangle + \mathcal{R}_{ik}(t_1)\mathcal{R}_{jl}(t_2)\langle h_k(t_1)h_l(t_2)\rangle\,, \tag{59}$$

where the stochastic average

$$\langle h_k(t_1)h_l(t_2)\rangle = \int_0^{t_1}\int_0^{t_2} \mathrm{e}^{-\gamma(t_1+t_2-t_1'-t_2')}\mathcal{R}^{-1}_{km}(t_1')\mathcal{R}^{-1}_{ln}(t_2')\langle f_m(t_1')f_n(t_2')\rangle\, dt_1' dt_2' \,, \tag{60}$$

is calculated from the correlation function Eq. (22) and the property of rotation matrix $\mathcal{R}(t)$. The time integration is done over t_2' and the integration over t_1' runs from 0 to the minimum between t_1 and t_2. So that

$$\langle h_k(t_1)\, h_l(t_2)\rangle = 2\lambda\delta_{kl}\int_0^{\min(t_1,t_2)} e^{-\gamma(t_1+t_2-2t_1')}\, dt_1' = \frac{\lambda}{\gamma}(e^{-\gamma|t_1-t_2|} - e^{-\gamma(t_1+t_2)})\delta_{kl} \,. \tag{61}$$

On the other hand, to calculate the correlation $\langle u_{0k}u_{0l}\rangle$ given in the first term of Eq. (59), we have to say something about the initial value of the velocity $\mathbf{u}(0)$. We assume that it is distributed according to a Maxwellian distribution function at equilibrium with the thermal bath at temperature T. In two-dimensions it is given by

$$P(\mathbf{u}_0) = \frac{m}{2\pi k_B T}\exp\left(-\frac{m|\mathbf{u}_0|^2}{2k_B T}\right), \tag{62}$$

and therefore $\langle u_{0k}u_{0l}\rangle = (k_B T/m)\delta_{kl}$. Upon substitution of this expression and the result of Eq. (61) into Eq. (59) we finally get

$$\begin{aligned} \langle u_i(t_1)u_j(t_2)\rangle &= \frac{k_B T}{m} e^{-\gamma(t_1+t_2)}\mathcal{R}_{ik}(t_1)\mathcal{R}_{jk}(t_2) \\ &+ \frac{\lambda}{\gamma}\mathcal{R}_{ik}(t_1)\mathcal{R}_{jk}(t_2)(e^{-\gamma|t_1-t_2|} - e^{-\gamma(t_1+t_2)}) \,. \end{aligned} \tag{63}$$

The equal time $t_1 = t_2 = t$ correlation function allows the calculation of the averaged kinetic energy

$$\langle E\rangle = \frac{m}{2}\langle u_i(t)u_i(t)\rangle = k_B T e^{-2\gamma t} + \frac{\lambda m}{\gamma}(1 - e^{-2\gamma t}) \,, \tag{64}$$

where the orthogonality properties of matrix $\mathcal{R}(t)$ have been taken into account. When we take $t \to \infty$, we obtain $\langle E\rangle \to \lambda m/\gamma$, which in a two-dimensional case (x-y plane) must satisfy the equipartition of energy property, hence $\langle E\rangle = k_B T$. The direct comparison between these two mean values drives to $\lambda = \gamma k_B T/m$, which is the usual fluctuation dissipation relation. It means that the presence of the magnetic field does not modify it, a result which sometimes is taken for granted. Notice that in this limiting case, $\langle E\rangle$ is independent of time.

4.1. Fokker-Planck Equation in a Magnetic Field

Let us now concentrate on the solution of the FP equation associated with Eq. (54). The drift and diffusion coefficients in this case are

$$D_i = -\Lambda_{ij}\, u_j, \qquad D_{ij} = \lambda\, \delta_{ij} \,, \tag{65}$$

and therefore, the FP equation for the transition probability density $P(\mathbf{u}, t|\mathbf{u}_0)$ reads

$$\frac{\partial P}{\partial t} = \nabla_{\mathbf{u}} \cdot (\Lambda \mathbf{u} P) + \lambda \nabla^2_{\mathbf{u}} P , \tag{66}$$

and satisfies the initial condition

$$\lim_{t \to 0} P(\mathbf{u}, t|\mathbf{u}_0) = \delta(\mathbf{u} - \mathbf{u}_0) = \delta(u_x - u_{0x})\delta(u_y - u_{0y}) . \tag{67}$$

According to Chandrasekhar [19], we require to solve the first-order equation of (66) without the Laplacian term, involving the Lagrangian subsidiary system

$$\dot{\mathbf{u}} = \Lambda \mathbf{u} , \tag{68}$$

together with the integrals

$$e^{\Lambda t} \mathbf{u} = \mathbf{I}_1 , \tag{69}$$

with $\mathbf{I}_1 = \mathbf{u}_0$. As far as we known, trying to solve Eq. (66) by this way or any other is not an easy task, due to the coupling derivative appearing in the first term of its right hand side (rhs). In 1990 this equation was solved explicitly by Ferrari [9], by means of a transformation allowing him to write Eq. (66) into another one similar to Eq. (24). Here, we use an alternative method of solution first proposed by Jiménez-Aquino and Romero-Bastida [13]. The strategy allows the transformation of the Langevin Eq. (54) into another which looks like as an ordinary Langevin equation. The transformation is the following

$$\mathbf{u}' = e^{-Wt} \mathbf{u} . \tag{70}$$

In the new velocity-space, the Langevin Eq. (54) is shown to be

$$\dot{\mathbf{u}}' = -\gamma \mathbf{u}' + \overline{\mathbf{f}}'(t) , \tag{71}$$

where $\overline{\mathbf{f}}'(t) = \mathcal{R}^{-1}(t)\overline{\mathbf{f}}(t)$, which accounts for a rotation of the noise vector $\overline{\mathbf{f}}(t)$. Indeed, Eq. (71) is now quite similar to that of an ordinary Brownian motion in which the drift and diffusion coefficients are given by

$$D'_i = -\gamma \, u'_i , \tag{72}$$

$$D'_{ij} = \lambda \mathcal{R}^{-1}_{ik}(t)\mathcal{R}^{-1}_{jk}(t) = \lambda \, \delta_{ij} . \tag{73}$$

The drift coefficient D'_i has the same structure as in Eq. (23), but it is written in terms of the new variables. It is remarkable the fact that the diffusion coefficient D'_{ij} is the same as in the usual case given in Eq. (23). This result comes from the statistical properties of the noise $\overline{\mathbf{f}}(t)$ and the fact that the transformed noise $\overline{\mathbf{f}}'(t)$ shares them, i.e. the transformed noise correspond to Gaussian white noise with the same intensity as the original one $\overline{\mathbf{f}}(t)$. It must be noticed that, the drift and diffusion coefficients (72) and (73) can also be derived from Eqs. (16) and (17) respectively. The associated FP equation for the transition probability density $P'(\mathbf{u}', t|\mathbf{u}'_0)$ is then

$$\frac{\partial P'}{\partial t} = \gamma \nabla_{\mathbf{u}'} \cdot (\mathbf{u}' P') + \lambda \nabla^2_{\mathbf{u}'} P' , \tag{74}$$

satisfying the initial condition

$$\lim_{t\to 0} P'(\mathbf{u}', t|\mathbf{u}'_0) = C_0\, \delta(\mathbf{u}' - \mathbf{u}'_0) = C_0\, \delta(u'_x - u'_{0x})\delta(u'_y - u'_{0y})\,, \tag{75}$$

where C_0 is an undetermined constant which will be obtained later on. Now, Eq. (74) is easy to solve because it is very similar to Eq. (24). Its solution is connected with the solution of the associated first-order equation involving the two integrals of the Lagrangian subsidiary system

$$\dot{\mathbf{u}}' = -\gamma \mathbf{u}'\,, \tag{76}$$

which are given by

$$e^{\gamma t}\mathbf{u}' = \mathbf{I}'_1\,, \tag{77}$$

where $\mathbf{I}'_1 = \mathbf{u}'_0$. Under these conditions, it can be seen that the solution of Eq. (74) is similar to the solution (33) but in the two-dimensional case, that is

$$P'(\mathbf{u}', t|\mathbf{u}'_0) = \frac{\gamma}{2\pi\lambda(1 - e^{-2\gamma t})} \exp\bigg(- \frac{\gamma|\mathbf{u}' - e^{-\gamma t}\mathbf{u}'_0|^2}{2\lambda(1 - e^{-2\gamma t})}\bigg)\,. \tag{78}$$

To return to the original variable $\mathbf{u}$, we further define the following variables

$$\mathbf{S} = \mathbf{u} - e^{-\Lambda t}\mathbf{u}_0, \qquad\qquad \mathbf{S}' = \mathbf{u}' - e^{-\gamma t}\mathbf{u}'_0\,, \tag{79}$$

which are a different way to write Eqs. (69) and (77) respectively. It can shown that $\mathbf{S}' = e^{-Wt}\mathbf{S}$ and therefore

$$|\mathbf{u}' - e^{-\gamma t}\mathbf{u}'_0|^2 = (\mathbf{u}' - e^{-\gamma t}\mathbf{u}'_0) \cdot (\mathbf{u}' - e^{-\gamma t}\mathbf{u}'_0) = |\mathbf{u} - e^{-\Lambda t}\mathbf{u}_0|^2\,. \tag{80}$$

Let us remember also that the transition probability P is transformed by $P' = J\,P$, where J is the Jacobian of the transformation given by Eq. (19). If we now define $P(\mathbf{S}) \equiv P(\mathbf{u}, t|\mathbf{u}_0)$ and $P'(\mathbf{S}') \equiv P'(\mathbf{u}', t|\mathbf{u}'_0)$ and take into account the transformation between $\mathbf{S}$ and $\mathbf{S}'$, it is easily shown that $J = 1$ and thus $P' = P$. Therefore, the quantity C_0 given in Eq. (75) will be $C_0 = 1$, and the solution of Eq. (66) is finally

$$P(\mathbf{S}) \equiv P(\mathbf{u}, t|\mathbf{u}_0) = \frac{\gamma}{2\pi\lambda(1 - e^{-2\gamma t})} \exp\bigg(- \frac{\gamma|\mathbf{u} - e^{-\Lambda t}\mathbf{u}_0|^2}{2\lambda(1 - e^{-2\gamma t})}\bigg)\,. \tag{81}$$

This transition probability density was also calculated by Ferrari [9] and Czopnik and Garbaczewski [10], using other methods of solution. For large times $\gamma t \gg 1$, it reduces to the Maxwellian distribution

$$P(\mathbf{u}) = \frac{m}{2\pi k_B T} \exp\bigg(- \frac{m|\mathbf{u}|^2}{2k_B T}\bigg)\,. \tag{82}$$

Since the process is Markovian, other consequences can be derived from Eq. (81). This solution can easily be extended to arbitrary $t_0 = t' \neq 0$. For this case, we substitute anywhere $t - t'$ instead of t and $\bar{\mathbf{u}} = (\bar{u}_1, \bar{u}_2)$ instead of $\mathbf{u}_0$, where $\bar{\mathbf{u}} = \mathbf{u}(t')$. In the stationary state the joint probability density $\mathcal{W}_2(\mathbf{u}, t; \bar{\mathbf{u}}, t')$ may be expressed by the product of Eqs. (81) and (82). For times $t \geq t'$ and $t \leq t'$ it can be shown that

$$\mathcal{W}_2(\mathbf{u}, t; \bar{\mathbf{u}}, t') = \frac{\gamma^2}{4\pi^2\lambda^2(1 - e^{-2\gamma t})} \exp\bigg(- \frac{\gamma[|\mathbf{u}|^2 - 2\mathbf{u}\cdot\bar{\mathbf{u}}e^{-\Lambda|t-t'|} + |\bar{\mathbf{u}}|^2]}{2\lambda(1 - e^{-2\gamma|t-t'|})}\bigg)\,. \tag{83}$$

This joint probability density determines a stationary Markovian stochastic process for which we can calculate various mean values. For instance, the mean values of the velocity components $u_i(t)$ for $i = x, y$, are

$$\langle u_i(t) \rangle = \int_{-\infty}^{\infty} u_i \, P(\mathbf{u}) \, d\mathbf{u} = 0 \,. \tag{84}$$

The matrix of velocity autocorrelation functions reads

$$\langle u_i(t) u_j(t') \rangle = \int_{-\infty}^{\infty} u_i \bar{u}_j \, \mathcal{W}_2(\mathbf{u}, t; \bar{\mathbf{u}}, t') \, d\mathbf{u} d\bar{\mathbf{u}} \,. \tag{85}$$

We can show after some algebra that

$$\langle u_x(t) u_x(t') \rangle = \langle u_y(t) u_y(t') \rangle = \frac{\lambda}{\gamma} e^{-\gamma |t-t'|} \cos \Omega |t - t'| \,. \tag{86}$$

On the other hand, the mean square displacement for the x and y velocity components can easily be calculated if the particle starts at time $t = 0$ at position $\mathbf{x} = \mathbf{x}_0$, where $\mathbf{x}_0 = (x_0, y_0)$. For each component, this mean value reads

$$\langle (x_i - x_{0i})^2 \rangle = \int_0^t \int_0^t \langle u_i(t_1) u_i(t_2) \rangle \, dt_1 dt_2 \,, \tag{87}$$

where $x_i = x, y$ and $x_{0i} = x_0, y_0$. By defining $(\Delta x_i)^2 \equiv (x_i - x_{0i})^2$, it can be shown after some calculations that

$$\begin{aligned} \langle (\Delta x)^2 \rangle = \langle (\Delta y)^2 \rangle = 2 D_e \, t \; &+ \; \frac{2\lambda(1 - C^2)}{\gamma^3 (1 + C^2)^2} [e^{-\gamma t} \cos \Omega t - 1] \\ &- \; \frac{4\lambda \, C}{\gamma^3 (1 + C^2)^2} \, e^{-\gamma t} \sin \Omega t \,, \end{aligned} \tag{88}$$

where $D_e = D/(1 + C^2)$ or $D_e = k_B T / \alpha_e$, with $C = \Omega / \gamma = qB / c\alpha$ a dimensionless parameter, and $\alpha_e = \alpha(1 + C^2)$ which accounts for an effective friction coefficient and depends on the magnetic field. Again, an analysis for large times, $\gamma t \gg 1$ shows that the mean square displacements reads

$$\langle (\Delta x)^2 \rangle = \langle (\Delta y)^2 \rangle = 2 D_e \, t \,, \tag{89}$$

The coefficient D_e accounts for an effective diffusion coefficient since the Einstein's diffusion constant D has been normalized by a dimensionless factor $1 + C^2$. It is clear that $D_e < D$, which means that the diffusion process in the presence of a magnetic field is slower than in the absence of this field. This effect looks like an increasing value of the friction coefficient by the factor $(1 + C^2) > 1$, when the magnetic field is present. The result (89) agrees with that calculated by Kurşunoğlu [5] in 1962. In Fig. 1, it is shown a comparison between the theoretical result of the MSD given by Eq. (89) and that calculated by numerical simulation [15]. According to that time scale, the ballistic regime is very short that it is not practically appreciated. The agreements are excellent.

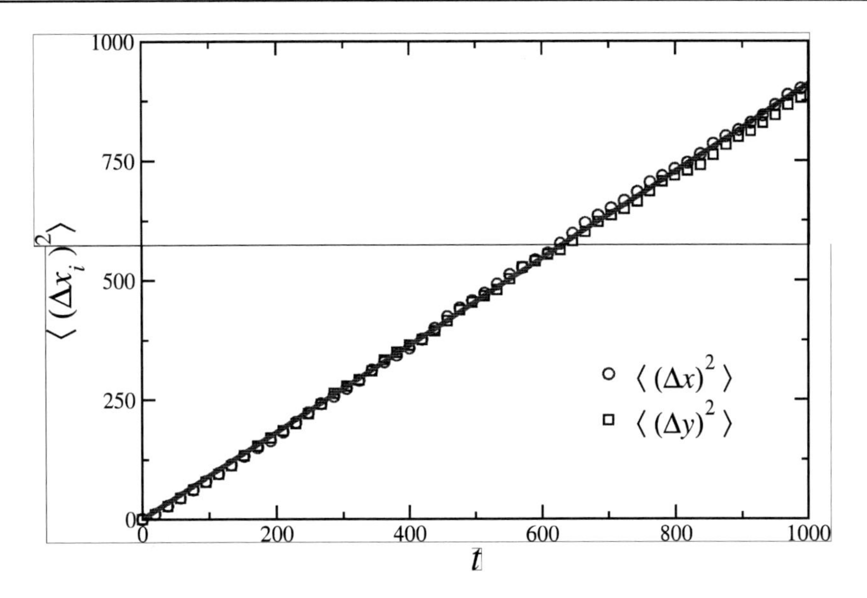

Figure 1. Comparison between the mean square displacement given by Eq. (89) [straight line] and the simulation results [circles and squares] for a charged Brownian particle in a magnetic field.

4.2. Fokker-Planck-Kramers Equationin a Magnetic Field

In the phase-space, the Langevin equation for a charged particle in a magnetic field and in the absence of other external forces can be written as

$$\dot{\mathbf{r}} = \mathbf{v}, \tag{90}$$

$$\dot{\mathbf{v}} = -\gamma \mathbf{v} + \frac{q}{mc}\mathbf{v} \times \mathbf{B} + \mathbf{f}(t). \tag{91}$$

We also consider a constant magnetic field pointing along the z-axis; in this case Eqs. (90) and (91) can also be split into two independent processes. One is described by a phase-space planar Langevin equation given by

$$\dot{\mathbf{x}} = \mathbf{u}, \tag{92}$$

$$\dot{\mathbf{u}} = -\gamma \mathbf{u} + W\mathbf{u} + \overline{\mathbf{f}}(t), \tag{93}$$

where W is the same as that given in Eq. (55) and $\overline{\mathbf{f}}(t) = (f_x, f_y)$ satisfies also the Gaussian white noise properties. The other process is described by the phase-space Langevin equation parallel to the magnetic field written as

$$\dot{z} = v_z, \tag{94}$$

$$\dot{v}_z = -\gamma v_z + f_z(t), \tag{95}$$

which coincides with the z-component of Eqs. (34) and (35) respectively, therefore the solution of the associated FPK equation corresponds to the same z-contribution of Eq. (49). So that, we only concentrate on the solution of the FPK equation associated with Eqs.

(92) and (93). The differential equation associated with the transition probability density $P(\mathbf{x}, \mathbf{u}, t|\mathbf{x}_0, \mathbf{u}_0)$ is now

$$\frac{\partial P}{\partial t} + \mathbf{u} \cdot \nabla_{\mathbf{x}} P = \nabla_{\mathbf{u}} \cdot (\Lambda \mathbf{u} P) + \lambda \nabla_{\mathbf{u}}^2 P \,, \tag{96}$$

satisfying the initial condition

$$\lim_{t \to 0} P(\mathbf{x}, \mathbf{u}, t|\mathbf{x}_0, \mathbf{u}_0) = \delta(\mathbf{u} - \mathbf{u}_0)\delta(\mathbf{x} - \mathbf{x}_0) \,. \tag{97}$$

Again the Chandrasekhar's proposal is related to the first-order solution of the Lagrangian subsidiary system

$$\dot{\mathbf{u}} = -\gamma \mathbf{u}, \qquad \dot{\mathbf{x}} = \mathbf{u} \,, \tag{98}$$

which is expressed in terms of the two-dimensional integrals $\mathbf{I}_1$, $\mathbf{I}_2$ as follows

$$e^{\Lambda t} \mathbf{u} = \mathbf{I}_1, \qquad \mathbf{x} + \Lambda^{-1} \mathbf{u} = \mathbf{I}_2 \,, \tag{99}$$

with $\mathbf{I}_1 = \mathbf{u}_0$, $\mathbf{I}_2 = \mathbf{x}_0 + \Lambda^{-1}\mathbf{u}_0$ [19], and the matrix

$$\Lambda^{-1} = \begin{pmatrix} \frac{\gamma}{\gamma^2+\Omega^2} & \frac{\Omega}{\gamma^2+\Omega^2} \\ -\frac{\Omega}{\gamma^2+\Omega^2} & \frac{\gamma}{\gamma^2+\Omega^2} \end{pmatrix} . \tag{100}$$

The complete solution of Eq. (96) is somewhat difficult to obtain due to the coupling in the term $\nabla_{\mathbf{u}} \cdot (\Lambda \mathbf{u} P)$. To proceed further, we also transform the phase-space $(\mathbf{x}, \mathbf{u})$ Langevin Eqs. (92), (93), into another phase-space $(\mathbf{x}', \mathbf{u}')$, where the Langevin equation can be seen as a phase-space Langevin equation for the ordinary Brownian motion. This can be achieved if we define the new variable $\dot{\mathbf{x}}' = \mathbf{u}'$, where we already know that $\mathbf{u}' = e^{-Wt}\mathbf{u}$ or $\dot{\mathbf{x}}' = e^{-Wt}\dot{\mathbf{x}}$, and $\dot{\mathbf{x}} = \mathbf{u}$. With these change of variables, Eqs. (92) and (93) transform into

$$\dot{\mathbf{x}}' = \mathbf{u}' \,, \tag{101}$$

$$\dot{\mathbf{u}}' = -\gamma \mathbf{u}' + \bar{\mathbf{f}}'(t) \,, \tag{102}$$

where again $\bar{\mathbf{f}}'(t) = \mathcal{R}^{-1}(t)\bar{\mathbf{f}}(t)$ satisfies the Gaussian white noise properties if the noise $\bar{\mathbf{f}}(t)$ does. Indeed, in the new phase-space, these equations are very similar to those of ordinary Brownian motion and therefore its associated FPK equation for the transition probability density $P'(\mathbf{x}', \mathbf{u}', t|\mathbf{x}_0', \mathbf{u}_0')$ reads

$$\frac{\partial P'}{\partial t} + \mathbf{u}' \cdot \nabla_{\mathbf{x}'} P' = \gamma \nabla_{\mathbf{u}'} \cdot (\mathbf{u}' P') + \lambda \nabla_{\mathbf{u}'}^2 P' \,, \tag{103}$$

subject to the initial condition

$$P'(\mathbf{x}', \mathbf{u}', 0|\mathbf{x}_0', \mathbf{u}_0') = K\delta(\mathbf{u}' - \mathbf{u}_0')\delta(\mathbf{x}' - \mathbf{x}_0') \,, \tag{104}$$

K being a quantity which will be determined later on. The Lagrangian subsidiary system associated with Eq. (103) is then

$$\dot{\mathbf{u}}' = -\gamma \mathbf{u}', \qquad \dot{\mathbf{x}}' = \mathbf{u}' \,, \tag{105}$$

and their corresponding four integrals are

$$e^{\gamma t}\mathbf{u}' = \mathbf{I}_1', \qquad \mathbf{x}' + \gamma^{-1}\mathbf{u}' = \mathbf{I}_2', \tag{106}$$

where $\mathbf{I}_1' = \mathbf{u}_0'$ and $\mathbf{I}_2' = \mathbf{x}_0' + \gamma^{-1}\mathbf{u}_0'$. It is now easy to find the solution of Eq. (103), because it is very similar to Eq. (36). The solution for the transition probability density $P'(\mathbf{R}', \mathbf{S}') \equiv P(\mathbf{x}', \mathbf{u}', t|\mathbf{x}_0', \mathbf{u}_0')$, can be written in the form of Eq. (41) but in the two-dimensional case, i.e.

$$P'(\mathbf{R}', \mathbf{S}') = \frac{1}{4\pi^2(FG - H^2)}\exp\left(- \frac{(F|\mathbf{S}'|^2 - 2H\mathbf{R}' \cdot \mathbf{S}' + G|\mathbf{R}'|^2)}{2(FG - H^2)}\right), \tag{107}$$

where F, G, and H are exactly the same as that defined before, and

$$\mathbf{R}' = \mathbf{x}' - \mathbf{x}_0' - \frac{\mathbf{u}_0'}{\gamma}(1 - e^{-\gamma t}), \tag{108}$$

$$\mathbf{S}' = \mathbf{u}' - \mathbf{u}_0' e^{-\gamma t}, \tag{109}$$

which correspond the solutions given in Eqs. (106). The transformation allowing us to return to the original phase-space $(\mathbf{x}, \mathbf{u})$, can be obtained from Eqs. (99) and (106), with the help of $\mathbf{u}' = e^{-Wt}\mathbf{u}$. In this case, it can be shown that

$$\mathbf{x}' - \mathbf{I}_2' = \gamma^{-1}e^{-Wt}\Lambda(\mathbf{x} - \mathbf{I}_2). \tag{110}$$

Let us now define the transition probability density $P(\mathbf{R}, \mathbf{S}) \equiv P(\mathbf{x}, \mathbf{u}, t|\mathbf{x}_0, \mathbf{u}_0)$ as the solution of Eq. (96), where the variables

$$\mathbf{R} = \mathbf{x} - \mathbf{x}_0 - \Lambda^{-1}(1 - e^{-\Lambda t})\mathbf{u}_0, \tag{111}$$

$$\mathbf{S} = \mathbf{u} - e^{-\Lambda t}\mathbf{u}_0, \tag{112}$$

which are another way of how to write the solutions given in Eq. (99). The transformation (110) yields to

$$\mathbf{R}' = \gamma^{-1} e^{-Wt}\Lambda\mathbf{R}, \tag{113}$$

$$\mathbf{S}' = e^{-Wt}\mathbf{S}. \tag{114}$$

According to the preceding section, the transformation between the transition probabilities P and P' is given through the corresponding Jacobian

$$d\mathbf{S}d\mathbf{R} = Jd\mathbf{S}'d\mathbf{R}'. \tag{115}$$

It is easy to show that $J = J_S J_R$, where

$$J_S \equiv |\mathrm{Det}(\partial S_i/\partial S_j')| = 1/J_S' = 1/|\mathrm{Det}(\partial S_i'/\partial S_j)|, \tag{116}$$

and

$$J_R \equiv |\mathrm{Det}(\partial R_i/\partial R'_j)| = 1/J'_R = 1/|\mathrm{Det}(\partial R'_i/\partial R_j)|\,, \tag{117}$$

and then $J = J_S J_R = 1/J'_S J'_R = 1/J'$. From Eqs. (113) and (114) we conclude that

$$J'_S = 1, \qquad J'_R = \frac{\gamma^2 + \Omega^2}{\gamma^2} = 1 + C^2\,, \tag{118}$$

where we already know that $C = \Omega/\gamma$. The transformation between the two transition probability densities is then $P = J'_R P'$ and the undetermined constant given in Eq. (104) is given by $K = 1 + C^2$. It is easy to show from the transformations (113) and (114) the following relations

$$|\mathbf{S}'|^2 = |\mathbf{S}|^2\,, \tag{119}$$

$$|\mathbf{R}'|^2 = K|\mathbf{R}|^2\,, \tag{120}$$

$$\mathbf{S}'\cdot\mathbf{R}' = \mathbf{S}\cdot\mathbf{R} + C(\mathbf{S}\times\mathbf{R})_z\,, \tag{121}$$

where $(\mathbf{S}\times\mathbf{R})_z = (S_x R_y - S_y R_x)$ is the z-component of the cross product. Upon substitution of Eqs. (119)-(121) into Eq. (107) we finally get the solution of Eq. (96) for the transition probability density $P(\mathbf{R},\mathbf{S})$, which can be written in the two-dimensional case as

$$P(\mathbf{R},\mathbf{S}) = K_0 \exp\left(- \frac{\left[F|\mathbf{S}|^2 - 2H\mathbf{R}\cdot\mathbf{S} + 2CH(\mathbf{S}\times\mathbf{R})_z + KG|\mathbf{R}|^2\right]}{2(FG - H^2)}\right), \tag{122}$$

where $K_0 = K/4\pi^2(FG - H^2)$. Equation (122), describes the charged particle's diffusion process taking place in the phase-space $(\mathbf{x}, \mathbf{u})$. Evidently, the phase-space transition probability density $P(R_z, S_z) \equiv P(z, v_z, t|z_0, v_{0z})$, according to Eq. (49), is given by

$$P(R_z, S_z) = \frac{1}{\sqrt{4\pi^2(FG - H^2)}} \exp\left(- \frac{\left[FS_z^2 - 2HR_zS_z + GR_z^2\right]}{2(FG - H^2)}\right). \tag{123}$$

which describes the diffusion process in the phase-space (z, v_z), and

$$R_z = z - z_0 - \gamma^{-1} v_{0z}(1 - e^{-\gamma t})\,, \tag{124}$$

$$S_z = v_z - e^{-\gamma t} v_{0z}\,. \tag{125}$$

Some consequences of the result in Eq. (122) can be analyzed:

(i) In the absence of magnetic field ($\Omega = 0$), we have $K = 1$ and then Eq. (122) reduces to the same solution given by Eq. (41) for the two-dimensional case if $\mathbf{r} = (x, y)$ and $\mathbf{v} = \mathbf{u} = (v_x, v_y)$.

(ii) We can calculate the velocity-space $\mathbf{S}$ and configuration-space $\mathbf{R}$ transition probability densities from the marginal integrations

$$P(\mathbf{S}) = \int P(\mathbf{R},\mathbf{S})\, d\mathbf{R}\,, \tag{126}$$

$$P(\mathbf{R}) = \int P(\mathbf{R},\mathbf{S})\, d\mathbf{S}\,. \tag{127}$$

From Eq. (126), we get after integration that

$$P(\mathbf{S}) \equiv P(\mathbf{u}, t|\mathbf{u}_0) = \frac{\gamma}{2\pi\lambda(1-e^{-2\gamma t})} \exp\bigg(- \frac{\gamma|\mathbf{u} - e^{-\Lambda t}\mathbf{u}_0|^2}{2\,\lambda(1-e^{-2\gamma t})}\bigg), \tag{128}$$

which is exactly the same as Eq. (81), as expected. Similarly, the marginal integration of Eq. (127) allows us to find

$$P(\mathbf{R}) \equiv P(\mathbf{x}, t|\mathbf{x}_0, \mathbf{u}_0) = N_{_R}(t) \exp\bigg(- \frac{\gamma\,|\mathbf{x} - \mathbf{x}_0 - \Lambda^{-1}(1-e^{-\Lambda t})\mathbf{u}_0|^2}{2\,D_e(2\gamma\,t - 3 + 4e^{-\gamma t} - e^{-2\gamma t})}\bigg), \tag{129}$$

where $N_{_R}(t) \equiv \gamma/2\pi\,D_e(2\gamma\,t - 3 + 4e^{-\gamma t} - e^{-2\gamma t})$, and $D_e = D/1 + C^2$ is the effective diffusion constant. It is worth noticing that the effective diffusion depends on the magnetic field and it causes its reduction. When we consider the large time limit $\gamma t \gg 1$, we get the stationary probability density

$$P(\mathbf{x}, t|\mathbf{x}_0) = \frac{1}{4\pi D_e t} \exp\bigg(- \frac{|\mathbf{x} - \mathbf{x}_0|^2}{2D_e t}\bigg), \tag{130}$$

which is also an expected result.

5. Brownian Motion in a Field of Force

To go further in the study of Brownian motion, we now will consider the presence of other forces like an electric field $\mathbf{E}(t)$ or mechanical forces $\mathbf{F}_m(t)$, both may be dependent on time but space independent. The sum of both forces per unit mass is defined as $\mathbf{a}(t) \equiv [q\mathbf{E}(t) + \mathbf{F}_{\mathrm{m}}(t)]/m$. In this case, the Langevin equation for the charged particle will be

$$\dot{\mathbf{v}} = -\gamma\mathbf{v} + \frac{q}{mc}\mathbf{v} \times \mathbf{B} + \mathbf{a}(t) + \mathbf{f}(t)\,. \tag{131}$$

The properties of Gaussian white noise for the internal noise of intensity λ are also assumed, and $\lambda = \gamma k_{_B}T/m$. As before, again the magnetic field is $\mathbf{B} = (0, 0, B)$. So that Eq. (131), again split into two independent processes: one is given on the x-y plane and the other along the z-axis. On the x-y plane we have the planar Langevin equation

$$\dot{\mathbf{u}} = -\gamma\mathbf{u} + W\mathbf{u} + \bar{\mathbf{a}}(t) + \bar{\mathbf{f}}(t)\,, \tag{132}$$

where W is the same as Eq. (55) and $\bar{\mathbf{a}}(t) = (a_x, a_y)$ is the two-dimensional time-dependent vector. Along the z-axis parallel to the magnetic field, the Langevin equation is

$$\dot{v}_z = -\gamma v_z + a_z(t) + f_z(t)\,, \tag{133}$$

which is not affected by the magnetic field but only by the external force $a_z(t)$. The solution of its associated Fokker-Planck equation has been given in [19, 25], so that we only pay attention to the planar Langevin Eq. (132).

5.1. Fokker-Planck Equation in a Field of Force

To construct the FP equation associated with Eq. (132), we need the corresponding drift and diffusion coefficients, which are given by

$$D_i = -\Lambda_{ij}\, u_j + a_i, \qquad D_{ij} = \lambda\, \delta_{ij}\,, \tag{134}$$

with $i, j = 1, 2$. In this case the FP equation for the transition probability density $P(\mathbf{u}, t|\mathbf{u}_0)$ is now

$$\frac{\partial P}{\partial t} + \bar{\mathbf{a}} \cdot \nabla_{\mathbf{u}} P = \nabla_{\mathbf{u}} \cdot (\Lambda \mathbf{u} P) + \lambda\, \nabla^2_{\mathbf{u}} P\,, \tag{135}$$

subject to the initial condition

$$P(\mathbf{u}, 0|\mathbf{u}_0) = \delta(\mathbf{u} - \mathbf{u}_0)\,. \tag{136}$$

The method to find the solution of Eq. (135) goes through the same steps as illustrated to solve Eq. (24). The associated two-dimensional integral is obtained from Eq. (135)

$$\dot{\mathbf{u}} = -\Lambda \mathbf{u} + \bar{\mathbf{a}}(t)\,, \tag{137}$$

whose solution is

$$e^{\Lambda t}\mathbf{u} - \bar{\mathbf{a}}_1(t) = \mathbf{I}_1, \qquad \bar{\mathbf{a}}_1(t) = \int_0^t e^{\Lambda s}\bar{\mathbf{a}}(s)\, ds\,, \tag{138}$$

and $\mathbf{I}_1 = \mathbf{u}_0$. Due to the mathematical difficulties mentioned in previous sections, we calculate its solution by means of two transformations taking into account the strategy of solution given before. The first transformation is the introduction of a new variable $\mathbf{U}$ defined as

$$\mathbf{U} = \mathbf{u} - \langle \mathbf{u} \rangle\,, \tag{139}$$

where $\langle \mathbf{u} \rangle$ is the deterministic solution of Eq. (132), i.e., it satisfies the deterministic equation

$$\frac{d\langle \mathbf{u} \rangle}{dt} = -\gamma \langle \mathbf{u} \rangle + W \langle \mathbf{u} \rangle + \bar{\mathbf{a}}(t) \tag{140}$$

and the new variable $\mathbf{U}$ satisfies the Langevin equation

$$\frac{d\mathbf{U}}{dt} = -\gamma \mathbf{U} + W\mathbf{U} + \bar{\mathbf{f}}(t) = -\Lambda \mathbf{U} + \bar{\mathbf{f}}(t)\,. \tag{141}$$

It is easy to show that the FP equation associated with Eq. (141) for the transition probability density $P(\mathbf{U}, t|\mathbf{U}_0)$ is

$$\frac{\partial P}{\partial t} = \nabla_{\mathbf{U}} \cdot (\Lambda \mathbf{U} P) + \lambda\, \nabla^2_{\mathbf{U}} P\,, \tag{142}$$

with the initial condition $P(\mathbf{U}, 0|\mathbf{U}_0) = \delta(\mathbf{U} - \mathbf{U}_0)$. Since Eq. (142) has the same structure of Eq. (66), we follow similar steps to find its solution. Hence, we introduce another transformation for the variable $\mathbf{U}$. This is given by

$$\mathbf{U}' = e^{-Wt}\mathbf{U}\,, \tag{143}$$

which leads to the Langevin equation

$$\frac{d\mathbf{U}'}{dt} = -\gamma \mathbf{U}' + \bar{\mathbf{f}}'(t)\,, \tag{144}$$

where $\bar{\mathbf{f}}'(t) = \mathcal{R}^{-1}(t)\mathbf{f}(t)$. Evidently the associated Fokker-Planck equation for the transition probability density $P'(\mathbf{U}', t|\mathbf{U}'_0)$ reads

$$\frac{\partial P'}{\partial t} = \gamma\, \nabla_{\mathbf{U}'} \cdot (\mathbf{U}'P') + \lambda\, \nabla^2_{\mathbf{U}'} P'\,, \tag{145}$$

subject to the initial condition

$$P'(\mathbf{U}', 0|\mathbf{U}'_0) = \widehat{C}_0 \delta(\mathbf{U}' - \mathbf{U}'_0)\,, \tag{146}$$

and $\widehat{C}_0$ is an undetermined constant. Again, the algebraic structure of Eq. (145) is very similar to Eq. (74) and therefore its solution is the same as Eq. (78) when written in terms of $\mathbf{U}'$ variable

$$P'(\mathbf{U}', t|\mathbf{U}'_0) = \frac{\gamma}{2\pi\lambda(1 - e^{-2\gamma t})} \exp\left(- \frac{\gamma|\mathbf{U}' - e^{-\gamma t}\mathbf{U}'_0|^2}{2\lambda(1 - e^{-2\gamma t})} \right). \tag{147}$$

To write this solution in the $\mathbf{U}$ variable, we consider the inverse transformation and the auxiliary variables

$$\widehat{\mathbf{S}} = \mathbf{U} - e^{-\Lambda t}\mathbf{U}_0, \qquad \widehat{\mathbf{S}}' = \mathbf{U}' - e^{-\gamma t}\mathbf{U}'_0\,, \tag{148}$$

and thus $\widehat{\mathbf{S}}' = e^{-Wt}\widehat{\mathbf{S}}$. Now, the transition probability density is written in terms of $\mathbf{S}$ and $\mathbf{S}'$, such that $P(\widehat{\mathbf{S}}) \equiv P(\mathbf{U}, t|\mathbf{U}_0)$ and $P'(\widehat{\mathbf{S}}') \equiv P'(\mathbf{U}', t|\mathbf{U}'_0)$. In this case the Jacobian corresponding to this transformation is $J = 1$, a fact which allows us to write that $P' = P$ and $\widehat{C}_0 = 1$. Besides $|\mathbf{S}'|^2 = |\mathbf{S}|^2$ and the transition probability density $P(\mathbf{U}, t|\mathbf{U}_0)$ looks like Eq. (147), just written in terms of $\mathbf{U}$. Thus, the solution of Eq. (142) then is

$$P(\mathbf{U}, t|\mathbf{U}_0) = \frac{\gamma}{2\pi\lambda(1 - e^{-2\gamma t})} \exp\left(- \frac{\gamma|\mathbf{U} - e^{-\Lambda t}\mathbf{U}_0|^2}{2\lambda(1 - e^{-2\gamma t})} \right). \tag{149}$$

Now we return to the original variable $\mathbf{u}$

$$\widehat{\mathbf{S}} = \mathbf{U} - e^{-\Lambda t}\mathbf{U}_0 = \mathbf{u} - \langle \mathbf{u} \rangle - e^{-\Lambda t}(\mathbf{u}_0 - \langle \mathbf{u}_0 \rangle) = \mathbf{u} - e^{-\Lambda t}(\mathbf{u}_0 + \bar{\mathbf{a}}_1(t))\,, \tag{150}$$

where the solution of Eq. (140) has been used, that is $\langle \mathbf{u} \rangle = e^{-\Lambda t}(\langle \mathbf{u}_0 \rangle + \bar{\mathbf{a}}_1(t))$. The solution of the FP equation (135) is finally given by

$$P(\mathbf{u}, t|\mathbf{u}_0) = \frac{\gamma}{2\pi\lambda(1 - e^{-2\gamma t})} \exp\left(- \frac{\gamma|\mathbf{u} - e^{-\Lambda t}(\mathbf{u}_0 + \bar{\mathbf{a}}_1(t))|^2}{2\lambda(1 - e^{-2\gamma t})} \right). \tag{151}$$

As the time gets large $\gamma t \gg 1$, this probability density can be approximated by

$$P(\mathbf{u}) = \frac{m}{2\pi k_B T} \exp\left(- \frac{m\,|\widehat{\mathbf{S}}_{as}|^2}{2\,k_B T} \right), \tag{152}$$

where $\widehat{\mathbf{S}}_{as}$ is asymptotic approximation of $\widehat{\mathbf{S}}$ which depends on the particular expression of the time-dependent external force $\bar{\mathbf{a}}(t)$, i.e. the explicit expressions of both electrical and mechanical forces.

5.2. The Probability Density for an Initially Maxwellian Velocity Distribution

The solution given in Eq. (151) along with the initial condition $P(\mathbf{u}, 0|\mathbf{u}_0) \equiv \delta(\mathbf{u} - \mathbf{u}_0)$, has been named to as the "fundamental" solution [9]. However, a more general probability density can be calculated by assuming an arbitrary initial condition $f(\mathbf{u}, 0)$. This can be calculated through the following integration

$$f(\mathbf{u}, t) = \int_{\mathbf{u}_0} f(\mathbf{u}_0, 0)\, P(\mathbf{u}, t|\mathbf{u}_0)\, d\mathbf{u}_0 \,. \tag{153}$$

When we consider the particular case of an initial Maxwellian distribution function given by

$$f(\mathbf{u}, 0) = \frac{m}{2\pi\, k_B T_0} \exp\left(- \frac{m|\mathbf{u} - \langle \mathbf{u} \rangle_0|^2}{2k_B T_0} \right), \tag{154}$$

after integration over the initial velocity $\mathbf{u}_0$, we obtain

$$f(\mathbf{u}, t) = \frac{m}{2\pi\, k_B T_t} \exp\left(\frac{m|\mathbf{u} - e^{-\Lambda t}((\langle \mathbf{u} \rangle_0 + \bar{\mathbf{a}}_1(t)|^2}{2k_B T_t} \right), \tag{155}$$

where T_0 corresponds to the temperature in the initial distribution function. Notice that it is different from the heat bath equilibrum temperature T. Also we have defined an equivalent time dependent temperature as follows

$$T_t = T\left[1 - \left(1 - \frac{T_0}{T}\right) e^{-2\gamma t}\right]. \tag{156}$$

Two limiting cases can be considered in Eq. (155):

(i) When $T_0 = 0$, the initial condition becomes $f(\mathbf{u}, 0) = \delta(\mathbf{u} - \mathbf{u}_0)$ with $\mathbf{u}_0 = \langle \mathbf{u} \rangle_0$, and the equivalent temperature $T_t = T(1 - e^{-2\gamma t})$, which obviously tends to the bath equilibrium temperature after times $t \gg 1/\gamma$. Therefore $f(\mathbf{u}, t)$ reduces to the fundamental solution (151).

(ii) For $T_0 = T$, the initial velocity distribution is the same as Eq. (154), i.e. a Maxwellian at the equilibrium temperature T around the initial mean velocity $\langle \mathbf{u} \rangle_0$. The probability density $f(\mathbf{u}, t)$ is in this case the same as Eq. (155), with $T_t = T$.

5.3. Fokker-Planck-Kramers Equation in a Field of Force

The next step in the study of Brownian motion in a field of force and in the presence of a constant magnetic field is done when we consider the FPK equation. It is constructed from the Langevin equation written in Eq. (132) along with $\dot{\mathbf{x}} = \mathbf{u}$ and given in the x-y plane. The description in (z, v_z) space is uncoupled from the planar problem and it is described with Eq. (133) along with $\dot{z} = v_z$, and we will not repeat their solution. Hence, the phase-space $(\mathbf{x}, \mathbf{u})$ FPK equation for the corresponding transition probability density $P(\mathbf{x}, \mathbf{u}, t|\mathbf{x}_0, \mathbf{u}_0)$ is given as [14, 19]

$$\frac{\partial P}{\partial t} + \mathbf{u}\cdot\nabla_{\mathbf{x}} P + \bar{\mathbf{a}} \cdot \nabla_{\mathbf{u}} P = \nabla_{\mathbf{u}} \cdot (\Lambda \mathbf{u} P) + \lambda\, \nabla^2_{\mathbf{u}} P\,, \tag{157}$$

subject to the initial condition

$$P(\mathbf{x}, \mathbf{u}, 0|\mathbf{x}_0, \mathbf{u}_0) = \delta(\mathbf{x} - \mathbf{x}_0)\delta(\mathbf{u} - \mathbf{u}_0)\,. \tag{158}$$

To solve Eq. (157) we define the variable $\dot{\mathbf{X}} \equiv \mathbf{U}$, where we know that $\mathbf{U} = \mathbf{u} - \langle \mathbf{u} \rangle$ and then $\dot{\mathbf{X}} = \dot{\mathbf{x}} - \langle \dot{\mathbf{x}} \rangle$ since $\dot{\mathbf{x}} = \mathbf{u}$, and express the Langevin equations in this new set of phase-space $(\mathbf{X}, \mathbf{U})$ coordinates. In this space, the Langevin equations read

$$\frac{d\mathbf{X}}{dt} = \mathbf{U}\,, \tag{159}$$

$$\frac{d\mathbf{U}}{dt} = -\gamma \mathbf{U} + W\mathbf{U}(t) + \bar{\mathbf{f}}(t)\,. \tag{160}$$

The FPK equation for the transition probability density $P(\mathbf{X}, \mathbf{U}, t|\mathbf{X}_0, \mathbf{U}_0)$, associated with them becomes

$$\frac{\partial P}{\partial t} + \mathbf{U}\cdot\nabla_{\mathbf{X}} P = \nabla_{\mathbf{U}} \cdot (\Lambda \mathbf{U} P) + \lambda \nabla^2_{\mathbf{U}} P\,, \tag{161}$$

subject to the initial condition

$$P(\mathbf{X}, \mathbf{U}, 0|\mathbf{X}_0, \mathbf{U}_0) = \delta(\mathbf{X} - \mathbf{X}_0)\delta(\mathbf{U} - \mathbf{U}_0)\,, \tag{162}$$

which is solved by the method presented in previous sections. In fact, Eqs. (161) and (162) have the same structure as Eqs. (96) and (97), and therefore the steps for the required solution are very similar. We also define the transformed variables $\dot{\mathbf{X}}' \equiv \mathbf{U}'$, where $\mathbf{U}' = \mathbf{u}' - \langle \mathbf{u}' \rangle = \dot{\mathbf{x}}' - \langle \dot{\mathbf{x}}' \rangle$, since $\dot{\mathbf{x}}' = \mathbf{u}'$. The Langevin equations satisfied by these new variables are as follows

$$\dot{\mathbf{X}}' = \mathbf{U}'\,, \tag{163}$$

$$\dot{\mathbf{U}}' = -\gamma \mathbf{U}' + \bar{\mathbf{f}}'(t)\,. \tag{164}$$

The FPK equation associated with these equations for the transiton probability density $P'(\mathbf{X}', \mathbf{U}', t|\mathbf{X}'_0, \mathbf{U}'_0)$ becomes

$$\frac{\partial P'}{\partial t} + \mathbf{U}'\cdot\nabla_{\mathbf{X}'} P' = \gamma \nabla_{\mathbf{U}'} \cdot (\mathbf{U}' P') + \lambda \nabla^2_{\mathbf{U}'} P'\,, \tag{165}$$

subject to the initial condition

$$P'(\mathbf{X}', \mathbf{U}', 0|\mathbf{X}'_0, \mathbf{U}'_0) = \widehat{K}\delta(\mathbf{U}' - \mathbf{U}'_0)\delta(\mathbf{X}' - \mathbf{X}'_0)\,, \tag{166}$$

where $\widehat{K}$ will be determined bellow. Equation (165) is also very similar to Eq. (103) and therefore we can conclude that the solution of Eq. (165), for the transition probability density $P'(\widehat{\mathbf{R}}', \widehat{\mathbf{S}}') \equiv P'(\mathbf{X}', \mathbf{U}', t|\mathbf{X}'_0, \mathbf{U}'_0)$ reads

$$P'(\widehat{\mathbf{R}}', \widehat{\mathbf{S}}') = \frac{1}{4\pi^2(FG - H^2)} \exp\left(- \frac{(F|\widehat{\mathbf{S}}'|^2 - 2H\widehat{\mathbf{R}}' \cdot \widehat{\mathbf{S}}' + G|\widehat{\mathbf{R}}'|^2)}{2(FG - H^2)} \right), \tag{167}$$

where

$$\widehat{\mathbf{R}}' = \mathbf{X}' - \mathbf{X}'_0 - \gamma^{-1}\mathbf{U}'_0(1 - e^{-\gamma t})\,, \tag{168}$$

$$\widehat{\mathbf{S}}' = \mathbf{U}' - \mathbf{U}'_0 e^{-\gamma t}\,. \tag{169}$$

The transformation which allows us to return to the space of coordinates $(\mathbf{X}, \mathbf{U})$, is then

$$\mathbf{X}' - \mathbf{J}_2' = \gamma^{-1} e^{-Wt} \Lambda(\mathbf{X} - \mathbf{J}_2)\,, \tag{170}$$

which is also very similar to Eq. (110). If we define the $\widehat{\mathbf{R}}$ and $\widehat{\mathbf{S}}$ variables as

$$\widehat{\mathbf{R}} = \mathbf{X} - \mathbf{X}_0 - \Lambda^{-1}(1 - e^{-\Lambda t})\mathbf{U}_0\,, \tag{171}$$

$$\widehat{\mathbf{S}} = \mathbf{U} - e^{-\Lambda t}\mathbf{U}_0\,, \tag{172}$$

then it can be shown that $\widehat{\mathbf{R}}' = \gamma^{-1} e^{-Wt} \Lambda \widehat{\mathbf{R}}$ and $\widehat{\mathbf{S}}' = e^{-Wt}\widehat{\mathbf{S}}$. On the other hand, the transformation between the probabilities is $P d\widehat{\mathbf{S}} d\widehat{\mathbf{R}} = P' d\widehat{\mathbf{S}}' d\widehat{\mathbf{R}}'$ where $d\widehat{\mathbf{S}} d\widehat{\mathbf{R}} = J\, d\widehat{\mathbf{S}}' d\widehat{\mathbf{R}}'$. Repeating the same algebraic steps given before, we conclude that

$$J'_{\widehat{S}} = 1, \qquad J'_{\widehat{R}} = \frac{\gamma^2 + \Omega^2}{\gamma^2} = 1 + C^2\,, \tag{173}$$

and $P = J'_{\widehat{R}} P'$. The constant given in Eq. (166) is then $\widehat{K} = J'_{\widehat{R}} = 1 + C^2$. Also we have that

$$|\widehat{\mathbf{S}}'|^2 = |\widehat{\mathbf{S}}|^2\,, \tag{174}$$

$$|\widehat{\mathbf{R}}'|^2 = \widehat{K}|\widehat{\mathbf{R}}|^2\,, \tag{175}$$

$$\widehat{\mathbf{S}}' \cdot \widehat{\mathbf{R}}' = \widehat{\mathbf{S}} \cdot \widehat{\mathbf{R}} + C(\widehat{\mathbf{S}} \times \widehat{\mathbf{R}})_z\,, \tag{176}$$

where $(\widehat{\mathbf{S}} \times \widehat{\mathbf{R}})_z = (\widehat{S}_x \widehat{R}_y - \widehat{S}_y \widehat{R}_x)$ is the z-component of the cross product. Upon substitution of Eqs. (174)-(176) into Eq. (167) we obtain the solution of Eq. (161), for the transition probability density $P(\widehat{\mathbf{R}}, \widehat{\mathbf{S}}) \equiv P(\mathbf{X}, \mathbf{U}, t|\mathbf{X}_0, \mathbf{U}_0)$, which is given by

$$P(\widehat{\mathbf{R}}, \widehat{\mathbf{S}}) = \widehat{K}_0 \exp\left(- \frac{\left[F|\widehat{\mathbf{S}}|^2 - 2H\widehat{\mathbf{R}} \cdot \widehat{\mathbf{S}} + 2CH(\widehat{\mathbf{S}} \times \widehat{\mathbf{R}})_z + \widehat{K}G|\widehat{\mathbf{R}}|^2\right]}{2(FG - H^2)} \right), \tag{177}$$

where $\widehat{K}_0 = \widehat{K}/4\pi^2(FG - H^2)$. The solution (177) is now written in terms of the original variables $\mathbf{x}$ and $\mathbf{u}$, to get the explicit solution of Eq. (157). First, we have shown in Eq. (150) how to relate the variable $\widehat{S}$ whith the velocity

$$\widehat{\mathbf{S}} = \mathbf{U} - e^{-\Lambda t}\mathbf{U}_0 = \mathbf{u} - e^{-\Lambda t}(\mathbf{u}_0 + \bar{\mathbf{a}}_1(t))\,. \tag{178}$$

Secondly, we need the definition $\dot{\mathbf{X}} = \dot{\mathbf{x}} - \langle \dot{\mathbf{x}} \rangle$, to show after integration that

$$\mathbf{X} - \mathbf{X}_0 = \mathbf{x} - \mathbf{x}_0 - \int_0^t \langle \mathbf{u} \rangle\, dt\,, \tag{179}$$

where $\langle \mathbf{u} \rangle$ is known directly from Eq. (140), and therefore

$$\mathbf{X} - \mathbf{X}_0 = \mathbf{x} - \mathbf{x}_0 - \Lambda^{-1}(1 - e^{-\Lambda t})\langle \mathbf{u}_0 \rangle - \bar{\mathbf{a}}_2(t)\,. \tag{180}$$

where we have defined $\bar{\mathbf{a}}_2(t) = \int_0^t \bar{\mathbf{a}}_1(s)ds$. Upon substitution of this equation into Eq. (171), we find that

$$\widehat{\mathbf{R}} = \mathbf{x} - \mathbf{x}_0 - \Lambda^{-1}(1 - e^{-\Lambda t})\mathbf{u}_0 - \bar{\mathbf{a}}_2(t)\,. \tag{181}$$

Finally, Eq. (177) along with Eqs. (178) and (181) constitute the solution of the FPK Eq. (157) for the transition probability density $P(\widehat{\mathbf{R}}, \widehat{\mathbf{S}}) \equiv P(\mathbf{x}, \mathbf{u}, t|\mathbf{x}_0, \mathbf{u}_0)$.

The transition probability density given in Eq. (177) allows, the calculation of the marginal probability densities $P(\widehat{\mathbf{S}})$ and $P(\widehat{\mathbf{R}})$, through the corresponding integrals.

$$P(\widehat{\mathbf{S}}) = \int P(\widehat{\mathbf{R}}, \widehat{\mathbf{S}})\, d\widehat{\mathbf{R}}\,, \tag{182}$$

$$P(\widehat{\mathbf{R}}) = \int P(\widehat{\mathbf{R}}, \widehat{\mathbf{S}})\, d\widehat{\mathbf{S}}\,. \tag{183}$$

The first integration leads to Eq. (151) as it should be, and the integration in Eq. (183) leads, after a long but straightforward algebra to

$$P(\mathbf{x}, t|\mathbf{x}_0, \mathbf{u}_0) = N_R(t)\exp\left\{ - \frac{\gamma|\mathbf{x} - \mathbf{x}_0 - \Lambda^{-1}(1 - e^{-\Lambda t})\mathbf{u}_0 - \bar{\mathbf{a}}_2(t)|^2}{2\, D_e(2\gamma\, t - 3 + 4e^{-\gamma t} - e^{-2\gamma t})} \right\}, \tag{184}$$

where $N_R(t)$ is the same as that given below Eq. (129). In the large time limit $\gamma t \gg 1$, Eq. (184) reduces to

$$P(\mathbf{x}, t|\mathbf{x}_0) \simeq \frac{1}{(4\pi\, D_e\, t)} \exp\left(- \frac{|\widehat{\mathbf{R}}_{as}|^2}{4D_e\, t} \right), \tag{185}$$

where $\widehat{\mathbf{R}}_{as}$ is the asymptotic approximation of $\widehat{\mathbf{R}}$ as given by Eq. (181).

For a more general probability density $f(\mathbf{x}, \mathbf{u}, t)$ satisfying an arbitrary initial condition $f(\mathbf{x}, \mathbf{u}, 0)$, we have

$$f(\mathbf{x}, \mathbf{u}, t) = \int_{\mathbf{x}_0} d\mathbf{x}_0 \int_{\mathbf{u}_0} d\mathbf{u}_0\, f(\mathbf{x}_0, \mathbf{u}_0, 0) P(\mathbf{x}, \mathbf{u}, t|\mathbf{x}_0, \mathbf{u}_0)\,, \tag{186}$$

where $P(\mathbf{x}, \mathbf{u}, t|\mathbf{x}_0, \mathbf{u}_0)$ is the fundamental solution given in Eq. (177).

6. Hall-Type Fluctuation Relation for a Charged Brownian Particle

In 2008 Roy and Kumar established two fluctuation relations for a two-dimensional system of noninteracting electrons under the influence of crossed electric and magnetic fields in the high friction limit with the surrounding medium [62]. One relation has been named the longitudinal or barotropic-type fluctuation relation while the other is named to as the transversal or Hall fluctuation relation. The Roy and Kumar's study was extended and generalized by Jiménez-Aquino *et al.* to a classical system of noninteracting harmonic oscillators in the presence of an electromagnetic field, in the high friction as well as in the complete cases by Jiménez-Aquino *et al.* [66,67]. It was shown that in the absence of the harmonic force, the Roy and Kumar's fluctuation relations were obtained as a particular case. Our purpose in this section is to show that it is also possible to obtain both aforementioned fluctuation relations, through another method than that proposed in [66]. This, consists in a two-dimensional system of noninteracting charged particles in the presence of an electromagnetic field as that governed by the Langevin Eq. (132) along with $\dot{\mathbf{x}} = \mathbf{u}$. Here, we assume that $\mathbf{F}_{\rm m} = 0$ and therefore $\mathbf{a}(t) = (q/m)\mathbf{E}(t)$.

6.1. Fluctuation Relations

The fluctuation relations can be obtained from the probability density $f(\mathbf{x}, t)$, which is calculated from the general integration

$$f(\mathbf{x}, t) = \int d\mathbf{x}_0 \int d\mathbf{u}_0 \, f(\mathbf{x}_0, \mathbf{u}_0, 0) P(\mathbf{x}, t|\mathbf{x}_0, \mathbf{u}_0) \,, \tag{187}$$

where $f(\mathbf{x}_0, \mathbf{u}_0, 0)$ is an arbitrary initial distribution function and $P(\mathbf{x}, t|\mathbf{x}_0, \mathbf{u}_0)$ is given by Eq. (184), which for further purposes can be written as

$$P(\mathbf{x}, t|\mathbf{x}_0, \mathbf{u}_0) = \frac{\gamma}{2\pi \, D_e F_0} \exp\left\{ - \frac{\gamma |\mathbf{x} - \mathbf{x}_0 - \Lambda^{-1}(1 - e^{-\Lambda t})\mathbf{u}_0 - \bar{\mathbf{a}}_2(t)|^2}{2 \, D_e F_0} \right\}, \tag{188}$$

where $D_e = D/\widehat{K}$, and $F_0 = (2\gamma \, t - 3 + 4e^{-\gamma t} - e^{-2\gamma t})$. The selection of the initial distribution is a matter of choice and, in principle, we can propose the simplest expression by means of choosing $f(\mathbf{x}_0, \mathbf{u}_0, 0) = \delta(\mathbf{x}_0)\delta(\mathbf{u}_0)$. However, if we recall that the system is embedded in a thermal bath at temperature T, it will be convenient to suggest a more general initial distribution. Hence, we will assume that it can be written as a product of functions in the configuration and velocity spaces. In order to compare with the results obtained in the high friction limit [62], we will assume the position dependence as a Gaussian function with a width determined by the temperature T and a delta function in the velocity $\mathbf{u}_0$, in such a way that

$$f(\mathbf{x}_0, \mathbf{u}_0, 0) = \frac{\gamma}{2\pi D_e} \exp\left(- \frac{\gamma \, |\mathbf{x}_0|^2}{2D_e} \right) \delta(\mathbf{u}_0) \,. \tag{189}$$

Substituting Eqs. (188) and (189) into Eq. (187), and after integration we get

$$f(\mathbf{x}, t) = \frac{m\gamma^2 \widehat{K}}{2\pi k_B T(1 + F_0)} \exp\left(- \frac{m\gamma^2 \widehat{K} \, |\mathbf{x} - \langle \mathbf{x} \rangle|^2}{2k_B T(1 + F_0)} \right), \tag{190}$$

where the mean value $\langle \mathbf{x} \rangle = \bar{\mathbf{a}}_2(t)$. From this probability density we can check that the variance for each component $\xi = x, y$, defined as $\sigma_\xi^2 \equiv \langle \xi^2 \rangle - \langle \xi \rangle^2$, is $\sigma_\xi^2 = k_B T(1 + F_0)/m\gamma^2 \widehat{K}$. The fluctuation relation for this probability density then is

$$\frac{f(\mathbf{x}, t)}{f(-\mathbf{x}, t)} = \exp\left(\frac{2m\gamma^2 \widehat{K} \, \mathbf{x} \cdot \langle \mathbf{x} \rangle}{k_B T(1 + F_0)} \right). \tag{191}$$

In order to understand the physical meaning contained in Eq. (191), notice that the exponent has a quotient of terms which have energy units and $(1 + F_0)$ should be dimensionless. We will show below that the exponent is related with total amount of dimensionless energy that the particle can exchange with the heat bath in the form of work $\mathbb{W}(t) = 2m\gamma^2 \widehat{K} \mathbf{x} \cdot \langle \mathbf{x} \rangle / k_B T(1 + F_0) = 2\mathbf{x} \cdot \langle \mathbf{x} \rangle / \sigma_\xi^2(t)$ performed along a single stochastic trajectory $\mathbf{x}(t)$ [47,49]. First, we can notice that the variance of this dimensionless work defined as $\sigma_{\mathbb{W}}^2 \equiv$

$\langle \mathbb{W}^2 \rangle - \langle \mathbb{W} \rangle^2$ satisfies $\sigma_{\mathbb{W}}^2 = 2\langle \mathbb{W} \rangle$. In principle, we can also formulate a fluctuation relation for $\mathbb{W}$ taking into account that it is a Gaussian random variable (GRV), since $\mathbf{x}$ is also a GRV. This is indeed the case, because the solution of equation $\dot{\mathbf{x}} = \mathbf{u}$ coming from Eq. (132) tell us that $\mathbf{x}$ is a GRV because $\mathbf{u}$ is Gaussian as well. Therefore probability density for $\mathbb{W}$ satisfies

$$P(\mathbb{W}) = \frac{1}{\sqrt{2\pi\sigma_{\mathbb{W}}^2}} \exp\left(- \frac{[\mathbb{W} - \langle \mathbb{W} \rangle]^2}{2\sigma_{\mathbb{W}}^2} \right). \tag{192}$$

It is possible to show that the fluctuation relation for the total dimensionless work is $P(\mathbb{W})/P(-\mathbb{W}) = e^{\mathbb{W}}$, consistently with the transient work-fluctuation theorem [33, 38, 40, 55].

Coming back to Eq. (191), we proceed to obtain an explicit expression for $\mathbb{W}(t)$ when the electric field is a constant $\mathbf{E} = (E_x, E_y)$. In this case the mean value of position reads

$$\langle \mathbf{x} \rangle = \frac{q}{m}(\Lambda^{-1}\mathbf{E}t + e^{-\Lambda t}\Lambda^{-2}\mathbf{E} - \Lambda^{-2}\mathbf{E}), \tag{193}$$

where Λ^{-1} is the same as Eq. (100). After a direct substitution we obtain

$$\mathbb{W}(t) = \frac{\Phi_1(t)}{k_B T} q\mathbf{E} \cdot \mathbf{x} + \frac{\Phi_2(t)}{k_B T} |\mathbf{x} \times q\mathbf{E}|, \tag{194}$$

where

$$\Phi_1(t) = \frac{2}{(1+F_0)} \left[\gamma t + \frac{(1-C^2)}{1+C^2}(e^{-\gamma t}\cos\Omega t - 1) - \frac{2C}{1+C^2} e^{-\gamma t} \sin\Omega t \right], \tag{195}$$

$$\Phi_2(t) = \frac{2}{(1+F_0)} \left[C\gamma t + \frac{2C}{1+C^2}(e^{-\gamma t}\cos\Omega t - 1) + \frac{1-C^2}{1+C^2} e^{-\gamma t} \sin\Omega t \right], \tag{196}$$

which are clearly dimensionless functions. We clearly see from Eq. (194) that $q\mathbf{E} \cdot \mathbf{x}$ represents what we can call as the *translational work* done on the charged particle by the constant electric field. The quantity $q\mathbf{E} \cdot \mathbf{x}\gamma t$ is the time increasing of this work, $e^{-\gamma t}\cos(\Omega t)\, q\mathbf{E} \cdot \mathbf{x}$ as well as $e^{-\gamma t}\sin(\Omega t)\, q\mathbf{E} \cdot \mathbf{x}$ are the oscillating dissipations of the work since $e^{-\tilde{\gamma} t} q\mathbf{E} \cdot \mathbf{x}$ is its dissipation, and $\Omega = qB/mc$ the cyclotron frequency. On the other hand, the factor $\tau = |\mathbf{x} \times q\mathbf{E}|$ is the modulus of the torque $\boldsymbol{\tau} = \mathbf{x} \times q\mathbf{E}$ done by the electric force. The second term of (194) contains a term of the form $\tau\, C\gamma t = \tau\, \Omega t$. If we now identify $\theta(t) = \Omega t$, thus $\tau\, \Omega t = \tau\theta(t)$ represents what we can call as the *rotational work* done on the charged particle by the electric torque. The other terms including the factors $e^{-\gamma t}\cos(\Omega t)$ and $e^{-\gamma t}\sin(\Omega t)$ can be identified as the oscillating dissipation of the rotational work. So, the quantity $\mathbb{W}(t)$ can be then understood as the total amount of dissipative work (translational and rotational) done on the charged particle as it is driven out of equilibrium by the constant electric field. In this case, the fluctuation relation (191), can shortly be written as $f(\mathbf{x}, t)/f(-\mathbf{x}, t) = e^{\mathbb{W}(t)}$, which quantifies the probability of observing a given trajectory at $(\mathbf{x}, t)$ during the forward process, and the probability of observing its backward counterpart at $(-\mathbf{x}, t)$. It is related with the total amount of dissipative work done by the external electric field [55].

To complete the analysis we will study some particular limits which help to disentangle the terms in the general expression. In the short time regime, characterized by $\gamma t \ll 1$ and $\Omega t \ll 1$, the exponent $\mathbb{W}(\gamma t \ll 1, \Omega t \ll 1) \to 0$. If we recall that this regime corresponds to the ballistic regime in the Brownian movement of the particle, we have that $f(\mathbf{x},t)/f(-\mathbf{x},t) \to 1$, and hence $f(\mathbf{x},t) = f(-\mathbf{x},t)$. In this case, the probability of observing trajectories during the forward processes is the same as their backward counterparts. This can be understood in the following way: at the very beginning (short times) of the process, the Brownian particle does not yet feel the presence of the surrounding medium, so that it moves as a free particle in the absence of friction and the process becomes reversible. The limit of short times is then equivalent to consider the zero friction case $\gamma \to 0$. On the other hand, in the large time regime $\gamma t \gg 1$, we obtain

$$\mathbb{W}(\gamma t \gg 1) = \frac{1}{k_B T}(q\mathbf{E}\cdot\mathbf{x} + C|\mathbf{x}\times q\mathbf{E}|)\,. \tag{197}$$

We have already commented above that the first and the second terms of this equation, represent the translational work done by the electric field and rotational work done by the electric torque respectively. It is interesting to notice that the second term of Eq. (197) can also be written as $C|\mathbf{x}\times q\mathbf{E}| = qC|\mathbf{x}\times\mathbf{E}| = (q/mc\gamma)\mathbf{x}\cdot(\mathbf{E}\times\mathbf{B})$, which shows in an explicit way how the magnetic and electric fields are coupled to give a contribution. In a similar way it can be written as $qC|\mathbf{x}\times\mathbf{E}| = (q/mc\gamma)\mathbf{E}\cdot(\mathbf{x}\times\mathbf{B})$, an expression which shows the torque caused by the magnetic field (this torque is on the x-y plane).

6.2. Hall-like Fluctuation Relation

Now we will consider the fluctuation relations associated to the marginal probability densities $f(x,t)$ and $f(y,t)$. To obtain $f(x,t)$ we take an integration over the y coordinate of the probability density (190) and then calculate the quotient

$$\frac{f(x,t)}{f(-x,t)} = \exp\left(\frac{2m\gamma^2\widehat{K}\,x\langle x\rangle}{k_B T(1+F_0)}\right). \tag{198}$$

In a similar way, we calculate $f(y,t)$ by integrating over the x coordinate of the probability density (190), this leads to

$$\frac{f(y,t)}{f(-y,t)} = \exp\left(\frac{2m\gamma^2\widehat{K}\,y\langle y\rangle}{k_B T(1+F_0)}\right). \tag{199}$$

Both expressions become interesting when we consider the particle as an electron $q = -e$, the electric field as a constant pointing along the x-axis, $E_x = E$, $E_y = 0$ and the large time limiting case $\gamma t \gg 1$. In this case $C = -eB/c\alpha$ and therefore Eq. (198) reduces to

$$\frac{f(x,t)}{f(-x,t)} = \exp\left(\frac{-eEx}{k_B T}\right), \tag{200}$$

which is exactly the same longitudinal or barotropic-type fluctuation relation first established by Roy and Kumar [62] for a two-dimensional system of noninteracting electrons under the action of crossed electromagnetic fields in a high friction limit. On the other hand, Eq. (199) allows us to write

$$\frac{f(y,t)}{f(-y,t)} = \exp\left(\frac{-e^2 E B y}{k_B T c \alpha}\right), \tag{201}$$

corresponding to the transversal or Hall-fluctuation relation established first by Roy and Kumar [62], which shows the fluctuations coming from the effect of both fields, the electric field along the x-axis and the magnetic field pointing along the z-axis. The crossed effect produces fluctuations along the y-axis, which is a signature of the usual Hall effect. Both fluctuation relations, Eqs. (200), (201) were also obtained as a particular case of two-dimensional system of noninteracting harmonic oscillators under the action of an electromagnetic field in a high friction limit [66]. It is clear that in the limit of short times, both marginal fluctuation relations satisfy $f(x,t) = f(-x,t)$ and $f(y,t) = f(-y,t)$ as expected.

It is also possible to consider that the electric field depends explicitly on time of the form $\mathbf{E}(t) = (\mathcal{E},\mathcal{E})\phi(t)$, where $\mathcal{E}$ is a constant and $\phi(t)$ a dimensionless function of time. Now, the exponent in Eq. (187) can be written as

$$\mathbb{W}(t) = \frac{2m\gamma^2 \widehat{K}\mathbf{x}\cdot\langle\mathbf{x}\rangle}{k_B T(1+F_0)} = \frac{2q\mathcal{E}\widehat{K}}{k_B T(1+F_0)}\left[xI_x(\gamma t) + yI_y(\gamma t)\right], \tag{202}$$

where $I_x(\gamma t)$ and $I_y(\gamma t)$ are the x and y components of the vectorial function

$$\mathbf{I}(\gamma t) = \frac{1}{\gamma^2}\,\hat{\mathbf{e}}\int_0^{\gamma t} e^{-\Lambda/\gamma\xi'}\,d\xi'\int_0^{\xi'} e^{\Lambda/\gamma\xi}\phi(\xi)\,d\xi\,, \tag{203}$$

and the vector $\hat{\mathbf{e}} = (1,1)$. Notice that this expression depends on the specific functionality contained in the function $\phi(t)$. Also, it has a dependence with the magnetic field which appears in the rotation matrix contained in Λ. The explicit expression can be calculated once the function $\phi(t)$ is specified.

7. Brownian Motion in Harmonic Traps in the Presence of an Electromagnetic Field

The study of fluctuation theorems from a theoretical as well as from experimental points of view has taken the Brownian motion of a particle in harmonic traps as a prototype system. In fact, the first experiment performed to demonstrate the fluctuation theorem was reported by Wang *et al.* [38], for a small system over short times. In this experiment the trajectory of a colloidal particle is followed when it is captured in an optical trap which is translated relative to the surrounding water molecules. The experiment confirms a theoretically predicted work relation associated to the integrated transient fluctuation theorem (ITFT), which gives an expression for the ratio between the probability to find positive and negative values of the fluctuations in the total work done on the system for a given time in a transient state. After the experiment by Wang *et al.* [38] others were continued, for example with colloidal particles in harmonic trap potentials [42, 43]. Inspired by Wang's *et al.* experiment [38], and using a model of a Brownian particle in a harmonic potential with a minimum moving arbitrarily, van Zon and Cohen (vZC) [40] showed theoretically that all quantities of interest for these theorems and the corresponding non-integrated ones (TFT and stationary state fluctuation theorem (SSTF)) hold. In this section, we will focus on the generalization of van Zon and Cohen's theorem to the case in which the Brownian harmonic oscillator is under the action of an electromagnetic field [68]. In our study, the magnetic field is again the vector $\mathbf{B} = (0, 0, B)$ and the electric field is also a time-varying vector which will be responsible for the arbitrary dragging of the potential minimum.

7.1. Langevin and Smoluchowski Equations for the Harmonic Oscillator

Due to the orientation of the magnetic field, the Langevin equation for the charged harmonic oscillator of mass m and charge q can be split into two independent differential equations: one is given on the x-y plane and the other along the z-axis. Along the z-axis, the Langevin equation is the same z-component of the Langevin equation studied in Ref. [40]. Hence, we just pay attention on the planar Langevin equation, for which we have already defined the two-dimensional vectors: $\mathbf{x} = (x, y)$, $\mathbf{u} = (v_x, v_y)$, $\bar{\mathbf{f}}(t) = (f_x, f_y)$, and now the planar electric field as $\bar{\mathbf{E}}(t) = (E_x, E_y)$. In this case, the two-dimensional harmonic trap $\bar{U}$ and its corresponding harmonic force $\bar{\mathbf{F}}$ are given by

$$\bar{U}(\mathbf{x}, t) = \frac{k}{2}|\mathbf{x} - \mathbf{x}^*|^2, \qquad \bar{\mathbf{F}}(\mathbf{x}, \mathbf{x}^*) = -k(\mathbf{x} - \mathbf{x}^*)\,, \tag{204}$$

where $\mathbf{x}^*$ the position of its minimum and k a constant. For $t = 0$, the potential minimum is located at the origin, $\mathbf{x}^*(0) \equiv \mathbf{x}_0 = 0$, whereas for $t > 0$, it moves with a velocity $\mathbf{u}^*(t)$. The electromagnetic field acts on the charged particle via the planar Lorentz force given by $F_L = q\bar{\mathbf{E}} + (q/c)\mathbf{u} \times \mathbf{B}$. Here, we are interested in study of the over-damped approximation of the Lagevin equation for the charged harmonic oscillator. This limiting case is satisfied when $\omega^2 \ll \gamma^2[1 + (\Omega/\gamma)^2]^2$, where $\gamma = \alpha/m$, $\omega^2 = k/m$, and again $\Omega = qB/mc$ is the cyclotron frequency. This condition is equivalent to $km \ll \alpha_e^2$, where $\alpha_e = \alpha(1 + C^2)$ and $C = \Omega/\gamma = qB/c\alpha$. It is clear that α_e plays the role of an effective friction coefficient which is magnetic field dependent. Thus, in the over-damped approximation the planar Langevin equation reads

$$\frac{d\mathbf{x}}{dt} = -\widetilde{\gamma}\,\mathbf{x} - \widetilde{W}\mathbf{x} + \tilde{\Lambda}\mathbf{x}^* + k^{-1}\tilde{\Lambda}\,\bar{\mathbf{f}}(t)\,, \tag{205}$$

where $\widetilde{W}$ and $\tilde{\Lambda} = \widetilde{\gamma}\mathbf{I} + \widetilde{W}$ are given by

$$\widetilde{W} = \begin{pmatrix} 0 & \widetilde{\Omega} \\ -\widetilde{\Omega} & 0 \end{pmatrix}, \quad \tilde{\Lambda} = \begin{pmatrix} \widetilde{\gamma} & \widetilde{\Omega} \\ -\widetilde{\Omega} & \widetilde{\gamma} \end{pmatrix}, \tag{206}$$

$\mathbf{I}$ being the unit matrix, $\widetilde{\gamma} = k/\alpha(1 + C^2) = k/\alpha_e$ and $\widetilde{\Omega} = kC/\alpha(1 + C^2)$. In Eq. (205), the position vector for the potential minimum is now $\mathbf{x}^* = (q/k)\bar{\mathbf{E}}(t)$ and $\dot{\mathbf{x}}^* = \mathbf{u}^*(t) = (q/k)\dot{\bar{\mathbf{E}}}(t)$, which tells us that the dragging of the potential minimum depends on the electric field rate of change. The fluctuating force $\bar{\mathbf{f}}(t)$ satisfies the properties of Gaussian white noise with zero mean value $\langle f_i(t)\rangle$=0 and a correlation function

$$\langle f_i(t) f_j(t')\rangle = 2\lambda\,\delta_{ij}\,\delta(t - t')\,, \tag{207}$$

where λ is related with the friction constant by $\lambda = \alpha k_B T$, k_B being the Boltzmann's constant and T the temperature of the bath. Next, we separate the average motion of the charged Brownian particle from the stochastic motion. For the average motion we denote by the variable $\langle \mathbf{x} \rangle$ as the deterministic solution of Eq. (205), that is

$$\frac{d\langle \mathbf{x} \rangle}{dt} = -\widetilde{\gamma} \langle \mathbf{x} \rangle - \widetilde{W} \langle \mathbf{x} \rangle + \tilde{\Lambda} \mathbf{x}^* , \tag{208}$$

with the initial condition $\langle \mathbf{x}(0) \rangle = \langle \mathbf{x} \rangle_0 = 0$. Now, we introduce the fluctuating variable $\mathbf{X} = \mathbf{x} - \langle \mathbf{x} \rangle$, for which the Langevin equation reads

$$\frac{d\mathbf{X}}{dt} = -\widetilde{\gamma}\, \mathbf{X} - \widetilde{W} \mathbf{X} + k^{-1} \tilde{\Lambda}\, \bar{\mathbf{f}}(t) . \tag{209}$$

As we will see later on, we require the stationary probability density for the fluctuating variable $\mathbf{X}$. However, we can go further because we can calculate not only this stationary probability but also the probability density for all times $t > 0$, through the explicit solution of the FP equation associated with that fluctuating variable. As we can see, Eq. (209) resembles to Eq. (141). Consequently, the solution of the FP equation associated with Eq. (209) must be similar to that given by Eq. (149). Let us briefly give the algebraic steps leading to the solution. We first introduce the transformation of variable

$$\mathbf{X}' = e^{\widetilde{W} t} \mathbf{X} = \mathbf{x}' - \langle \mathbf{x}' \rangle , \tag{210}$$

where $\mathbf{x}' = e^{\widetilde{W} t} \mathbf{x}$, $\langle \mathbf{x}' \rangle = e^{\widetilde{W} t} \langle \mathbf{x} \rangle$, and $\mathcal{R}(t) = e^{\widetilde{W} t}$ being an orthogonal rotation matrix given by

$$\mathcal{R}(t) = \begin{pmatrix} \cos \widetilde{\Omega}\, t & \sin \widetilde{\Omega}\, t \\ -\sin \widetilde{\Omega}\, t & \cos \widetilde{\Omega}\, t \end{pmatrix} , \tag{211}$$

and its inverse is given by $\mathcal{R}^{-1}(t) = e^{-\widetilde{W} t}$. Clearly in the new space of coordinates we have

$$\frac{d\langle \mathbf{x}' \rangle}{dt} = -\widetilde{\gamma} \langle \mathbf{x}' \rangle + \tilde{\Lambda} \mathbf{x}'^* , \tag{212}$$

$$\frac{d\mathbf{X}'}{dt} = -\widetilde{\gamma}\, \mathbf{X}' + \bar{\mathbf{f}}'(t) , \tag{213}$$

with $\bar{\mathbf{f}}'(t) = k^{-1} \tilde{\Lambda}\, \mathcal{R}(t) \bar{\mathbf{f}}(t)$. Again, the transformed noise $\bar{\mathbf{f}}'(t)$ satisfies the same statistical properties of Gaussian white noise as the original noise $\bar{\mathbf{f}}(t)$. The associated Fokker-Planck equation for the transition probability density $P'(\mathbf{X}', t|\mathbf{X}'_0)$ is [19,21,24],

$$\frac{\partial P'}{\partial t} = \widetilde{\gamma}\, \nabla_{\mathbf{X}'} \cdot (\mathbf{X}' P') + \widetilde{\lambda}\, \nabla^2_{\mathbf{X}'} P' , \tag{214}$$

with $\widetilde{\lambda} = \lambda / \alpha^2 (1 + C^2)$. The solution to Eq. (214) is well known [19,21,24], and reads

$$P'(\mathbf{X}', t|\mathbf{X}'_0) = \frac{\beta k}{2\pi(1 - e^{-2\widetilde{\gamma} t})} \exp\left(- \frac{\beta k\, |\mathbf{X}' - e^{-\widetilde{\gamma} t} \mathbf{X}'_0|^2}{2(1 - e^{-2\widetilde{\gamma} t})} \right) , \tag{215}$$

where $P'(\mathbf{X}',0|\mathbf{X}'_0) = \delta(\mathbf{X}' - \mathbf{X}'_0)$, and $\beta = 1/k_B T$. According to the transformation (210), the transition probability density for the fluctuating variable $\mathbf{X}$ for all times $t > 0$ can be readily shown to be

$$P(\mathbf{X}, t|\mathbf{X}_0) = \frac{\beta k}{2\pi(1 - e^{-2\tilde{\gamma} t})} \exp\left(- \frac{\beta k \, |\mathbf{X} - e^{-\Lambda t}\mathbf{X}_0|^2}{2(1 - e^{-2\tilde{\gamma} t})} \right). \tag{216}$$

Clearly the corresponding stationary distribution is given by $P_{\text{eq}}(\mathbf{X}) = (\beta k/2\pi)\, e^{-(\beta k/2)|\mathbf{X}|^2}$. In a similar way, it is possible to consider the initial condition as given through a Gaussian distribution $P(\mathbf{X}_0) = (\beta k/2\pi)e^{-(\beta k/2)|\mathbf{X}_0|^2}$.

7.2. Transient Fluctuation Theorem for the Total Work

In this paper, we adopt the definition of the dimensionless total work done on a system during a time τ as given in reference [40],

$$\mathbb{W}_\tau = \frac{1}{k_B T} \int_0^\tau \mathbf{u}^* \cdot \mathbf{F}(\mathbf{x}, \mathbf{x}^*) \, dt \,. \tag{217}$$

According to this definition and the developments done above, we calculate the statistical properties of the total dimensionless work for the harmonic force defined in Eq. (204), which in terms of the $\mathbf{X}$ variable (defined above Eq. (209)) can be written as

$$\mathbb{W}_\tau = -\beta k \int_0^\tau [\mathbf{u}^* \cdot \mathbf{X} + \mathbf{u}^* \cdot (\langle \mathbf{x} \rangle - \mathbf{x}^*)] \, dt \,, \tag{218}$$

Equation (218) shows that the total work done on the system is a linear function of the stochastic variable $\mathbf{X}$. Accordingly, the statistical properties of the dimensionless total work corresponds to a Gaussian process. Therefore the probability distribution P_T of the total work can be written as follows

$$P_T(\mathbb{W}_\tau) = \frac{1}{\sqrt{2\pi V_T(\tau)}} \exp\left(- \frac{[\mathbb{W}_\tau - M_T(\tau)]^2}{2V_T(\tau)} \right), \tag{219}$$

where we have defined $M_T(\tau) \equiv \langle \mathbb{W}_\tau \rangle$ as the mean value of the work and $V_T(\tau) \equiv \langle \mathbb{W}_\tau^2 \rangle - \langle \mathbb{W}_\tau \rangle^2$ as its variance. The probability distribution written in Eq. (219) contains the time evolution of the total work from the initial time up to time τ. This fact means that we are studying the distribution corresponding to the transient situation. We will use the subscript T for all quantities in the transient case. Taking into account that $\langle \mathbf{X} \rangle = 0$, the work mean value reads

$$M_T(\tau) = -\beta k \int_0^\tau \mathbf{u}^* \cdot (\langle \mathbf{x} \rangle - \mathbf{x}^*) \, dt \,, \tag{220}$$

and the variance becomes affected just by the first term in Eq. (218), so that

$$V_T(\tau) = (\beta k)^2 \int_0^\tau dt_1 \int_0^\tau \mathbf{u}^*(t_1) \cdot \langle \mathbf{X}(t_1)\mathbf{X}(t_2) \rangle \cdot \mathbf{u}^*(t_2) \, dt_2 \,. \tag{221}$$

In order to calculate the work mean value we need the solution for the $\langle \mathbf{x} \rangle$ variable, which can be calculated from Eq. (212)

$$\langle \mathbf{x}'(t) \rangle = e^{-\widetilde{\gamma}t} \langle \mathbf{x}_0' \rangle + \int_0^t e^{-\widetilde{\gamma}(t-t')} \tilde{\Lambda} \mathbf{x}'^*(t')\, dt' = e^{-\widetilde{\gamma}t} \langle \mathbf{x}_0' \rangle + e^{-\widetilde{\gamma}\, t} \int_0^t e^{\Lambda\, t'} \tilde{\Lambda} \mathbf{x}^*(t')\, dt' , \tag{222}$$

where the relations $\tilde{\Lambda} = \widetilde{\gamma} \mathbf{I} + \widetilde{W}$ and $\mathcal{R}(t) = e^{\widetilde{W}\, t}$ have been used. Since $\langle \mathbf{x}_0 \rangle = 0$ then $\langle \mathbf{x}_0' \rangle = 0$ and after integration by parts, Eq. (222) reduces to

$$\langle \mathbf{x}'(t) \rangle = \mathbf{x}'^*(t) - \int_0^t e^{-\widetilde{\gamma}(t-t')} \mathcal{R}(t') \mathbf{u}^*(t')\, dt' , \tag{223}$$

and therefore the solution for the $\langle \mathbf{x}(t) \rangle$ variable reads

$$\langle \mathbf{x}(t) \rangle = \mathbf{x}^*(t) - \mathcal{R}^{-1}(t) \int_0^t e^{-\widetilde{\gamma}(t-t')} \mathcal{R}(t') \mathbf{u}^*(t')\, dt' . \tag{224}$$

The direct substitution of Eq. (224) into Eq. (220) leads to the work mean value which can be written as

$$M_{\scriptscriptstyle T}(\tau) = \beta k \int_0^{\tau} dt' \int_0^{t'} e^{-\widetilde{\gamma}(t'-t'')} \mathbf{U}^*(t') \cdot \mathbf{U}^*(t'')\, dt'' , \tag{225}$$

where we have defined $\mathbf{U}^*(t) = \mathcal{R}(t) \mathbf{u}^*(t)$, which accounts for a rotation of the velocity $\mathbf{u}^*$. To evaluate the variance we take into account the symmetry of the time-correlation function $\langle \mathbf{X}(t_1) \mathbf{X}(t_2) \rangle$ under interchange of t_1 and t_2, hence Eq. (221) can be written as

$$V_{\scriptscriptstyle T}(\tau) = 2(\beta k)^2 \int_0^{\tau} dt' \int_0^{t'} \mathbf{u}^*(t') \cdot \langle \mathbf{X}(t') \mathbf{X}(t'') \rangle \cdot \mathbf{u}^*(t'')\, dt'' . \tag{226}$$

To continue the calculation we need the time-correlation $\langle \mathbf{X}(t') \mathbf{X}(t'') \rangle$, which corresponds to a stationary process, so that this time-correlation can also be written as $\langle \mathbf{X}(t' - t'') \mathbf{X}_0 \rangle$. Also, the solution for this variable can be obtained from Eqs. (209) and (213) with the properties of matrix $\mathcal{R}(t)$, from which for all $t \geq 0$, it can be written as

$$\mathbf{X}(t) = e^{-\widetilde{\gamma}t} \mathcal{R}^{-1}(t) \mathbf{X}_0 + \mathcal{R}^{-1}(t) \int_0^t e^{-\widetilde{\gamma}(t-t')}\, \overline{\mathbf{f}}'(t')\, dt' . \tag{227}$$

Since $\langle \mathbf{X}_0 \mathbf{X}_0 \rangle = (k_{\scriptscriptstyle B} T/k) \mathbf{I}$ and if $\langle \overline{\mathbf{f}}(t) \mathbf{X}_0 \rangle = 0$, we can show that for $t \geq 0$

$$\langle \mathbf{X}(t) \mathbf{X}_0 \rangle = (\beta k)^{-1}\, e^{-\widetilde{\gamma}t} \mathcal{R}^{-1}(t)\, \mathbf{I} , \tag{228}$$

which implies that for $t' \geq t''$

$$\langle \mathbf{X}(t' - t'') \mathbf{X}_0 \rangle = (\beta k)^{-1}\, e^{-\widetilde{\gamma}(t'-t'')} \mathcal{R}^{-1}(t') \mathcal{R}(t'') \mathbf{I} . \tag{229}$$

With the direct substitution of Eq. (229) into Eq. (226) it can be verified that the variance yields to

$$V_{\scriptscriptstyle T}(\tau) = 2\beta k \int_0^{\tau} dt' \int_0^{t'} e^{-\widetilde{\gamma}(t'-t'')} \mathbf{U}^*(t') \cdot \mathbf{U}^*(t'')\, dt'' , \tag{230}$$

and the comparison with Eq. (225) allows us to conclude that

$$V_T(\tau) = 2\,M_T(\tau)\,. \tag{231}$$

Then, according to Eqs. (219) and (231), we can write the ratio of the probability distributions $P_T(\mathbb{W}_\tau)$ and $P_T(-\mathbb{W}_\tau)$ as follows

$$\frac{P_T(\mathbb{W}_\tau)}{P_T(-\mathbb{W}_\tau)} = e^{2M_T(\tau)\mathbb{W}_\tau/V_T(\tau)} = e^{\mathbb{W}_\tau}\,. \tag{232}$$

Therefore, the transient work-fluctuation theorem (TFT) for the dragging of an electrically Brownian charged harmonic oscillator in the presence of a uniform magnetic field and a time-varying electric field, is also satisfied in this case.

7.3. Stationary State Fluctuation Theorem for the Total Work

In this section we will consider the stationary state fluctuation theorem (SSFT) when an electromagnetic field is present. The SSFT is formulated for the dimensionless total work done on the system, during the time interval $[t_i, t_i + \tau]$ along a single trajectory in the stationary state. First of all, we consider the total work given in Eq. (217) for the plane perpendicular to the magnetic field. In this case the total dimensionless work done on the system over the time interval $[t_i, t_i + \tau]$ is given as

$$\mathbb{W}_{\tau,t_i} = \beta \int_{t_i}^{t_i+\tau} \mathbf{F}(\mathbf{x}, \mathbf{x}^*) \cdot \mathbf{u}^*\, dt\,. \tag{233}$$

We emphasize that this work will be calculated for a sequence of initial times t_i and all segments correspond to a time interval τ, besides we consider this work evaluated along a single stationary state trajectory ($i = 1, 2, 3, \ldots$). As a second step we recall that the stochastic variable $\mathbf{X} = \mathbf{x} - \langle \mathbf{x} \rangle$ being a Gaussian process has a Gaussian stationary state. This fact allows us to assure that the total work will have the same property, therefore the distribution of $\mathbb{W}_{\tau,t_i}$ for each t_i is also Gaussian and it is given by

$$P_{t_i}(\mathbb{W}_{\tau,t_i}) = \frac{1}{\sqrt{2\pi V_{t_i}(\tau)}} \exp\left(- \frac{[\mathbb{W}_{\tau,t_i} - M_{t_i}(\tau)]^2}{2V_{t_i}(\tau)} \right), \tag{234}$$

where the mean value and the variance are respectively given by

$$M_{t_i}(\tau) = -\beta k \int_{t_i}^{t_i+\tau} \mathbf{u}^* \cdot (\langle \mathbf{x} \rangle - \mathbf{x}^*)\, dt\,, \tag{235}$$

$$V_{t_i}(\tau) = 2\beta k \int_{t_i}^{t_i+\tau} dt' \int_{t_i}^{t'} e^{-\tilde{\gamma}(t'-t'')} \mathbf{U}^*(t') \cdot \mathbf{U}^*(t'')\, dt''\,. \tag{236}$$

We also assume that for each t_i large enough, M_{t_i}, and V_{t_i} will reach steady state values, and consequently become independent of t_i. If in addition, the correlation between different segments $[t_i, t_i + \tau]$ and $[t_j, t_j + \tau]$, decays sufficiently fast as $|t_i - t_j|$ gets larger, the distribution of W_{τ,t_i} along a trajectory in the stationary state is given by

$$P_S(\mathbb{W}_{\tau,S}) = \frac{1}{\sqrt{2\pi V_S(\tau)}} \exp\left(- \frac{[\mathbb{W}_{\tau,S} - M_S(\tau)]^2}{2V_S(\tau)} \right), \tag{237}$$

where the subscript S denotes the distribution of $\mathbb{W}_{\tau,t_i} \to \mathbb{W}_{\tau,S}$ over segments along the stationary state trajectory. Thus for large time and according to Eqs. (225) and (235), the mean value is

$$M_S(\tau) = \lim_{t\to\infty} \beta k \int_t^{t+\tau} dt' \int_0^{t'} e^{-\widetilde{\gamma}(t'-t'')} \mathbf{U}^*(t') \cdot \mathbf{U}^*(t'')\, dt'' , \tag{238}$$

and according to Eq. (236), the variance reads as

$$V_S(\tau) = \lim_{t\to\infty} 2\beta k \int_t^{t+\tau} dt' \int_t^{t'} e^{-\widetilde{\gamma}(t'-t'')} \mathbf{U}^*(t') \cdot \mathbf{U}^*(t'')\, dt'' . \tag{239}$$

In Eqs. (238) and (239), the integration limits make a difference when compared with Eqs. (225) and (230). This difference is manifested when we realized that the equality between V_S and $2M_S$ is not satisfied, in contrast with the transient case where we have obtained that $V_T = 2M_T$. The corresponding deviation can be calculated from

$$\frac{P_S(\mathbb{W}_\tau)}{P_S(-\mathbb{W}_\tau)} = \exp\left(\frac{\mathbb{W}_\tau}{1-\varepsilon(\tau)}\right) , \tag{240}$$

where $\varepsilon(\tau)$ represents the deviation between the mean value and the variance, and it is given by

$$\varepsilon(\tau) = \frac{2M_S(\tau) - V_S(\tau)}{2M_S(\tau)} . \tag{241}$$

The direct substitution of Eqs. (238-239) in Eq. (241) leads us to the following expression for the quantity $\varepsilon(\tau)$,

$$\varepsilon(\tau) = \frac{1}{M_S(\tau)} \lim_{t\to\infty} \beta k \int_t^{t+\tau} dt' \int_0^{t} e^{-\widetilde{\gamma}(t'-t'')} \mathbf{U}^*(t') \cdot \mathbf{U}^*(t'')\, dt'' , \tag{242}$$

which according to Eq. (224) can be written as

$$\varepsilon(\tau) = \frac{1}{M_S(\tau)} \lim_{t\to\infty} \beta k (\mathbf{x}^* - \langle \mathbf{x} \rangle) \cdot \int_0^{\tau} e^{-\widetilde{\gamma} t'} \mathbf{U}^*(t+t')\, dt' . \tag{243}$$

In Eq. (243) the denominator corresponds to the total mean value of the dimensionless work done of the system in the stationary state in time τ. In the numerator, the exponential in the integral will make the integral bounded for large time τ, provided that $\mathbf{U}^*(t)$ does not grow exponentially in time with an exponent bigger than the relaxation time $\widetilde{\tau}_r = \widetilde{\gamma}^{-1}$. We notice that this relaxation time depends on the magnetic field. Then $\varepsilon(\tau)$ approaches to zero proportionally to $1/\tau$ as τ goes to infinity, hence

$$\varepsilon(\tau) \to 0 \;\text{ as }\; \tau \to \infty . \tag{244}$$

As a consequence we obtain that

$$V_S(\tau) \to 2M_S(\tau) \;\text{ as }\; \tau \to \infty , \tag{245}$$

and in such a case the SSFT holds, that is

$$\frac{P_S(\mathbb{W}_{\tau,S})}{P_S(-\mathbb{W}_{\tau,S})} \to e^{\mathbb{W}_{\tau,S}} \quad \text{as } \tau \to \infty . \tag{246}$$

Once we have proved the TFT and SSFT given respectively by Eqs. (232) and (246), we can prove the integrated version of these theorems following similar steps as established in Ref. [40]. It is important to notice that the magnetic field can drive to different effects, though the formal structure is similar.

7.4. Applications

Here we will study two cases in which the two-dimensional harmonic trap potential is moving. In the first case, we assume that the electric field depends linearly with time $\mathbf{E}(t) = (\mathcal{E}t, \mathcal{E}t)$, so the potential minimum is dragged with uniform velocity (linear motion). This is the same kind of motion studied by Jayannavar and Sahoo [53] to verify the Bohr-van Leeuwen theorem [69] using the Jarzynski equality. As a second application, we will consider an oscillating electric field which produces a circular motion in the minimum of the harmonic trap [40]. Both physical models can be used in principle by experimentalists to corroborate the work-fluctuation theorems, as has been done in the absence of a magnetic field [38].

Linear electric field in the transient case. In this first case, the position vector for the potential minimum can be written as $\mathbf{x}^* = (q/k)\mathcal{E}t = (ut, ut)$, meaning that the electric field drags the minimum of the trap with a constant velocity $\mathbf{u}^* = u_{opt}(1,1)$, where $u_{opt} = (q/k)\mathcal{E}$ is the so called optical speed and $\mathcal{E}$ the amplitude of the electric field per unit time. After a long but straightforward algebra (see Appendix A of Ref. [68]), it can be shown that the work mean value (225) and its variance (230) are given respectively by

$$M_T(\tau) = 2\varpi\left\{\tau - \tau_r(1 - C^2)[1 - e^{-\tau/\tilde{\tau}_r}\cos(\tilde{\Omega}\,\tau)] - 2\tau_r\,C\,e^{-\tau/\tilde{\tau}_r}\sin(\tilde{\Omega}\,\tau)\right\}, \tag{247}$$

$$V_T(\tau) = 2M_T(\tau), \tag{248}$$

where $\varpi = \alpha\beta\,u_{\text{opt}}^2 = \alpha\beta(q\mathcal{E})^2/k^2$ stands as the work delivered to the system per unit time. Eq. (247) multiplied by the factor $k_B T$ is the total work mean value defined in Eq. (217), and it is exactly the same as the one calculated by Jayannavar and Sahoo in Ref. [53], by an alternative method. In the absence of the magnetic field ($C = 0$), the transient dimensionless work mean value reduces to $M_T(\tau) = 2\varpi\{\tau - \tau_r\,[1 - e^{-k\tau/\gamma}]\}$, which is the same expression calculated by van Zon and Cohen [40] and Mazonka and Jarzynski [32] except by the factor 2 which comes from the planar character of the vector $\mathbf{u}^*$.

Linear electric field in the stationary case.

In this case, it can be shown that the work mean value (238) and its variance (239) become (see Appendix A of Ref. [68])

$$M_S(\tau) = 2\varpi\tau, \quad (249)$$

$$V_S(\tau) = V_T(\tau) = 2M_T(\tau). \quad (250)$$

On the other hand, the deviation $\varepsilon(\tau)$ defined in Eq. (241) is given by

$$\varepsilon(\tau) = \frac{1}{\tau}\left\{\tau_r(1-C^2)[1-e^{-\tau/\widetilde{\tau}_r}\cos(\widetilde{\Omega}\,\tau)] + 2C\,\tau_r\,e^{-\tau/\widetilde{\tau}_r}\sin(\widetilde{\Omega}\,\tau)\right\}, \quad (251)$$

which is proportional to $1/\tau$ and it is clear that $\varepsilon(\tau) \to 0$ as $\tau \to \infty$. Thus, according to Eq. (245) we have that $V_S(\tau) = 2M_S(\tau)$ in this limit and the SSFT holds. In addition, this equality can be obtained from Eqs. (249) and (250), because from Eq. (247) we see that for large τ, $M_T(\tau)$ is proportional to τ and therefore $M_T(\tau) = M_S(\tau)$.

Oscillating electric field in the transient case.

As a second example, let us consider an oscillating electric field driving the potential minimum into a circular motion. Now, its position vector is given as $\mathbf{x}^*(t) = \mathrm{r}(\sin\Omega_0 t, (1-\cos\Omega_0 t))$ for $t \geq 0$, $\mathrm{r} = qE/k$ is the radius of the circle and, E is the amplitude of the electric field taken as a constant. Thus, the dragging velocity is $\mathbf{u}^*(t) = u_{\mathrm{opt}}(\cos\Omega_0 t, \sin\Omega_0 t)$, and $u_{\mathrm{opt}} = \mathrm{r}\,\Omega_0$ is the corresponding optical speed. In this case, the transient mean value of the total dimensionless work (225) and its variance (230) lead to the following expressions (see Appendix B of Ref. [68])

$$M_T(\tau) = \varpi_e\left\{\tau - \widetilde{\tau}_r\frac{2\,\widetilde{\tau}_r\,\widehat{\Omega}\,e^{-\tau/\widetilde{\tau}_r}\sin(\widehat{\Omega}\tau)}{1+\widetilde{\tau}_r^2\widehat{\Omega}^2} - \widetilde{\tau}_r\frac{[1-\widetilde{\tau}_r^2\,\widehat{\Omega}^2][1-e^{-\tau/\widetilde{\tau}_r}\cos(\widehat{\Omega}\tau)]}{1+\widetilde{\tau}_r^2\,\widehat{\Omega}^2}\right\}, \quad (252)$$

$$V_T(\tau) = 2\,M_T(\tau), \quad (253)$$

where ϖ_e is the work per unit time given by

$$\varpi_e = \frac{\beta\alpha_e\,u_{\mathrm{opt}}^2}{(1+\widetilde{\tau}_r^2\,\widehat{\Omega}^2)}, \quad (254)$$

and $\widehat{\Omega} = \Omega_0 - \widetilde{\Omega}$. It is very interesting to notice that in the special case where $\Omega_0 = \widetilde{\Omega}$, the mean value of the total work has a maximum given by $M_T(\tau) = \beta\alpha_e u_{opt}^2[\tau - \widetilde{\tau}_r(1-e^{-\tau/\widetilde{\tau}_r})]$. In fact, in Figs. 2 and 3, we show the symmetric behavior around the maximum of a reduced value of (252), that is $M_T \equiv M_T(\tau)/\widetilde{\tau}_r\beta\alpha_e\,u_{\mathrm{opt}}^2$, as a function of the dimensionless variables $x = \tau/\widetilde{\tau}_r$ and $y = \widehat{\Omega}\,\widetilde{\tau}_r$. Thus, the maximum means that we have a resonance caused by the tuning in the electric field with the Larmor frequency. In other words, we have a case where both the magnetic and the oscillating electric field are tuned giving place to a resonant situation which can be explored by an experimental device.

On the other hand, we can compare our result with that calculated by van Zon and Cohen [40] in the absence of the magnetic field. In this case $\widetilde{\Omega} = 0$ and hence $\widehat{\Omega} = \Omega_0$,

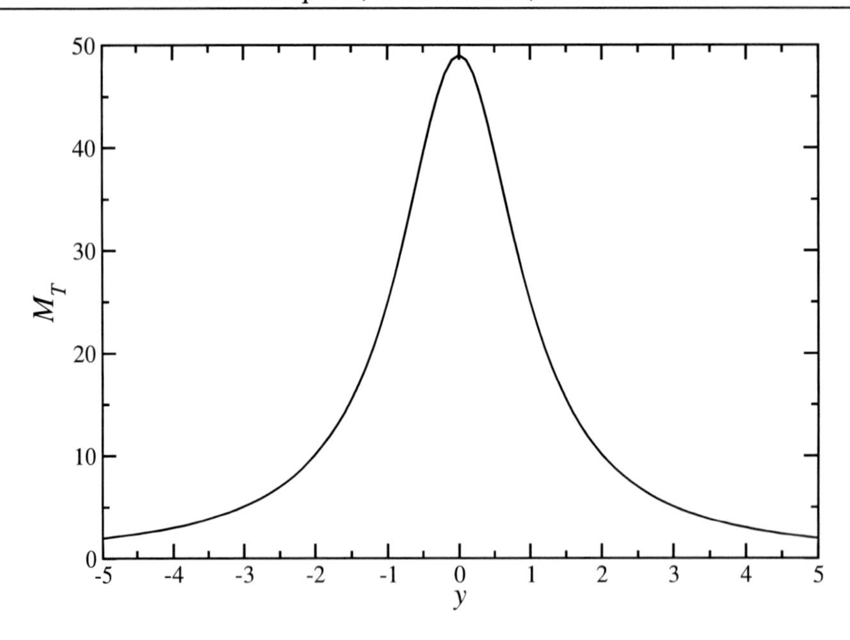

Figure 2. The reduced value of the total work M_T as a function of y for a fixed value of $x = 50$

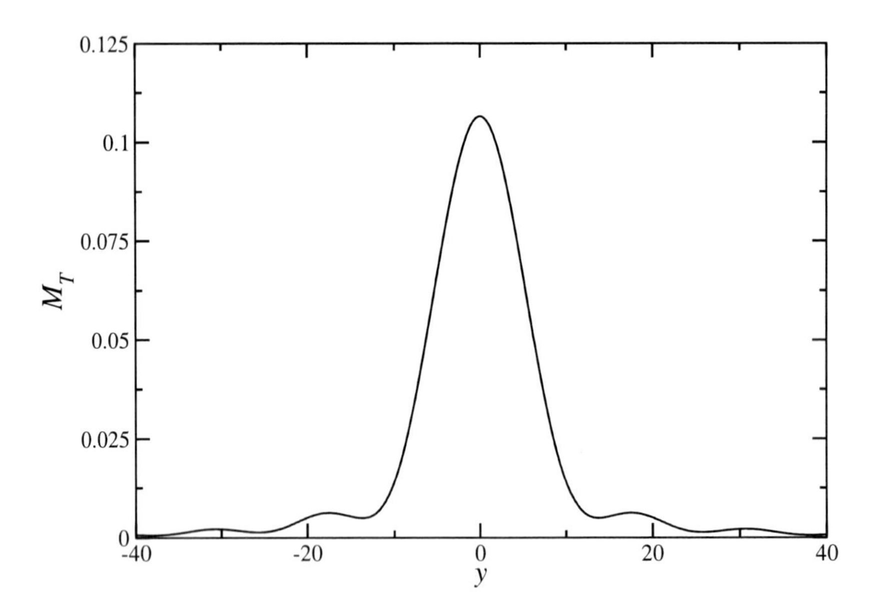

Figure 3. The reduced value of the total work M_T as a function of y for a fixed value of $x = 0.5$

$\alpha_e = \alpha$, $\widetilde{\tau}_r = \tau_r = \alpha/k$, and therefore Eq. (252) reduces to

$$M_T(\tau) = \psi\left\{\tau - \tau_r \frac{2\,\tau_r\,\Omega_0\,e^{-\tau/\tau_r}\sin(\Omega_0\tau)}{1+\tau_r^2\,\Omega_0^2} - \tau_r \frac{[1-\tau_r^2\,\Omega_0][1-e^{-\tau/\tau_r}\cos(\Omega_0\tau)]}{1+\tau_r^2\,\Omega_0^2}\right\}, \tag{255}$$

where

$$\psi = \frac{\alpha\beta\,\mathrm{r}^2\,\Omega_0^2}{(1+\tau_r^2\,\Omega_0^2)}\,. \tag{256}$$

which agree with the corresponding results of Ref. [40].

Oscillating electric field in the stationary case.

In this case, the explicitly expressions for the work mean value (238) and its corresponding variance (239) are (see Appendix B of Ref. [68])

$$M_S(\tau) = 2\varpi_e\,\tau\,, \tag{257}$$

$$V_S(\tau) = V_T(\tau) = 2M_T(\tau)\,. \tag{258}$$

Now, the $\varepsilon(\tau)$ parameter is in this case

$$\varepsilon(\tau) = \frac{\widetilde{\tau}_r}{\tau\,(1+\widetilde{\tau}_r^2\,\widehat{\Omega}^2)}\{2\,\widetilde{\tau}_r\,\widehat{\Omega}\,e^{-\tau/\widetilde{\tau}_r}\sin(\widehat{\Omega}\,\tau)+[1-\widetilde{\tau}_r^2\,\widehat{\Omega}^2][1-e^{-\tau/\widetilde{\tau}_r}\cos(\widehat{\Omega}\,\tau)]\}\,, \tag{259}$$

which also vanishes like $1/\tau$ as $\tau \to \infty$, therefore $V_S = 2M_S$ consistently with the SSFT. Also, this identity can be obtained from Eq. (258), in fact taking into account Eq. (252) it can be checked that $M_T(\tau) \to 2\varpi_e\tau = M_S(\tau)$ as $\tau \to \infty$.

Jarzynski equality.

It is well known that Jarzynski equality [30, 31] relates nonequilibrium work done $\mathbb{W}$ don a system to the equilibrium free energy differences ΔF, that is, $\langle e^{-\beta\mathbb{W}}\rangle = e^{-\beta\Delta F}$. In the transient case, and according to Eqs. (219) and (231), we can show that $\langle e^{-\beta\mathbb{W}}\rangle = 1$, from which we can conclude that $\Delta F = 0$, consistently with the Bohr-van Leeuwen theorem on the absence of orbital diamagnetism in a classical system of charged particles in thermodynamic equilibrium [69].

8. Detection of Weak Signals in an Electromagnetic Field

In this last section, we are going to study the decay process of the unstable state of a charged Brownian particle in the presence of a constant electromagnetic field, as a mechanism to detect weak amplitudes of the applied electric field. The decay process initiates when the particle embedded in a thermal bath of temperature T, is located around the unstable state of a two-dimensional bistable potential (this is a potential characterized by a maximum representing the unstable state, and two minimums representing the two stable states). In the absence of the electromagnetic field, the particle decays down-hill in the potential due to the presence of thermal noise. When an electric field is present, the decay process is accelerated due to the presence of this additional electric field. When both, electric and magnetic fields are taken into account, the particle decays following a stochastic rotational trajectory. We show that this decay process can be characterized by means of the mean first passage time (MFPT) distribution, through the quasi-deterministic (QD) approach [70, 72, 77]. In a similar way as done in the laser system [72], the detection process takes place around the initial unstable state, so that a linear approximation of the nonlinear bistable potential will be enough. Thus, the decay process will be stopped by a potential barrier located in a prescribed reference value, which is chosen as a quantity proportional to the steady-state value of the bistable potential. For the detection of the weak signal, we use a similar criterion established in Ref. [72, 75] for a laser system. This is related to the statistic of the MFPT. As before, we use again the same vectorial notation for the two-dimensional vectors: $\mathbf{x}$ being the position, $\mathbf{u}$ the velocity, $\bar{\mathbf{E}} = (E_x, E_y)$ the electric field, and $\bar{\mathbf{f}}(t) = (f_x, f_y)$ for the Gaussian white noise.

8.1. Langevin Equation in an External Potential in the Presence of an Electromagnetic Field

Consider the two-dimensional Brownian motion of a charged particle of mass m and charge q, in a two-dimensional bistable potential and in the presence of a uniform electromagnetic field, in the high friction limit with a thermal bath at temperature T. The bistable potential is given by $V(x,y) = -(a/2)(x^2+y^2)+(b/4)(x^2+y^2)^2$ with $a, b > 0$, and its associated force reads $\mathbf{F} = -\nabla V = a\mathbf{x} - b\,r^2\mathbf{x}$, where $r^2 = x^2+y^2$ is the square modulus of the vector $\mathbf{x}$. Due to the presence of the electromagnetic field, a Lorentz force $\mathbf{F}_L = (q/c)\mathbf{u}\times\mathbf{B} + q\bar{\mathbf{E}}$ will be present. For the case $\mathbf{B} = (0, 0, B)$ the over-damped approximation of the two-dimensional Langevin equation can be written as

$$\frac{d\mathbf{x}}{dt} = \widetilde{a}\mathbf{x} + \mathfrak{W}\,\mathbf{x} - b\,r^2\mathfrak{L}\,\mathbf{x} + q\mathfrak{L}\,\bar{\mathbf{E}} + \mathfrak{L}\,\bar{\mathbf{f}}(t)\,, \tag{260}$$

where $\mathfrak{W}$ and $\mathfrak{L}$ are matrices given by

$$\mathfrak{W} = \begin{pmatrix} 0 & \widetilde{\Omega} \\ -\widetilde{\Omega} & 0 \end{pmatrix}, \quad \mathfrak{L} = \frac{1}{\alpha_e}\begin{pmatrix} 1 & C \\ -C & 1 \end{pmatrix}, \tag{261}$$

with $\widetilde{a} = a/\alpha_e$, $\widetilde{\Omega} = \widetilde{a}C$ such that $C = qB/c\alpha$ is a dimensionless parameter, and $\alpha_e = \alpha(1 + C^2)$, is an effective, magnetic-field dependent friction coefficient. The noise term $\bar{\mathbf{f}}(t)$, is Gaussian with zero mean value $\langle f_i(t)\rangle = 0$ and correlation function $\langle f_i(t)f_j(t')\rangle = 2\lambda\delta_{ij}\delta(t-t')$. Here λ is the noise intensity which, according to the fluctuation-dissipation relation in the absence of time-dependent external forces, satisfies $\lambda = \alpha k_B T$, with k_B the Boltzmann's constant. Our scheme will be formulated in the transformed space of coordinates $\mathbf{x}' = e^{-\mathfrak{W}t}\mathbf{x}$, where the Langevin equation (260) transforms into

$$\frac{d\mathbf{x}'}{dt} = \widetilde{a}\,\mathbf{x}' - b\,r'^2\mathfrak{L}\,\mathbf{x}' + q\mathfrak{L}\,\mathfrak{R}^{-1}(t)\,\bar{\mathbf{E}} + \mathfrak{L}\,\mathfrak{R}^{-1}(t)\,\bar{\mathbf{f}}(t)\,, \tag{262}$$

where again $\mathfrak{R}(t) = e^{\mathfrak{W}t}$ is an orthogonal rotation matrix, and thus $\mathfrak{R}^{-1}(t) = e^{-\mathfrak{W}t}$ such that

$$\mathfrak{R}(t) = \begin{pmatrix} \cos\widetilde{\Omega}\,t & \sin\widetilde{\Omega}\,t \\ -\sin\widetilde{\Omega}\,t & \cos\widetilde{\Omega}\,t \end{pmatrix}. \tag{263}$$

The third and fourth terms of Eq. (262) represent the time-dependent rotating electric field and the fluctuating forces respectively. Also, $r'^2 = x'^2 + y'^2$ with $r'^2 = r^2$, which means that the modulus of vector $\mathbf{x}$ remains invariant under the transformation $\mathfrak{R}^{-1}(t)$, as it should be.

8.2. Quasi-deterministic Approach

We are interested in the calculation of the probability distribution of the passage time required by the charged particle to reach a prescribed reference value R', in its decay process from the initial unstable state caused by both the internal fluctuations and the external electric field. In a similar way as done in the laser system of Refs. [72, 73, 75], the detection process takes place around the initial unstable state of the bistable potential; thus the decay process of the charged particle must stop at a fixed value, representing the absorbing potential barrier, taken as $R^2 = C_0 r_{st}^2$, where $0 < C_0 < 1$ and $r_{st}^2 = a/b$. The constant C_0 is a quantity which must be determined by the experiment (as an example, in the laser system this value is taken as 2% of the intensity steady-sate value). In our case, the steady-state value can be calculated from the deterministic evolution of Eq. (262) without the electric field in terms of the variable $\mathfrak{r} \equiv r^2$, and it satisfies $d\mathfrak{r}/dt = (2\tilde{a}/\mathfrak{r}_{st})\mathfrak{r}(\mathfrak{r}_{st} - \mathfrak{r})$, where $\mathfrak{r}_{st} = a/b$ is the stationary-state value. In a similar way, in the transformed coordinate space the variable $\mathfrak{r}' \equiv r'^2$ satisfies the deterministic equation $d\mathfrak{r}'/dt = (2\tilde{a}/\mathfrak{r}'_{st})\mathfrak{r}'(\mathfrak{r}'_{st} - \mathfrak{r}')$, and also $\mathfrak{r}'_{st} = a/b$. To calculate the MFPT when the strength of the electric field is less or of the same order than the intensity of the noise, we use the formalism of the QD approach [72,77] which relies upon the linear approximation of Eq. (262) that reads as

$$\frac{d\mathbf{x}'}{dt} = \tilde{a}\,\mathbf{x}' + q\mathfrak{L}\,\mathfrak{R}^{-1}(t)\bar{\mathbf{E}} + \mathfrak{L}\,\mathcal{R}^{-1}(t)\bar{\mathbf{f}}(t)\,. \tag{264}$$

The solution of this last equation for the initial conditions $\mathbf{x}'_0 = (x'_0, y'_0) = (0,0)$ can be written as

$$\mathbf{x}' = \mathbf{h}'(t)\, e^{\tilde{a}t}\,, \tag{265}$$

where

$$\mathbf{h}'(t) = \int_0^t e^{-\tilde{a}s}\mathfrak{L}\,\mathfrak{R}^{-1}(s)[q\bar{\mathbf{E}} + \bar{\mathbf{f}}(s)]\, ds\,, \tag{266}$$

or written in terms of its components we have

$$h'_i(t) = \int_0^t e^{-\tilde{a}s}\mathfrak{L}_{ij}(\mathfrak{R}^{-1})_{jk}(s)[qE_k + f_k(s)]\, ds\,. \tag{267}$$

We recall that the QD approach has as a main goal to take advantage of the quasi-deterministic behavior of a stochastic process in the limit of large times [70,72,77] such that, for $t \gg 1/2\tilde{a}$, the stochastic process $h'_i(t)$ plays the role of an effective initial condition, in other words $h'_i(\infty)$ behaves as a Gaussian random variable. This property is guaranteed taking into account that for small noise intensity and weak electric field, the process $h'_i(t)$ satisfies the following condition

$$\lim_{t\to\infty} \frac{dh'_i(t)}{dt} = \lim_{t\to\infty} e^{-\tilde{a}t}\mathfrak{L}_{ij}(\mathfrak{R}^{-1})_{jk}(t)[qE_k + f_k(t)] \to 0\,, \tag{268}$$

and therefore $h'_i(\infty)$ becomes a Gaussian random variable, i.e. $h'_i(\infty) = h'_i$, which implies that in this limiting case the stochastic process described by Eq. (265) becomes a quasi-deterministic one. In terms of the modulus r'^2 it reads as

$$r'^2(t) = h'^2\, e^{2\tilde{a}t}\,, \tag{269}$$

where $h'^2 = h_1'^2 + h_2'^2$. The random first passage time required for the charged particle to reach a prescribed reference value R' is then

$$t^\star = (1/2\widetilde{a}) \ln(R'^2/h'^2) \,. \tag{270}$$

Clearly, the randomness of this passage time is due to the randomness of the h' variable, therefore, the statistical properties of the passage time must be determined through the marginal probability density $P(h')$. The latter can be calculated from the joint probability density given by the Gaussian distribution [21, 24]

$$P(h_1', h_2') = N\exp\Bigg[-\frac{1}{2}\sum_{i,j=1}^{2} (\sigma^{-1})_{ij}(h_i' - \langle h_i'\rangle)(h_j' - \langle h_j'\rangle)\Bigg] \,, \tag{271}$$

where $N = 1/2\pi(\det\sigma_{ij})^{1/2}$ is the normalization factor and $\sigma_{ij} = \langle h_i' h_j'\rangle - \langle h_i'\rangle\langle h_j'\rangle$ the correlation matrix. From Eq. (267) we have

$$\langle h_i'\rangle = q\int_0^\infty e^{-\widetilde{a}s}\mathfrak{L}_{ik}(\mathfrak{R}^{-1})_{kl}(s)\,E_l\,ds \,. \tag{272}$$

$$\langle h_i' h_j'\rangle = \langle h_i'\rangle\langle h_j'\rangle + \int_0^\infty\int_0^\infty e^{-\widetilde{a}(s+s')}\mathfrak{L}_{ik}\mathfrak{L}_{jl}\mathfrak{R}_{km}^{-1}(s)\mathfrak{R}_{ln}^{-1}(s')\langle f_m(s) f_n(s')\rangle\,ds ds' \,. \tag{273}$$

After some algebra Eq. (273) reduces to $\langle h_i' h_j'\rangle = \langle h_i'\rangle\langle h_j'\rangle + (\lambda/\alpha\, a)\,\delta_{ij}$, which tells us that the variables h_i' are independent and that $\sigma_{ij} = (\lambda/\alpha\, a)\,\delta_{ij}$ is a diagonal matrix with elements $\sigma_{ii} \equiv \sigma^2 = \lambda/\alpha\, a$. Therefore, the joint probability density (271) now reduces to

$$P(h_1', h_2') = \frac{1}{2\pi\sigma^2}\, e^{-(1/2\sigma^2)[(h_1' - \langle h_1'\rangle)^2 + (h_2' - \langle h_2'\rangle)^2]} \,. \tag{274}$$

The mean values $\langle h_i'\rangle$ can be calculated by assuming the particular case in which $\mathbf{E} = (E, E)/\sqrt{2}$, E being the modulus of this vector. In this case it is easy to show that $\langle h_1'\rangle = \langle h_2'\rangle = q\,E/\sqrt{2}\;a$. The marginal probability density $P(h')$ can be calculated using the following transformation

$$P(h_1', h_2') dh_1' dh_2' \to e^{-(1/2\sigma^2)[h' - 2\mathbf{p}'\cdot\mathbf{h}' + p'^2]}\; J(h', \theta)\, dh' d\theta' \,, \tag{275}$$

where $J(h', \theta)$ is the Jacobian of the transformation from the (h_1', h_2') space of coordinates to the new (h', θ') space of coordinates. The Jacobian in this case is $J(h_1', h_2') = h'$. The vector $\mathbf{p}' = (\langle h_1'\rangle, \langle h_2'\rangle)$ and its modulus is given by $p'^2 = \langle h_1'\rangle^2 + \langle h_2'\rangle$. After integration of the right hand side of Eq. (275) over θ' variable, we get the normalized marginal probability density $P(h')$ [72, 76]

$$P(h') = (h'/\sigma^2)\, I_0(p'\, h'/\sigma^2)\, e^{-(1/2\sigma^2)(h'^2 + p'^2)} \,, \tag{276}$$

with $p'^2 = (qE)^2/a^2$ and $I_0(x)$ is the modified Bessel's function of zeroth order [96]. The statistical properties of first passage time (FPT) distribution can be calculated through the moment generating function defined as $G(2\widetilde{a}\nu) \equiv \langle e^{-2\widetilde{a}\nu t}\rangle$, and thus $G(2\widetilde{a}\nu) =$

$\langle (R'^2/h'^2)^{-\nu} \rangle$. This generating function is calculated from the marginal probability $P(h')$ given by Eq. (276), giving as a result [76,77]

$$G(2\tilde{a}\nu) = (R'^2/\sigma^2)^{-\nu}\, e^{-\beta'^2} \sum_{m=0}^{\infty} \frac{\Gamma(m+\nu+1)}{(m!)^2} \beta'^{2m} = G_0(2\tilde{a}\nu)\, e^{-\beta'^2}\, M(\nu{+}1, 1, \beta'^2)\,, \tag{277}$$

and $G_0(2\tilde{a}\nu) = (R'^2/\sigma^2)^{-\nu}\Gamma(\nu+1)$ is the moment generating function in the absence of the external electric field, $M(\nu+1, 1, \beta'^2)$, is the Kummer confluent hypergeometric function [96] with $\beta'^2 = p'^2/2\sigma^2 = \alpha(qE)^2/2a\,\lambda$. The MFPT is then calculated from $\langle 2\tilde{a}\, t^\star \rangle = [-dG(2\tilde{a}\nu)/d\nu]_{\nu=0}$ and, after some algebra, leads to

$$\langle 2\tilde{a}\, t^\star \rangle = \langle 2\tilde{a}\, t^\star \rangle_0 + \sum_{m=1}^{\infty} \frac{(-1)^m\, \beta'^{2m}}{m m!}\,, \tag{278}$$

where

$$\langle 2\tilde{a}\, t^\star \rangle_0 = \ln(a\, R'^2/2\lambda) - \psi(1)\,, \tag{279}$$

is the MFPT in the absence of the external electric field ($\beta' = 0$) and $\psi(1) = -\gamma = -0.577$ the Euler's constant. The variance of the passage time distribution, defined as $\langle (\Delta t^\star)^2 \rangle = \langle t^{\star 2} \rangle - \langle t^\star \rangle^2$, can be calculated from $\langle (2\tilde{a}\, t^\star)^2 \rangle = [-d^2G(2\tilde{a}\nu)/d\nu^2]_{\nu=0}$, and, again after some algebra, we can show that [76,77]

$$\langle (2\tilde{a}\,\Delta t^\star)^2 \rangle = \psi'(1) + 2\sum_{m=2}^{\infty} \left(\sum_{k=1}^{m-1} \frac{1}{k} \right) \frac{(-1)^m \beta'^{2m}}{m m!} - \left[\sum_{m=1}^{\infty} \frac{(-1)^m \beta'^{2m}}{m m!} \right]^2 . \tag{280}$$

For practical purposes, if the parameter satisfies $\beta'^2 \leq 1$, which implies that the strength of the electric field is less or of the same order than the intensity of the noise, the MFPT and its variance can be approximated by

$$\langle 2\tilde{a}\, t^\star \rangle = \langle 2\tilde{a}\, t \rangle_0 - \beta'^2 + (\beta'^4/4)\,, \tag{281}$$

$$\langle (2\tilde{a}\,\Delta t^\star)^2 \rangle = \psi'(1) - (\beta'^4/2)\,, \tag{282}$$

8.3. Detection of Weak Electric Fields

The detection of weak electric fields can be undertaken by means of Eqs. (281) and (282) according to the following procedure. First of all, we recall that β'^2 is a measure of the electric field in relation to the intensity of the noise. Secondly, Eqs. (281) and (282) are valid under the assumption $\beta'^2 < 1$ which means that the electric field is less than the noise intensity, in other words, those equations can be used for weak electric fields. As a third element, we need a criteria allowing us to distinguish in a clear way how the electric field intensity can be detected. To choose such a criteria we notice that Eqs. (281) and (282) are very similar to Eqs. (22) and (26) in Ref. [72] with $\beta' = \beta/2$, where a criteria was established to detect weak signals in a laser system. Herein we adopt it and take the following

$$[\langle t^\star \rangle_{\beta'_c} - \langle t^\star \rangle_{\beta'=0}]^2 \geq (\Delta t^\star)^2_{\beta'=0}\,, \tag{283}$$

in which the difference between the time scales in the presence and in the absence of the electric field is greater or equal than the maximum variance. Using Eqs. (281) and (282) we obtain the value of $\beta'_c \geq [\psi'(1)]^{1/4} \approx 1.13$, which is the critical value for which the weak electric field can be detected. However, if we choose the parameter $\beta'/2$ instead of $\beta\prime$ and thus employing again the above equations, we get $\beta'_c \geq [4\psi'(1)]^{1/4} \approx 1.6$, which is the same value calculated in Ref. [72]. Below this value the electric field cannot be detected. On the other hand, if $\beta' > \beta'_c$, the amplitude of the electric field can be efficiently detected because it dominates over the noise intensity and the dynamics is dominated by the deterministic evolution. In this case it can be shown that the MFPT and its variance are approximated by [72,76,77]

$$\langle 2\tilde{a}\, t^\star \rangle \approx \ln(aR'/qE)^2 \tag{284}$$

$$\langle (\Delta t^\star)^2 \rangle \approx 1/2\tilde{a}^2\, \beta'^2 \tag{285}$$

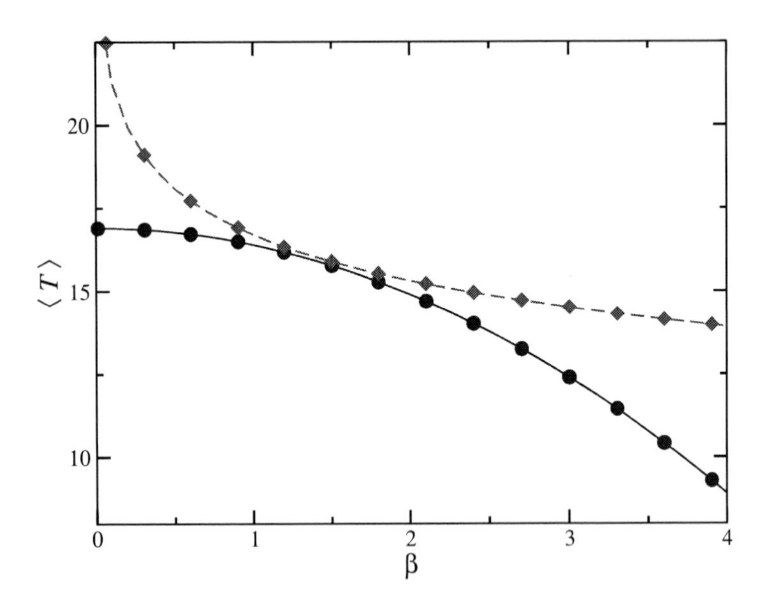

Figure 4. Results for the MPT $\langle T \rangle = \langle 2\tilde{a}t^\star \rangle$ as a function of β obtained from Eq. (281) [solid line] and from the deterministic solution Eq. (284) [dashed line] for system parameters such that $2a\alpha R'^2/\lambda = 1.8 \times 10^7$.

In Fig. 4 we show that the critical value of $\beta'_c \approx 1.6$ corresponds to the match between the two approximations (281) and (284). Here it is important to comment that Eqs. (281), (282), (284), and (285) have been derived from the QD approach. However, they are not actually appropriate to describe the rotational effects appearing in a natural way in the dynamical evolution of the charged particle in a magnetic field [see Eq. (260)] along its decay process to reach the value R'. This process is better understood in the transformed space of coordinates $\mathbf{x}' = (x', y')$, where the trajectory of the charged particle can be seen as rotational or not, depending on which force, electric or fluctuating, is greater. In fact, if $\beta' \leq 1$, it is shown in Fig 5(a) that the trajectory is practically a straight line, with no rotational effect associated whatsoever. This is the reason why the QD approach is better understood in the transformed space of coordinates.

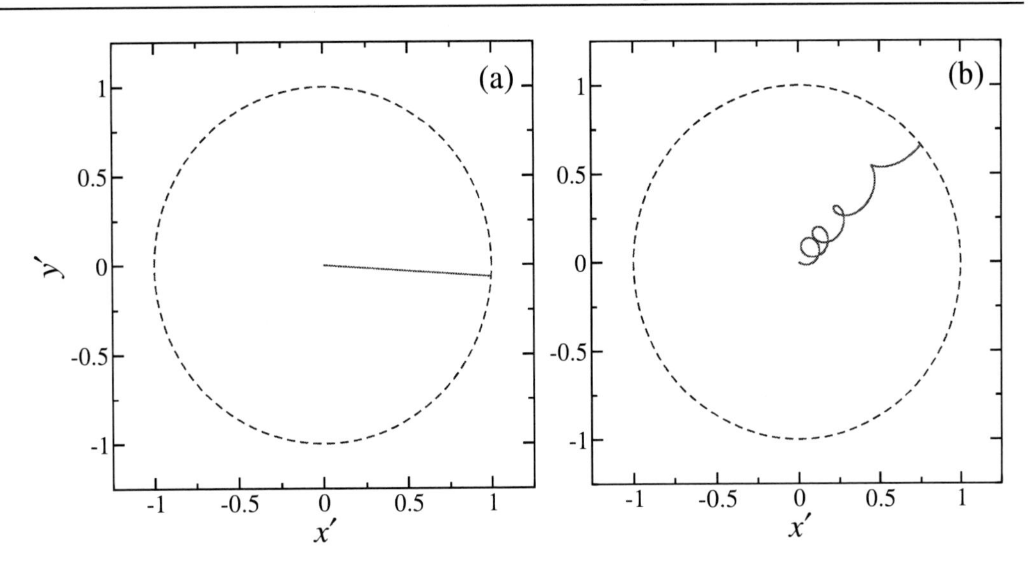

Figure 5. Dynamical evolution of a single trajectory of the system given by linear approximation of Eq. (264) in the (x', y') space for values $a = 300$, $\alpha = 270$, $C = 10$, $R' = 1.0$, (a) $qE = 10^{-3}$ and (b) $qE = 20.0$.

If $\beta' \gg 1$, the strength of the electric field is greater than the intensity of the noise and the dynamical evolution of the charged Brownian particle is rotational, as can be seen in Fig. 5(b). To characterize this rotational evolution, another approach has been required as shown in [75, 77].

References

[1] Einstein, A. *Ann. Phys.* 1905, 17, 549-540.

[2] Langevin, P. *Am. J. Phys.* 1908, 146, 530-533.

[3] Perrin, J. B. *Ann. Chim. Phys.* 1909, 18, 5-114.

[4] Taylor, J. B. *Phys. Rev. Lett.* 1961, 6, 262-263.

[5] Kurşunoğlu, B. *Ann. Phys.* 1962, 17, 259-268.

[6] Kurşunoğlu, B. *Phys. Rev.* 1963, 132, 21-26.

[7] Sturrock, P. A. *Phys. Rev.* 1966, 141, 186-191.

[8] Karmeshu, *Phys. Fluid* 1974, 17, 1828-1830.

[9] Ferrari, L. *Physica A* 1990, 163, 596-614.

[10] Czopnik, R.; Garbaczewky, P. *Phys. Rev. E* 2001, 63, 021105 1-9.

[11] Simões, T. P.; Lagos, R. E. *Physica A* 2005, 355, 274-282.

[12] Holod, I.; Zagorodny, A.; Weiland, J. *Phys. Rev. E* 2005 71, 046401 1-11

[13] Jiménez-Aquino, J. I.; Romero-Bastida, M. *Phys. Rev. E* 2006, 74, 041117 1-10

[14] Jiménez-Aquino, J. I.; Romero-Bastida, M. *Phys. Rev. E* 2007, 76, 021106 1-9.

[15] Jiménez-Aquino, J. I.; Romero-Bastida, M. *Rev. Mex.de Fis. E* 2006, 52, 182-187.

[16] Jiménez-Aquino, J. I.; Romero-Bastida, M.; Pérez-Guerrero Noyola, A. C. *Rev. Mex. Fis. E* 2008, 54, 81-86.

[17] Jiménez-Aquino, Velasco, R. M.; Uribe, F. J. *Phys. Rev. E* 2008, 77, 051105 1-13.

[18] Hou, L. J.; Miskovic, Z. L.; Piel, A.; Shukla, P. K. *Phys. Plasmas* 2009 16, 053705 1-12.

[19] Chandrasekhar, S. *Rev. Mod. Phys.* 1943, 15, 1-89.

[20] Uhlenbeck, G. E.; Ornstein, L. S. *Phys. Rev.* 1930, 36, 823-841

[21] Risken, H. *The Fokker-Planck equation: Methods of solution and Applications*; Springer-Verlag, Berlin, 1984.

[22] Mazo, R. M. *Brownian motion, Fluctuations, Dynamics and Applications*; Clarendon Press, Oxford, 2002.

[23] Coffey, W. T.; Kalmykov, Yu. P.; Waldron, J. T. *The Langevon Eqaution, with application to Stochastic problems in Physics and Chemistry and Electrical Engineering*; 2nd Ed. World Scientific, 2004.

[24] Van Kampen, N. G. *Stochastic Processes in Physics and Chemistry*; 3rd Ed., Elsevier, Amsterdam, 2007.

[25] Ferrari, L. *J. Chem. Phys.* 2003, 118, 11092-11099.

[26] Piña, E. *Dinámica de Rotaciones*; Universidad Autónoma Metropolitana, México, 1996.

[27] Evans, D. J.; Cohen, E. G. D.; Morriss, G. P. *Phys. Rev. Lett.* 1993, 71 2401-2404.

[28] Evans, D. J.; Searles, D. J. *Phys. Rev. E* 1994, 50, 1645-1648.

[29] Gallavotti, G.; Cohen, E. G. D. *Phys. Rev. Lett.* 1995, 74 2694-2697.

[30] Jarzynski, C. *Phys. Rev. Lett.* 1997, 78, 2690-2693.

[31] Jarzynski, C. *Phys. Rev. E.* 1997, 56, 5018-5035.

[32] Mazonka, O.; Jarzynski, C. *e-print cond-mat/* 1999, 9912121

[33] Crooks, G. E. *Phys. Rev. E* 1999, 60, 2721-2726

[34] Crooks, G. E. *Phys. Rev. E* 2000, 61, 2361-2366.

[35] Lepri, S.; Rondoni, L.; Benettin, G. *J. Stat. Phys.* 2000, 99, 857-872

[36] Searles, D. J.; Evans, D. J. *J. Chem. Phys.* 2000, 113, 3503-3509.

[37] Hatano, T.; Sasa, S. I. *Phys. Rev. Lett.* 2001, 86, 3463-3466.

[38] Wang, G. M.; Sevick, E. M.; Mittag, E.; Searles, D. J.; Evans, D. J. *Phys. Rev. Lett.* 2002, 89, 050601 1-4.

[39] Liphardt, J.; Dumont, S.; Smith, S. B.; Tinoco, I. Jr.; Bustamente, C. *Science* 2002, 296, 1832-

[40] van Zon, R.; Cohen, E. G. D. *Phys. Rev. E* 2003, 67, 046102 1-10.

[41] van Zon, R.; Cohen, E. G. D. *Phys. Rev. E* 2004, 69 056121 1-14.

[42] Carberry, D. M.; Reid, J. C.; Wang, G. M.; Sevick, E. M.; Searles, D. J.; Evans, D. J. *Phys. Rev. Lett* 2004, 92, 140601 1-4.

[43] Trepagnier, E. H.; Jarzynski, C.; Ritort, F.; Crooks, G. E.; Bustamante, C. J.; Liphardt, J. *Proc. Natl. Acad. Sciences* 2004, 101, 15038

[44] Douarche, F.; Ciliberto, S.; Patrosyan, A.; Rabbiosi, I. *Europhys. Lett.* 2005, 70, 593-599

[45] Collin, D.; Ritort, F.; Jarzynski, C. Smith, S. B.; Tinoco, I. Jr.; Bustamante, C. *Nature* 2005, 437, 231-234

[46] Bustamante, C.; Liphardt, J.; Ritort, F. *Phys. Today* 2005, 58, 43

[47] Seifert, U. *Phys. Rev. Lett.* 2005, 95, 040602 1-4.

[48] Dhar, A. *Phys. Rev. E* 2005, 71, 036126 1-5.

[49] Speck, T.; Seifert, U. *Eur. Phys. J. B* 2005, 43, 529

[50] Blickle, V.; Speck, T.; Helden, L.; Seifert, U.; Bechinger, C. *Phys. Rev. Lett.* 2006, 96, 070603 1-4.

[51] Tietz, C.; Schuler, S.; Speck, T.; Seifert, U.; Wrachtrup, J. *Phys. Rev. Lett.* 2006, 97, 050602 1-4.

[52] Lechner, W.; Oberhofer, H.; Dellago, C.; Geissler, P. L. *J. Chem. Phys.* 2006, 124, 044113 1-12

[53] Jayannavar, A. M.; Sahoo, M. *Phys. Rev. E* 2007, 75, 032102 1-4.

[54] Joubaud, S.; Garnier, N. B.; Ciliberto, S. *J. Stat. Mech.* 2007, P09018 1-28

[55] Horowitz, J,; Jarzynski, C. *J. Stat. Mech.* 2007, P11002 1-14

[56] Mai, T.; Dhar, A. *Phys. Rev. E* 2007, 75, 061101 1-7.

[57] Touchette, H.; Cohen, E. G. D. Phys. Rev. E 2007, 76, 020101 1-4

[58] Harris, R.; Schuetz, G. M. *J. Stat. Mech. Theory Exp.* 2007, P07020

[59] Saito, K.; Dhar, A. *Phy. Rev. Lett.* 2007 99 180601 1-4.

[60] Vilar, J. M. G.; Rubi, J. M. *Phys. Rev. Lett.* 2008, 100, 020601 1-4.

[61] Peliti, L. *Phys. Rev. Lett.* 2008, 101, 098903 1.

[62] Roy, D.; Kumar, N. *Phys, Rev. E* 2008, 78, 052102 1-4.

[63] Consolini, G.; De Michelis, P.; Tozzi, R. *J. Geophys. Res.* 2008, 113, A08222 1-11.

[64] Saha, S.; Jayannavar, A. M. *Phys. Rev. E* 2008, 77, 022105 1-4.

[65] Jiménez-Aquino, J. I.; Velasco, R. M.; Uribe, F. J. *Phys. Rev. E* 2008, 78, 032102 1-4.

[66] Jiménez-Aquino, J. I.; Velasco, R. M.; Uribe, F. J. *Phys. Rev. E* 2009, 79, 061109 1-6.

[67] Jiménez-Aquino, J. I.; Velasco, R.M.; Uribe, F. J. *New Trends in Statistical Physics*; Eds. Macías, A.; Dagdug, L. World Scientific, 2010.

[68] Jiménez-Aquino, J. I.; Uribe, F. J.; Velasco, R. M. *J. Phys A: Math. Theor.* 2010, 43, 255001 1-17.

[69] van Leeuwen, J. H. *J. Phys.* 1921, 2, 3619-

[70] De Pasquale, F.; Sancho, J. M.; San Miguel, M.; Tartaglia, P. *Phys. Rev. Lett.* 1986, 56, 2473-2476.

[71] Vemuri, G.; Roy, R. *Phys. Rev. A* 1989, 39, 2539-2543.

[72] Balle, S.; De Pasquale, F.; San Miguel, M. *Phys. Rev. A* 1990, 41, 5012-5015.

[73] Littler, I.; Balle, S.; Bergmann, K.; Vemuri, G.; Roy, R. *Phys. Rev. A* 1990 41, 4131-4134.

[74] Jiménez-Aquino, J. I.; Sancho, J. M. *Phys. Rev. A* 1991, 43, 589-590.

[75] Dellunde, J.; Torrent, M. C.; Sancho, J. M. *Opt. Commun.* 1993, 102, 277-280.

[76] Jiménez-Aquino, J. I.; Romero-Bastida, M. *Phys. Rev. E* 2002, 66, 061101 1-15.

[77] Jiménez-Aquino, J. I.; Romero-Bastida, M. *Phys. Rev. E* 2010, 81, 031128 1-7.

[78] Benzi, R.; Sutera, A.; Vulpiani, A. *J. Phys. A* 1981, 14, L453

[79] McNamara, B.; Wiesenfeld, K. *Phys. Rev. A* 1989, 39, 4854-4869.

[80] Klik I.; Gunther, L. *J. Stat. Phys.* 1990 60, 473-

[81] Jung, P.; Hänggi, P. *Phy. Rev. A* 1991, 44, 8032-8042.

[82] Douglass, J. K.; Wilkens, L.; Pantazelou, E.; Moss, F. *Nature* 1993, 365, 337-340.

[83] Pérez-Madrid, A.; Rubí, J. M. *Phys. Rev. E* 1995, 51, 4159-4164.

[84] Weisenfeld, K.; Moss, F. *Nature*, 1995, 373, 33-36

[85] Bezrukov, S. M.; Vodyanoy, I. *Nature* 1995, 378, 362-364.

[86] Collins, J. J.; Imhoff, T. T.; Grigg, P. *Nature* 1996, 383, 770-770.

[87] Gammaitoni, L.; Hänggi, P.; Jung, P.; Marchesoni, F. *Rev. Mod. Phys.* 1998, 70, 223-287.

[88] Pei, X.; Wilkens, L.; Moss, F. *J. Neurophysiol* 1996, 76, 3002-3011

[89] Yang, L. F.; Hou, Z. H.; Xin, H. W. *J. Chem. Phys.* 1999, 110, 3591-3595.

[90] Goychuk, I.; Hänggi, P. *PNAS* 2002, 99, 3552-3556.

[91] Miyakawa, K.; Tanaka, T.; Isikawa, H. *Phys. Rev. E* 2003, 67, 066206 1-4.

[92] Kitajo, K.; Nozaki, D.; Ward, L. M.; Yamamoto, Y. *Phys. Rev. Lett.* 2003, 90, 218103 1-4.

[93] Li, H.; Hou, Z.; Xin, H. *Chem. Phys. Lett.* 2005, 402, 444-449.

[94] Engel, T. A.; Helbig, B.; Russell, D. F.; Schimansky-Geier, L.; Neiman, A. B. *Phys. Rev. E* 2009, 80, 021919 1-7.

[95] Jianghai, D.; Aiguo, S.; Yiqing, W.; Xiulan, W. *IEEE Int. Conf. Neural Networks & Signal Processing*, 2003, Nanjing, China, December 14-17.

[96] M. Abramowitz and I. A. Stegun, *Handbook of Mathematical Functions* (Dover, New York, 1972).

In: Brownian Motion: Theory, Modelling and Applications ISBN: 978-1-61209-537-0
Editors: R.C. Earnshaw and E.M. Riley

Chapter 14

BROWNIAN MOTION: FROM QUANTUM DECOHERENCE TO DIFFUSION MRI MEDICAL IMAGING

*Nicolás F. Lori**
IBILI, Faculty of Medicine, Coimbra University, Coimbra, Portugal
Brain Imaging Network in Portugal

ABSTRACT

We develop an analysis of Brownian motion representation from the quantum to the diffusion MRI level. We then present an example of an application of Brownian motion computer simulations to diffusion MRI. The purpose of this work is to provide a general description of the mathematical and physical tools involved, and is aimed at graduate student level and above. The focus of the text is on the physics of Brownian motion, and so there is a large amount of equations.

Keywords: Brownian Motion, Diffusion MRI, Quantum Decoherence.

BRIEF HISTORY OF BROWNIAN MOTION

In 1905, on the same issue where Einstein proposed special relativity and the explanation of the photoelectric effect, he also proposed an explanation for the movement of pollen in water surfaces that can be observed in a microscope (1). This movement of pollen in water surfaces was first studied by Robert Brown, and is hence called Brownian motion. In Brownian motion, the pollen is observed as suffering shifts in velocity at very short time-intervals, and the accumulation of those shifts in velocity make the pollen have a rather seemingly arbitrary trajectory. Einstein was able to obtain an extremely satisfactory

* Dr. Nicolás F. Lori, Visual Neuroscience Laboratory, IBILI-Faculdade de Medicina, Azinhaga de Santa Comba, 3000-548 Coimbra Portugal, Phone +351 239480261, Fax +351 239480280/480217, Email: nflori@fmed.uc.pt

description of that motion, by simply associating the shifts in velocity to the random impact with neighboring atoms. This concept of motion can be generalized to the motion of water molecules in water, a.k.a. self-diffusion, and this will be the type of Brownian motion to be considered here. The description made by Einstein remained satisfactory until the development of quantum mechanics, where it became clear that the impact between molecules cannot be treated using the classical mechanics perspective. It was then necessary to develop a quantum formalism that was capable of representing such molecular interactions. Those mechanisms were quickly developed, but they had some difficulties in accurately representing the decoherence process the molecular impact would generate. Present day quantum mechanics has no such problems, and it is possible to represent Brownian motion using fully quantum processes. We present below a fully quantum treatment of Brownian motion that uses the quantum Darwinism approach, a recently developed approach to quantum mechanics.

BROWNIAN MOTION: FROM QUANTUM TO CLASSICAL

The quantum Brownian motion model consists of an environment made of a collection of harmonic oscillators interacting with an oscillating system of mass *M*, position *x*, and harmonic potential $V(x) = \frac{1}{2} M\omega x^2$, where ω is an oscillation frequency whose square-route is proportional to the water molecule's indeterminacy in momentum and inversely proportional to the indeterminacy in position (Eqs. (5.2) and (5.52) of Ref. (2)). The total Lagrangian representing the quantum system appears in Eqs. (5.1) to (5.3) of Ref. (2).

If the temperature *T* has a energy *KT* higher than all other relevant energy scales, including the energy content of the initial state and energy cutoff in the spectral density of the environment; then the Wigner quasi-distribution representation of the high temperature density matrix master equation can be expressed as a parameter *W* of an equation described in Eq. (5.40) of Ref. (2):

$$\dot{W} = -\frac{p}{M}\frac{\partial}{\partial \mathbf{x}}W + \frac{\partial V}{\partial x}\frac{\partial}{\partial \mathbf{p}}W + \varsigma\frac{\partial}{\partial \mathbf{p}}[pW] + \zeta MKT\frac{\partial^2}{\partial \mathbf{p}^2}W, \tag{1}$$

Where $\boldsymbol{x}$ represents the position vector and $\boldsymbol{p}$ represents the momentum vector.

The Eq. (1) is identical to the Chandrasekhar phase space distribution generalization of the Focker-Planck equation representing a classical molecular Brownian motion obeying the Langevin equation as described in Ref. (1):

$$\dot{\mathbf{u}} = -\zeta\mathbf{u} + \mathbf{a} \tag{2}$$

where $\boldsymbol{u}$ is the velocity, ζ is the friction coefficient, and **a** is the random acceleration. If *C(v)* describes the spectral environment with *v* an integration variable representing frequencies of the environment's harmonic oscillators, then using Eqs. (5.15) and (5.23b) of Ref. (2):

$\zeta = \frac{4}{M\omega}\int_0^\infty dl \int_0^\infty dv\, C(v)\sin(\omega l)\sin(vl)$. Taking into account the possible interactions of spatial asymmetries of molecular interactions, the friction coefficient becomes the friction tensor:

$$\zeta_{ij} = \frac{4}{M\omega_i}\int_0^\infty dl \int_0^\infty dv\, C_j(v)\sin(\omega_i l)\sin(vl) \tag{3}$$

Starting from the Langevin equation, Eq. (2), it is straightforward to obtain that: $\mathbf{u} = e^{-\zeta t}\mathbf{u}_0 + e^{-\zeta t}\int_0^t dl\, e^{\zeta l}\mathbf{a}(l)$, where the *0* index in vectors means "at time *t=0*". Integrating the previous expression, we obtain: $\mathbf{r} - \mathbf{r}_0 = \zeta^{-1}\left[\mathbf{1} - e^{-\zeta t}\right]\mathbf{u}_0 + \int_0^t dl\, \zeta^{-1}\left[\mathbf{1} - e^{-\zeta l}\right]\mathbf{a}(l)$. If $\otimes$ is the symbol for a Cartesian product between a dual vector and a vector, then the average-displacement covariance matrix is:

$$\begin{aligned}\left\langle [\mathbf{r}-\mathbf{r}_0]^T \otimes [\mathbf{r}-\mathbf{r}_0]\right\rangle &= \left\langle \mathbf{u}^T{}_0 \otimes \left[\mathbf{1} - e^{-\zeta^T t}\right]\zeta^{T-1}\zeta^{-1}\left[\mathbf{1} - e^{-\zeta t}\right]\otimes \mathbf{u}_0\right\rangle + \left\langle \int_0^t df\, \mathbf{a}^T(f)\left[\mathbf{1} - e^{-\zeta^T f}\right]\zeta^{T-1} \otimes \int_0^t dl\, \zeta^{-1}\left[\mathbf{1} - e^{-\zeta l}\right]\mathbf{a}(l)\right\rangle \\ &+ \left\langle \mathbf{u}^T{}_0 \otimes \left[\mathbf{1} - e^{-\zeta^T t}\right]\zeta^{T-1} \otimes \int_0^t dl\, \zeta^{-1}\left[\mathbf{1} - e^{-\zeta l}\right]\mathbf{a}(l)\right\rangle + \left\langle \int_0^t df\, \mathbf{a}^T(f)\left[\mathbf{1} - e^{-\zeta^T f}\right]\zeta^{T-1} \otimes \zeta^{-1}\left[\mathbf{1} - e^{-\zeta t}\right]\otimes \mathbf{u}_0\right\rangle\end{aligned} \tag{4}$$

Considering that **a** and $\mathbf{u_0}$ oscillate much faster than ζ as expressed in Eqs. (20.6) to (20.8) of Ref. [1], we can show that:

$$\left\langle \mathbf{u}^T{}_0 \otimes \left[\mathbf{1} - e^{-\zeta^T t}\right]\zeta^{T-1} \otimes \int_0^t dl\, \zeta^{-1}\left[\mathbf{1} - e^{-\zeta l}\right]\mathbf{a}(l)\right\rangle \approx \left\langle \int_0^t df\, \mathbf{a}^T(f)\left[\mathbf{1} - e^{-\zeta^T f}\right]\zeta^{T-1} \otimes \zeta^{-1}\left[\mathbf{1} - e^{-\zeta t}\right]\otimes \mathbf{u}_0\right\rangle \approx 0 \tag{5}$$

Using linear algebra transformations:

$$\left\langle [\mathbf{r}-\mathbf{r}_0]^T \otimes [\mathbf{r}-\mathbf{r}_0]\right\rangle = \left[\mathbf{1} - e^{-\zeta t}\right]\zeta^{-1}\left\langle \mathbf{u}^T{}_0 \otimes \mathbf{u}_0\right\rangle \zeta^{T-1}\left[\mathbf{1} - e^{-\zeta^T t}\right] + \int_0^t df\, \left[\mathbf{1} - e^{-\zeta f}\right]\zeta^{-1}\left\langle \mathbf{a}^T(f) \otimes \mathbf{a}(l)\right\rangle \int_0^t dl\, \zeta^{T-1}\left[\mathbf{1} - e^{-\zeta^T l}\right] \tag{6}$$

Again using that **a** and $\mathbf{u_0}$ oscillate much faster than ζ and a linear algebra transformation:

$$|\mathbf{u}|^2 = e^{-\zeta t}|\mathbf{u}_0|^2 e^{-\zeta^T t} + e^{-\zeta t}\left[\int_0^t df \int_0^t dl\, e^{\zeta f}\langle \mathbf{a}(f)\cdot \mathbf{a}(l)\rangle e^{\zeta^T l}\right]e^{-\zeta^T t} \tag{7}$$

Equipartition of energy requires that at thermal equilibrium: $|\mathbf{u}|^2 = 3\frac{KT}{m}$ in accordance with Eq. (13.17) of Ref. (1), and in the long time limit thermal equilibrium must be reached. So, the fast oscillation of **a** allows us to consider that

$e^{-\zeta t}\left[\int_0^t df \int_0^f dl\, e^{\zeta f} \langle \mathbf{a}(f)\cdot\mathbf{a}(l)\rangle e^{\zeta^T l}\right] e^{-\zeta^T t} \approx e^{-\zeta t}\left[\int_0^t dl e^{[\zeta+\zeta^T]l}\right] e^{-\zeta^T t}\left[\int_{-\infty}^{\infty}\langle \mathbf{a}^T(l)\cdot\mathbf{a}(l+s)\rangle ds\right]$. Using Eq. (7) obtains the relation between average behavior of random acceleration $\mathbf{a}$ and the symmetric component of the friction tensor $\hat{\zeta} = \frac{\zeta+\zeta^T}{2}$:

$$\int_0^\infty \langle \mathbf{a}(l)\otimes\mathbf{a}(l+s)\rangle ds = \frac{KT}{m}\hat{\zeta} \tag{8}$$

Using Eq. (8) in Eq. (6), calculations similar to those obtained in Eq. (20.14) of Ref. (1), and linear algebra we obtain:

$$\langle[\mathbf{r}-\mathbf{r}_0]^T\otimes[\mathbf{r}-\mathbf{r}_0]\rangle = \zeta^{-1}[\mathbf{1}-e^{-\zeta t}]\langle\mathbf{u}^T{}_0\otimes\mathbf{u}_0\rangle[\mathbf{1}-e^{-\zeta^T t}]\zeta^{T-1} + \frac{KT}{m}\zeta^{-1}\zeta^{T-1}\left[[\zeta+\zeta^T]t - 3 + 2[e^{-\zeta t}+e^{-\zeta^T t}] - e^{-[\zeta+\zeta^T]t}\right] \tag{9}$$

The relation between the average-displacement covariance matrix and symmetric component of ζ^{-1}, $\hat{\zeta}^{-1}$, in the long t limit is:

$$\langle[\mathbf{r}-\mathbf{r}_0]^T\otimes[\mathbf{r}-\mathbf{r}_0]\rangle = 2\frac{KT}{m}\hat{\zeta}^{-1}t\ , \tag{10}$$

which implies that the diffusion tensor in the Brownian motion perspective is:

$$D_{Brown} = \frac{KT}{m}\hat{\zeta}^{-1} \tag{11}$$

Equation (2) implies that D is symmetric even if ζ is asymmetric. This means that the anti-symmetric component of the diffusion tensor, is always zero for systems in thermal equilibrium.

If the medium where the molecule moves is asymmetric, then it might be no longer possible to represent the molecule's movement by a single diffusion tensor. Nevertheless, it is always possible to use a single tensor in representing a molecule moving between molecular impacts by use of Eq. (8) to obtain that the diffusion tensor for a time-correlation t is (3):

$$D_{\mathbf{a}}(t) = \int_0^t \langle \hat{\zeta}^{-1}\mathbf{a}(l)\otimes\mathbf{a}(l+s)\hat{\zeta}^{-1}\rangle ds \tag{12}$$

Another definition for the diffusion tensor is $D_{\mathbf{r}}(t) = \frac{\langle[\mathbf{r}-\mathbf{r}_0]^T\otimes[\mathbf{r}-\mathbf{r}_0]\rangle}{2[t-t_0]}$ which obtains:

$$D_{\mathbf{r}}(t) = \zeta^{-1}[\mathbf{1}-e^{-\zeta t}]\frac{\langle\mathbf{u}^T{}_0\otimes\mathbf{u}_0\rangle}{2t}[\mathbf{1}-e^{-\zeta^T t}]\zeta^{T-1} + \frac{KT}{2tm}\zeta^{-1}\zeta^{T-1}\left[[\zeta+\zeta^T]t - 3 + 2[e^{-\zeta t}+e^{-\zeta^T t}] - e^{-[\zeta+\zeta^T]t}\right] \tag{13}$$

Since equipartition of energy must apply on the long term limit, it's obtained that:

$$\lim_{t\to\infty} D_{\mathbf{a}}(t) = \lim_{t\to\infty} D_{\mathbf{r}}(t) = D_{Brown} \tag{14}$$

LOCAL AND GLOBAL DIFFUSION EQUATIONS

Another approach to dealing with the description of diffusion processes is to define the diffusion tensor based on the mass conservation equation. The consequence of the mass conservation equation is that if the only source of average molecular self-movement is along a concentration gradient then Fick's law of diffusion is obtained (1). The relationship between the concentration gradient and the flow of mass is defined through a diffusion tensor, D_{Fick}, which does not necessarily coincide with the previously mentioned definitions of diffusion tensor. Ficks' law states that the time-variation of the mass density ρ is equal to:

$$\frac{\partial \rho}{\partial t} = \nabla \cdot [D_{Fick} \cdot \nabla[\rho]] \tag{15}$$

From Eq. (15) it is obtained that the relationship between mass density in location $\mathbf{r}$ at time t, $\rho(\mathbf{r},t)$, and at time t=0 when $\rho(\mathbf{r},0) = c_0\delta(\mathbf{r})$ where δ is Dirac's delta function in 3D, is expressed as the Gaussian (1):

$$\frac{\rho(\mathbf{r},t)}{c_0} = \frac{1}{(2\pi)^{\frac{3}{2}}\left|2D_{Fick}t\right|^{\frac{1}{2}}} e^{-\frac{1}{2}\left[\mathbf{r}^T\cdot[2D_{Fick}t]^{-1}\cdot\mathbf{r}\right]} \tag{16}$$

The Wigner quasi-distribution, W, in Eq. (1) is in the classical limit equal to the displacement probability $P(\mathbf{r},t)$, which is equal to $\frac{\rho(\mathbf{r},t)}{c_o}$. The solution to Eq. (1) in Ref. (1), $W(\mathbf{r},t;\mathbf{u}_0)$, is nevertheless dependent of the velocity at time $t=0$, $\mathbf{u}_0$, which is associated to the local temperature T. The Wigner quasi-distribution is also a Gaussian, and so can be expressed identically to Eq. (16) but it is in that case obtained that D_{Fick} is the one described below and that $|\mathbf{r}|$ must be replaced by an effective $|\mathbf{r}|_{Fick}$ also described below:

$$D_{Fick}(t) = \frac{KT}{2tm}\zeta^{-1}\zeta^{T^{-1}}\left[\left[\zeta + \zeta^T\right]t - 3 + 2\left[e^{-\zeta t} + e^{-\zeta^T t}\right] - e^{-\left[\zeta+\zeta^T\right]t}\right] \tag{17}$$

$$|\mathbf{r}|_{Fick} = \left|\mathbf{r} - \zeta^{-1}\left[1 - e^{-\zeta t}\right]\mathbf{u}_0\right| \quad (18)$$

Like it occurred in Eq. (14), it is obtained that D_{Fick} also converges to D_{Brown} as t grows to infinity. But in the derivation of Fick's law it is assumed that D_{Fick} does not explicitly depend on time. Thus, the derivation of Fick's law obtains a result that contradicts its assumptions. It is for that reason that it is said that Fick's law is only valid in the case that the gradient of the mass density is small, that there are no boundaries, and where the times considered are large enough so that the displacements considered are large enough so that it makes sense to replace the discrete particle locations by a continuous field of mass density.

This implies that Fick's law of diffusion is realistic in the $t \to \infty$ limit, but not in the $t \to 0$ limit. In the $t \to 0$ limit one needs first to move into a Brownian motion perspective, and then to a quantum perspective. In the case of diffusion MRI the diffusion time being considered is a lot bigger than the average molecular impact time, we are thus able to use the $t \to \infty$ limit and simply use Fick's law of diffusion.

DISCRETE REPRESENTATION OF BROWNIAN MOTION

In the case that the time-scale we are considering is a lot bigger than the average time between molecular impacts, it is possible to consider a time-duration ε much bigger than the average time between molecular impacts and much shorter than the time-scale we are considering. In the case of diffusion MRI studies the time scale is on the order of the 10^{-3} seconds while the average time between molecular impacts is 10^{-12} seconds, which allows for an ε ranging between 10^{-9} and 10^{-6} seconds. During the time ε the acceleration $\mathbf{a}$ undergoes many fluctuations, which together with the initial velocity at the beginning of the time-lapse ε define a net unit-size vector pointing in an arbitrary direction, $\hat{\mathbf{r}}$. If $P(\hat{r})$ is the probability that the after-impact trajectory of the molecule is in the $\hat{\mathbf{r}}$ direction; then this corresponds to the probability of a displacement $\sqrt{2\hat{r}^T \cdot D \cdot \hat{r}\varepsilon}$ in that direction, where $D = D_{Brown}$. If $\wedge$ stands for the logic AND, and $\boldsymbol{g}$ for the motion-sensitizing magnetic field gradient, the MRI signal for a group of N molecules that move during a time equal to K times the time ε is:

$$E\left(\left[\mathbf{r}_{1[1]} \wedge \mathbf{r}_{2[1]} \wedge \cdots \wedge \mathbf{r}_{K-1[1]} \wedge \mathbf{r}_{K[1]}\right], \cdots, \left[\mathbf{r}_{1[m]} \wedge \mathbf{r}_{2[m]} \wedge \cdots \wedge \mathbf{r}_{K-1[m]} \wedge \mathbf{r}_{K[m]}\right]\right) = \sum_{m=1}^{N} \prod_{n=1}^{K} e^{i\gamma\varepsilon \sqrt{2\hat{r}_{n[m]}^T \cdot D \cdot \hat{r}_{n[m]}\varepsilon}\, g \cdot \hat{r}_{n[m]}} \quad (19)$$

For each molecule, we will have a combined effect of displacement and phase change effects that result in:

$$E\left(\mathbf{r}_1 \wedge \mathbf{r}_2 \wedge \cdots \wedge \mathbf{r}_{K-1} \wedge \mathbf{r}_K\right) = e^{i\gamma\varepsilon\left[\mathbf{g} \cdot \sqrt{2\hat{\mathbf{r}}_1^T \cdot D \cdot \hat{\mathbf{r}}_1 \varepsilon}\,\hat{r}_1 + \mathbf{g} \cdot \sqrt{2\hat{\mathbf{r}}_2^T \cdot D \cdot \hat{\mathbf{r}}_2 \varepsilon}\,\hat{\mathbf{r}}_{21} + \cdots \mathbf{g} \cdot \sqrt{2\hat{\mathbf{r}}_{K-1}^T \cdot D \cdot \hat{\mathbf{r}}_{K-1}\varepsilon}\,\hat{\mathbf{r}}_{K-1} + \mathbf{g} \cdot \sqrt{2\hat{\mathbf{r}}_K^T \cdot D \cdot \hat{\mathbf{r}}_K \varepsilon}\,\hat{\mathbf{r}}_K\right]} \quad (20)$$

Defining $\vec{r}_i = \sqrt{2\hat{\mathbf{r}}_i^T \cdot D \cdot \hat{\mathbf{r}}_i \varepsilon}\,\hat{\mathbf{r}}_i$, averaging over all the molecules (assuming they are in large enough number) one obtains:

$$E = \sum_{\substack{\text{All K steps} \\ \text{Trajectories}}} P\left(\mathbf{r}_1 \wedge \mathbf{r}_2 \wedge \cdots \wedge \mathbf{r}_{K-1} \wedge \mathbf{r}_K\right) e^{i\gamma\varepsilon\left[\mathbf{g}\cdot\mathbf{r}_1 + \mathbf{g}\cdot\mathbf{r}_2 + \cdots + \mathbf{g}\cdot\mathbf{r}_{K-1} + \mathbf{g}\cdot\mathbf{r}_K\right]} \tag{21}$$

In the case where a pulsed gradient spin-echo sequence is used, and if both gradients have constant amplitudes lasting each a time δ and having a time between the start of each of them equal to Δ (an example of motion-sensitizing the signal), if λ is a very small displacement amplitude, and p is a probability density function, then Eq. (21) becomes:

$$E = \sum_{\substack{\text{All } \vec{\mathbf{R}} \\ \text{that differ} \\ \text{less than } \lambda}} \lambda\, p\left(\left|\frac{\mathbf{r}_1 + \cdots + \mathbf{r}_{l+\frac{\delta}{\varepsilon}}}{\frac{\delta}{\varepsilon}} - \frac{\mathbf{r}_{l+\frac{\Delta}{\varepsilon}} + \cdots + \mathbf{r}_{l+\frac{\Delta+\delta}{\varepsilon}}}{\frac{\delta}{\varepsilon}} + \mathbf{R}\right| \le \lambda\right) e^{i\gamma\delta\mathbf{g}\cdot\left[\frac{\mathbf{r}_1 + \cdots + \mathbf{r}_{l+\frac{\delta}{\varepsilon}}}{\frac{\delta}{\varepsilon}} - \frac{\mathbf{r}_{l+\frac{\Delta}{\varepsilon}} + \cdots + \mathbf{r}_{l+\frac{\Delta+\delta}{\varepsilon}}}{\frac{\delta}{\varepsilon}}\right]} \tag{22}$$

The set of all $\vec{\mathbf{R}}$ that differ less than λ defines a cubic grid of size λ.

If $E(\mathbf{g})$ is the MRI signal when the applied magnetic field gradient is value of $\mathbf{g}$, then the ratio $\frac{E(\mathbf{g})}{E(0)} = E$ in the case where the medium has a homogeneous diffusion-tensor D (e.g. white matter with all the axons pointing in the same orientation) is given by the Stejskal-Tanner equation of Ref. (4):

$$\frac{E(\mathbf{g})}{E(0)} = e^{-[\gamma\delta\mathbf{g}][\gamma]^T \cdot \left[D\left[\Delta - \frac{\delta}{3}\right]\right] \cdot [\gamma\delta\mathbf{g}]} \tag{23}$$

Where ^{T} stands for the transpose of a vector.

Model-Based Brownian Motion Generation of Diffusion MRI Signal

As in Ref. (3), the Brownian motion of the water molecules will be approximated by a random walk of discrete step size. The size of the discrete step will depend on the time increment we choose for a particular simulation, and the characteristics of the local diffusion tensor. In living organisms, the temperature is usually about 37 °C, same as 310K, and the diffusion MRI scale we are interested in is much larger than the nanometer. Therefore, it is reasonable to approximate the water molecule's Brownian motion by a random walk.

In a 3D random walk, each step occurs in 3D, so there must be a probability of advancement in each of the three directions. In our case, we consider the simple model that diffusion is locally isotropic everywhere except where the water molecule interacts with domain borders. Because diffusion is locally isotropic in this model, the diffusion tensor is always in its diagonal form, which makes it easier to define the 3D steps of the random walk. Because the intracellular and extracellular parts of white matter have different diffusion properties, we consider that the intra-cylinder and the extra-cylinder components of the cylinder array have different isotropic diffusion coefficients.

Although it could be said that the axons in white mater are very similar to cylinders, modeling the axons using cylinders is simply a model. Moreover, when modeling, there are a lot of parameters about which authors disagree; and we will be focusing on the case where there are no crossing axons in the white matter fiber bundle being modeled. Thus, this modeling of the effect of Brownian motion on the motion-sensitized MRI signal is model-based. After the explanation of this model, we will go back to Eq. 22 for a description of a model-free approach to the treatment of the motion-sensitized MRI signal.

The random walk simulation used in this work is identical to the simulation of completely restricted random walks in Ref. (3), unless otherwise noted, and with the exception that the domain walls are now partially permeable, random walks are seeded in the extracellular as well as intracellular space, and the domains are cylinders rather than spheres. When a random walk nears a cylinder's wall, two things can happen. Either the next step intersects the wall, or it doesn't. If the step does not make contact with the wall, then nothing different from a free-diffusion random walk occurs (3). If the initial point is inside the cylinder and the step trajectory intersects the wall, then either the step passes through the wall without altering the step size or direction, or the step moves only in the component direction parallel to the cylinder axis. These two stepping conditions are randomized according to the probability P_{in-ex} that the step will pass through the wall. If the initial point is outside the cylinder, then the probability of stepping through the wall is P_{ex-in}. When the random walk moves from inside to outside the cylinder (or vice-versa), a corresponding instantaneous change in the diffusion coefficient is applied to the subsequent step, changing from the diffusion coefficient inside the cylinder D_{in} to the diffusion coefficient outside the cylinder D_{ex} (or vice-versa), and the random walk then proceeds in the extracellular (or intracellular) space. For a cylinder of radius a, this exchange time corresponds to an intracellular-extracellular exchange probability P_{in-ex} (Eq. (20) of Ref. (5)):

$$P_{in-ex} = \frac{2a\sqrt{\frac{dt}{6D_{in}}}}{T_{in-ex} - \frac{a^2}{8D_{in}}} \tag{24}$$

The extra-intra exchange probability is $P_{ex-in} = \sqrt{D_{in}/D_{ex}}\,P_{in-ex}$, from Eq. (18) of Ref. (5).

Several studies have indicated that the Stejskal-Tanner equation does not hold in all situations (e.g., Refs. (5,6)), but those studies also indicate that for the gradient $|g|=40.0*10^{-9}$ Tesla/micron or below, the Stejskal-Tanner equation is obeyed in the direction of the fiber. The exchange time from intracellular to extracellular space ms (T_{in-ex}) is assumed to be 600 ms as in Refs. (5,6). Replicating the simulations in Ref. (3) with 10000 simulated trajectories reproduces the experimental results obtained in Ref. (7) for the Splenium of the Corpus Callosum both for the direction parallel to the WM tract and the direction orthogonal to the WM tract, in a form that agrees with the Stejskal-Tanner equation.

The parameter values used for the simulations in this work were based in the experimental pulse sequence parameters of Ref. (7) for Figures 1a and 1b, and of Ref (8) for Figure 1c. The variation of the diffusion coefficient with the angle θ was also analyzed, where θ is the angle between **g** and the fiber axis. The angle θ is made to vary between 0 and 90 degrees.

If the Stejskal-Tanner equation is valid and the Apparent Diffusion Coefficient (ADC) is $D_{||}$ when **g** is parallel to the permeable cylinder and $D_{\perp}$ when **g** is orthogonal to the permeable cylinder, then the measured ADC will depend on θ. Since all the fibers in the simulation are pointing in the same direction, the θ-dependent ADC is simply the diffusion coefficient measured in that direction, and it is represented as $D(\theta)$. Using the Stejskal-Tanner equation, one would obtain "$D(\theta) = D_{||}\cos^2(\theta) + D_{\perp}\sin^2(\theta)$". The simulation results indicate that $D_{||}=1.1\pm0.1$ *micron²/millisecond* and $D_{\perp}=0.13\pm0.01$ *micron²/ms*, and the "$D(\theta) = D_{||}\cos^2(\theta) + D_{\perp}\sin^2(\theta)$" line matches the simulated data well. In these simulations it is therefore obtained that it is correct to adopt the Stejskal-Tanner for these experimental parameters.

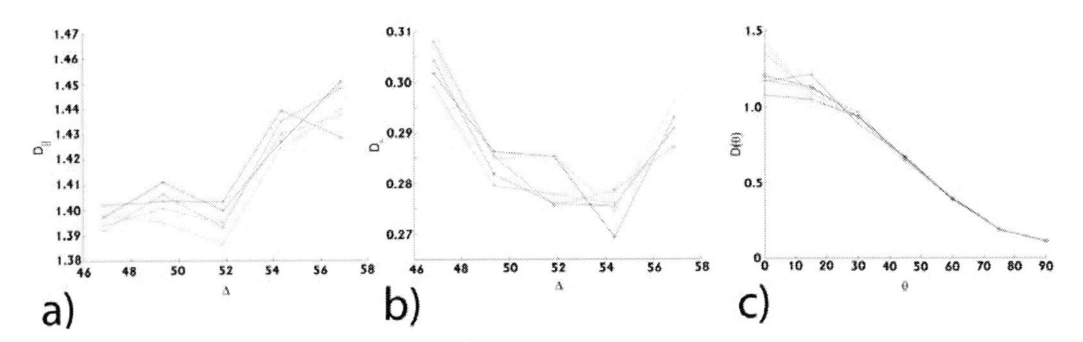

Figure 1. a) Comparison of the simulated observed diffusion coefficient for a *g* orientation parallel to the fiber for different fiber radius; for the $|g|$, Δ and δ values used in Ref. (7); at the most left are the values used for the tetrahedral orientations and at the most right are the values used for the orthogonal orientations. The tetrahedral orientation parameters were $|g|=34.6*10^{-9}$ *Tesla/micron*, $\Delta=46.85$ *millisecond* and $\delta=17.0$ *millisecond*, and for the orthogonal orientations they were $|g|=20.0*10^{-9}$ *Tesla/micron*, $\Delta=56.85$ *millisecond* and $\delta=27.0$ *millisecond*. The cylinder radius are 1 *micron* (red), 2 *micron* (green), 3 *micron* (blue), 5 *micron* (magenta), and 7 *micron* (cyan). The values of the diffusion coefficient are 1.2 *micron²/millisecond* if the simulated water molecule is inside the cylinder (D_{in}), and 2.5 *micron²/millisecond* if the simulated water molecule is outside the cylinder (D_{ex}). b) Same as (a), but here for a *g* orientation orthogonal to the fiber. c) Same as (a) except that $|g|=22.4*10^{-9}$ *Tesla/micron*, $\Delta=56.0$ *millisecond* and $\delta=55.0$ *millisecond* for all points, as in Ref. (8), and the only parameter that changes from left to right is θ, the angle between *g* and the axis of the fiber. The black line is the "$D(\theta) = D_{||}\cos^2(\theta) + D_{\perp}\sin^2(\theta)$" curve, with $D_{||}=1.1\pm0.1$ *micron²/millisecond* and $D_{\perp}=0.13\pm0.01$ *micron²/millisecond*. For all simulated lines appearing in this figure, the number of random walks was 10000.

These simulated results obtain that for the experimental conditions typically used in gaussian decomposition (9), Q-Ball Imaging (QBI) (8), and Diffusion Spectrum Imaging (DSI) (10), the Stejskal-Tanner equation is obeyed if the axons of the fiber bundles are all parallel. It is clear from Figure 1c that the black line matches the simulated data well. If white matter axons in a voxel point in different orientations, then the single tensor approximation is no longer valid and it is necessary to use more elaborate diffusion MRI data treatments such as QBI or DSI, and it is not necessarily valid that the Stejskal-Tanner equation would be valid for each of the axons if they are not parallel.

Model-Free Brownian Motion Generation of Diffusion MRI Signal

The relationship between displacement probability and MR signal intensity for a motion-sensitizing pulsed-gradient spin-echo appears in Eq. (22), which can be greatly simplified by defining $\mathbf{q} \equiv \gamma\delta\mathbf{g}$ and using:

$$\Delta\mathbf{r}_{[\Delta,\delta]} = \frac{\mathbf{r}_1 + \cdots + \mathbf{r}_{l+\frac{\delta}{\varepsilon}}}{\frac{\delta}{\varepsilon}} - \frac{\mathbf{r}_{l+\frac{\Delta}{\varepsilon}} + \cdots + \mathbf{r}_{l+\frac{\Delta+\delta}{\varepsilon}}}{\frac{\delta}{\varepsilon}} \tag{25}$$

Using Eq. (25) and the definition of $\boldsymbol{q}$, the continuous approximation of Eq. (22) becomes:

$$E(\mathbf{q}) = \int_{-\infty\mathbf{1}^3}^{\infty\mathbf{1}^3} p\left(\Delta\mathbf{r}_{[\Delta,\delta]}\right) e^{i\mathbf{q}\cdot\Delta\mathbf{r}_{[\Delta,\delta]}} d^3\Delta\mathbf{r}_{[\Delta,\delta]} \tag{26}$$

,

with $\int_{-\infty\mathbf{1}^3}^{\infty\mathbf{1}^3}$ symbolizing the integral over a 3D space. Since Eq. (26) indicates that the MR signal intensity reduction $E(\mathbf{q})$ is simply the inverse Fourier transformation of the displacement probability $P\left(\Delta\mathbf{r}_{[\Delta,\delta]}\right)$, then the forward Fourier transformation of $E(\mathbf{q})$ obtains:

$$P\left(\Delta\mathbf{r}_{[\Delta,\delta]}\right) = \int_{-\infty\mathbf{1}^3}^{\infty\mathbf{1}^3} E(\mathbf{q}) e^{-i\mathbf{q}\cdot\Delta\mathbf{r}_{[\Delta,\delta]}} d^3\mathbf{q} \tag{27}$$

,

The result in Eq. (27) is a model-free result, as opposed to the model-based results of the previous section. Starting from any point in space it is possible to define the probability that the molecule will be at another point at a different time. When the only displacement is caused by diffusion, the displacement probability is symmetric, but the occurrence of a net flow will make the displacement distribution become asymmetric.

The probability of a trajectory t_A^B between points A and B made of steps $\mathbf{\Delta l}\{t_A^B\}$ is equal to the integration of $P(\mathbf{\Delta l}\{t_A^B\})$ along the trajectory t_A^B:

$$\mathrm{P}(t_A^B) = \prod_{j=1}^{L} \left|\mathbf{\Delta l}_j\{t_A^B\}\right| P(\mathbf{\Delta l}\{t_A^B\}_j) \tag{28}$$

,

The probabilities $P(\mathbf{\Delta l}\{t_A^B\})$ for a certain step are obtained by interpolating the displacement probabilities from nearby voxels, with those displacement probabilities obtained from experimental MRI results mimicking Eq. (27), meaning that $E(\mathbf{q})$ values must be acquired for a large-enough 3-dimensional cubic grid of $\mathbf{q}$ vectors (typically made of about 500 different vectors). The most typical interpolation technique uses spline functions, but both nearest neighbor and linear interpolation approaches are also used (11). The definition of trajectories in white matter based on Eq. (28) is the principle behind DSI fiber tracking (10), and it allows the correct mimicking of the orientation and location of white matter fiber bundles even in the case where there are crossing with matter fiber bundles within a single voxel. But if we assume that all the steps $\mathbf{\Delta l}\{t_A^B\}$ have the same amplitude $|\mathbf{\Delta l}|$, then the probability of step orientation, $\varphi\{t_A^B\}$, is the only thing separating the different possible steps that can occur at each point. The probability of an orientation is called the Orientation Distribution Function (ODF) Φ and it is equal to the integral of the probability $P(\mathbf{\Delta l})$ over all the steps that have that same orientation φ, corresponding to the unit vector $\hat{\mathbf{e}}_\varphi$ pointing in the φ orientation:

$$\Phi(\hat{\mathbf{e}}_\varphi) = \int_{-\infty}^{\infty} p(\alpha\hat{\mathbf{e}}_\varphi)d\alpha \tag{29}$$

The normalized ODF $\Psi(\hat{\mathbf{e}}_\varphi)$ is obtained by dividing the non-normalized ODF $\Phi(\hat{\mathbf{e}}_\varphi)$ by the integral of the non-normalized ODF $\Phi(\hat{\mathbf{e}}_\varphi)$ over a unit-sphere:

$$\Psi(\hat{\mathbf{e}}_\varphi) = \frac{\Phi(\hat{\mathbf{e}}_\varphi)}{\int_{SPHERE} \Phi(\hat{\mathbf{e}}_\varphi)} \tag{30}$$

To experimentally reproduce Eq. (29) one needs to obtain about 500 different values of $E(\mathbf{q})$, but then a lot of that extra information would be lost by the integration expressed in Eq. (29). The goal of QBI (8) is to obtain the ODF without using that many **q** vectors. The QBI-based ODF obtained with a q-vector amplitude of q' for orientation $\hat{\mathbf{e}}_{\varphi}$, $\Phi_{q'}(\hat{\mathbf{e}}_{\varphi})$, is the Funk-Radon transform of the motion-sensitized MR signal reduction (8). If we consider a cylindrical coordinate system with the z-axis pointing in the orientation $\hat{\mathbf{e}}_{\varphi}$, then in that coordinate system: $\mathbf{q}\equiv(q,\vartheta,\zeta)$ and $\Delta\mathbf{r}\equiv(r,\theta,z)$. Thus implying that if $\delta(x)$ is the Dirac-delta function, the described Funk-radon transform is (8):

$$\Phi_{q'}(\hat{\mathbf{e}}_{\varphi})=\int_0^{\pi} q\,d\vartheta_{\mathbf{u}}\int_0^{\infty} dq\int_{-\infty}^{\infty} d\zeta\, E(q,\vartheta,\zeta)\delta(q'-q)\delta(\zeta) \tag{31}$$

Using Eq. (26) obtains the expression below:

$$\Phi_{q'}(\hat{\mathbf{e}}_{\varphi})=\int_0^{\pi} q\,d\vartheta\int_0^{\infty} dq\int_{-\infty}^{\infty} d\zeta\int_0^{\pi} r\,d\theta\int_0^{\infty} dr\int_{-\infty}^{\infty} dz\,P(r,\theta,z)e^{i[z\zeta+rq\cos(\theta-\vartheta)]}\delta(q'-q)\delta(\zeta) \tag{32}$$

Defining $p(r,\theta)=\int_{-\infty}^{\infty} dz\,P(r,\theta,z)$ with corresponding symmetric $\hat{p}(r,\theta)=\frac{1}{2}(p(r,\theta)+p(r,-\theta))$, and anti-symmetric $\tilde{p}(r,\theta)=\frac{1}{2}(p(r,\theta)-p(r,-\theta))$ components: together with $e^{iX\cos(\alpha)}=\sum_{n=-\infty}^{\infty} i^n J_n(X)e^{in\alpha}$ with J_n being the nth order Bessel function of the 1st kind; $\int_0^{\pi}\cos(n\theta_{\mathbf{u}}-n\vartheta_{\mathbf{u}})\,d\vartheta_{\mathbf{u}}=\left(\frac{2}{n}\right)\sin(n\theta_{\mathbf{u}})$ for n>0 and even; and $\int_0^{\pi}\cos(n\theta_{\mathbf{u}}-n\vartheta_{\mathbf{u}})\,d\vartheta_{\mathbf{u}}=0$ for n>0 and odd; obtains that integrating Eq. (32) over z, ζ, q and ϑ defines the result:

$$\Phi_{q'}(\hat{\mathbf{e}}_{\varphi})=q'\int_0^{\infty} dr\int_0^{\pi} r\,d\theta\left[J_0(q'r)\hat{p}(r,\theta)-\sum_{n=1}^{\infty}\frac{2}{n}J_{2n}(q'r)\sin(2n\theta)\tilde{p}(r,\theta)\right]. \tag{33}$$

In physical situations where there is only diffusion and no net flow, we have $\tilde{p}(r,\theta)=0$ (8)(10), and so we obtain the result of Ref. (8):

$$\Phi_{q'}(\hat{\mathbf{e}}_{\varphi})=q'\int_0^{\infty} dr\int_0^{\pi} r\,d\theta\, J_0(q'r)\hat{p}(r,\theta) \tag{34}$$

The relationship expressed in Eq. (34) implies that for large q', the QBI-based approximation to the ODF is approximately equal to the ODF, $\Phi_{q'}(\hat{\mathbf{e}}_{\varphi}) \approx \Phi(\hat{\mathbf{e}}_{\varphi})$. The QBI-based experimentally obtained ODF values for different orientations in different voxels allows for the use of a $\Phi(\varphi\{t_A^B\})$ in the same way as $P(\mathbf{\Delta l}\{t_A^B\})$ was used, but using the probability of an orientation instead of the probability; the probability of a trajectory is then:

$$\Phi(t_A^B) = \prod_{j=1}^{L} \Phi(\varphi\{t_A^B\}_j) \tag{35}$$

The use of Eq. (35) to select lines connecting points in the brain that have the highest values of $\Phi(t_A^B)$ is the basis of QBI fiber-tracking, and allows for the recovery of lines that have orientations and locations that mimic anatomical white matter connections in living humans without any need for anesthesia, radiation or contrast agents; even in the case where there are crossing white matter fiber bundles within a single voxel, and with the advantage over DSI of only needing about 200 different **q** vectors. The disadvantage in relation to DSI is that QBI allows a much less thorough analysis of the values of $P(\mathbf{\Delta l}\{t_A^B\})$. Below is an example of the type of experimental results that can be obtained using QBI.

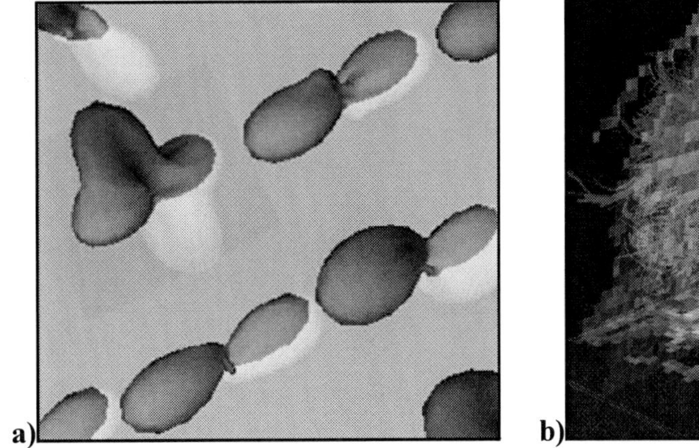

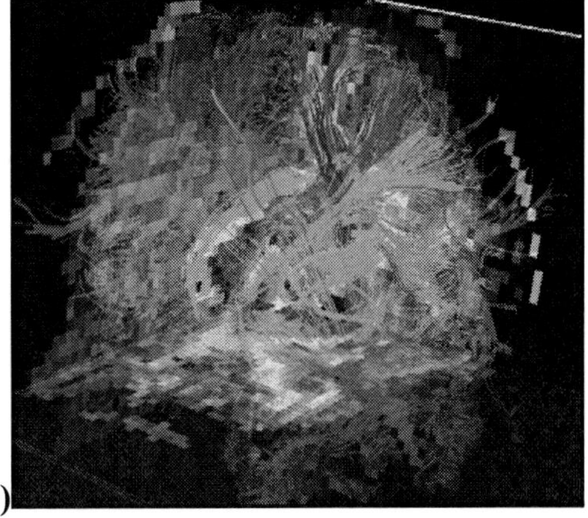

Figure 2. QBI experimental results in living human using a 3 Tesla MRI scanner and the TrackVis software. The color code for both the lines and the ODF surfaces indicates their preferred orientation, with the following color-code: Green->anterior-posterior, Red->left-right, Blue-superior-inferior. a) ODF surfaces in a transverse slice. b) QBI white matter fiber tracking experimental results, brain is being observed from the left-anterior perspective as the cube in the lower right corner indicates.

CONCLUSION

We describe, in a few pages, the relationship between quantum diffusion processes and diffusion MRI. It is clear from the description that a lot of the quantum aspects of the diffusion process are dissipated when we are at the diffusion MRI level. But it is not unlikely, that a bright mind would find a way of highlighting such a correlation between quantum phenomenon and diffusion MRI. Such an application might occur at the level of brain imaging, or at the level of quantum computing. But the possibilities are definitely there. The difficulties are also there, amongst them, the decoherence process that is very strong at very high temperatures (by very high meaning a lot higher than absolute zero) such as they occur in biological systems.

ACKNOWLEDGMENTS

Thanks are due to FCT for funding the PTDC/SAU-BEB/100147/2008 project; and to both Van J. Wedeen and Miguel Castelo-Branco for helpful comments.

REFERENCES

[1] D.A. McQuarrie, *Statistical Mechanics*, New York:Harper & Row, 1975.

[2] W.H. Zurek, Decoherence, einselection, and the quantum origins of the classical. *Reviews of Modern Physics*, 75 (2003), pp. 715-774.

[3] N.F. Lori, T.E. Conturo, D. Le Bihan, Definition of displacement probability and diffusion time in q-space magnetic resonance measurements that use finite-duration diffusion-encoding gradients, *J. Magn. Reson.* 165 (2003) 185-195.

[4] E. O. Stejskal, J. E. Tanner. Spin diffusion measurements: spin echoes in the presence of a time dependent field gradient. *J Chem Phys* 1965;42:288-292.

[5] C. Meier, W. Dreher, D. Leibfritz. Diffusion in compartmental systems. I. A comparison of an analytical model with simulations. *Magn Reson Med* 2003;50(3):500-509.

[6] C. Meier, W. Dreher, D. Leibfritz. Diffusion in compartmental systems. II. Diffusion-weighted measurements of rat brain tissue in vivo and postmortem at very large b-values. *Magn Reson Med* 2003;50(3):510-514.

[7] J. S. Shimony, R. C. McKinstry, E. Akbudak, J. A. Aronovitz, A. Z. Snyder, N. F. Lori, T. S. Cull, and T. E. Conturo. Quantitative diffusion-tensor anisotropy brain MR imaging: normative human data and anatomic analysis. *Radiology* 1999; 212:770-784.

[8] D. S. Tuch, T. G. Reese, M. R. Wiegell, V. J. Wedeen. Diffusion MRI of complex neural architecture. *Neuron* 2003;40(5):885-895.

[9] D. C. Alexander. Multiple-fibre reconstruction algorithms for diffusion MRI. *Ann N Y Acad Sci.* 2005 Dec;1064:113-33.

[10] V. J. Wedeen, P. Hagmann, W. I. Tseng, T. G. Reese, and R. M. Weisskoff. 2005. Mapping complex tissue architecture with diffusion spectrum magnetic resonance imaging, *Magn Reson Med*; 54, 1377-86.

[11] N. F. Lori, E. Akbudak, J. S. Shimony, T. S. Cull, A. Z. Snyder, R. K. Guillory, and T. E. Conturo. 2002. Diffusion tensor fiber tracking of brain connectivity: acquisition methods, reliability analysis and biological results. *NMR in Biomedicine*; 15, 494-515.

INDEX

#

A

B

C

D

E

F

G

H

I

J

K

L

M

N

O

P

Q

R

S